NEURAL PATHWAYS RESEARCH

NEURAL PATHWAYS RESEARCH

FLORIAN L. PICHLER
EDITOR

Nova Biomedical Books
New York

For permission to use material from this book please contact us:
Telephone 631-231-7269; Fax 631-231-8175
Web Site: http://www.novapublishers.com

Library of Congress Cataloging-in-Publication Data

Neural pathways research / Florian L. Pichler, editor.
p. ; cm.
Includes bibliographical references and index.
ISBN 978-1-60456-214-9 (hardcover)
1. Neural circuitry. I. Pichler, Florian L.
[DNLM: 1. Neural Pathways--physiology. 2. Brain--physiology. 3. Neuronal Plasticity. 4. Synaptic Transmission. WL 102 N49344 2008]
QP363.3.N443 2008
612.8'2--dc22

2007052111

Published by Nova Science Publishers, Inc. New York

Contents

Preface

A neural pathway is a neural tract connecting one part of the nervous system with another, usually consisting of bundles of elongated, myelin-insultated neurons, known collectively as white matter. Neural pathways serve to connect relatively distant areas of the brain or nervous system, compared to the local communication of grey matter. This new book presents the latest research in the field.

Chapter I - The main olfactory bulb (MOB) is the first relay station in the CNS for processing sensory information that comes from olfactory receptor neurons in the nasal epithelium. Neural signaling from olfactory nerve terminals to the principal MOB projection neurons, mitral and tufted cells (M/TCs), is in part mediated by glutamate acting at AMPA and NMDA ionotropic glutamate receptor (iGluR) subtypes. However, neuroanatomical studies demonstrate that M/TCs as well as inhibitory interneurons such as GABAergic granule cells (GCs) in the MOB express high levels of metabotropic glutamate receptors (mGluRs), suggesting that these receptors also play a role in the function of the MOB network. Recent experiments that employed a combination of patch-clamp electrophysiology, voltage-sensitive dye imaging, microsurgery, and use of mGluR gene knockout mice have illuminated the role of mGluRs in the MOB and point toward novel and potent regulatory roles of these receptors in shaping olfactory output from the MOB to higher olfactory centers in the brain.

Chapter II - Carbocyanine dye tracing was introduced into the neuroanatomical studies in the late 1980s (Godement et al.,1987, Honig and Hume, 1989) and has been applied later for numerous central and peripheral axonal systems. The main feature of the fluorescent lipophylic carbocyanine dyes as perinatal tracers is that the markers can be applied on the prefixed brain tissue and thus avoid in vivo stereotaxic insertions specific for most of the axonal brain markers. Even though there are three known dyes the 1,1'-dioctadecyl-3,3,3',3'-tetramethylindocarbocyanine perchlorate (DiI) is widely used as a retrograde and sometimes anterograde tracer.

This review reevaluates advantages and disadvantages of the DiI tracing on the basis of our own experience in using this method in the studies of the development of afferent and efferent hypothalamic connections and literature data. The authors have studied the time schedule of the formation of the several hypothalamic axonal systems: hypothalamic connections with the posterior and intermediate pituitary lobes and median eminence, mammillary body tracts and reciprocal septohypothalamic connections. Some of the hypothalamic fiber systems develop unexpectedly early in embryogenesis but other forms only postnatally.

Principal facts and questions raised in these studies are discussed: 1) Characteristics of the DiI distribution in the place of application; 2) Specificity of DiI labeling; 3) Comparison of the DiI and DiA; 4) Advantages of the confocal microscopic analysis of the material with carbocianine tracers; 5) What cells of the neuroepithelium are labeled with DiI; 6) Controls in the DiI experiments; 7) Advantages and limitations of the method; 8) Suggestions for new studies.

The authors are sure that even though DiI tracing needs long exposition time and extremely fine hand manipulations with the fetal brain material, it should be used as a very important tool for studying the early steps of the formation of brain connections. Such data will provide the basic knowledge concerning the main principles of brain network development.

Chapter III - Zinc is released with glutamate from neuron terminals in the hippocampus. Zinc may serve as a negative-feedback factor of presynaptic activity and negatively modulate postsynaptic calcium mobilization. On the other hand, the hippocampus is vulnerable to glutamate excitotoxicity, a final common pathway for numerous pathological processes such as Alzheimer's disease and amyotrophic lateral sclerosis, in addition to stroke/ischemia, temporal lobe epilepsy. The excitotoxicity is linked to the excessive influx of zinc and calcium. The crosstalk between zinc and calcium via calcium channels may play a role in both synaptic plasticity and excitotoxicity. This paper summarizes the involvement of zinc in functional and toxic aspects in the hippocampus focused on the crosstalk. The enhanced excitotoxicity in the hippocampus in zinc deficiency is also summarized.

Chapter IV - Among precocious Aves, day-old chicks show a preference for the moving object to which they were first exposed. This behavior, called "imprinting," is known as a form of learning. Visual and acoustic clues are naturally used by chicks as guides to their mothers. This imprinting phenomenon can be reproduced in the laboratory using such clues as an imprinting stimulus. The authors established a system of inducing visual imprinting behavior in chicks (*Gallus gallus domesticus*) using moving 2-dimensional images on a thin-film transistor liquid crystal display. Since this system enables control of the components of the image presented to the chicks, they were now able to discriminate those components to which the chicks direct their attention from among all the components of the stimulus, such as shapes, colors, textures and directions of movement.

By applying this system, the authors examined the neural pathways related to color recognition and memory. They adopted two methods of visualizing these pathways: the optical imaging of the intrinsic signal, which is suitable for mapping gross changes in activity on the surface of the brain, and c-FOS immunohistochemistry, which is suitable not only for examining the deep structure of the brain but for determining what types of neurons respond to specific stimuli within the layers and nuclei of the brain. Using the first approach, the authors found that the visual wulst, one of the higher visual centers in Aves, is related to both color recognition and the acquisition of memory. Using the second approach, they found that the cholecystokinin neurons in the visual wulst and adjacent areas are indispensable for the acquisition of memory.

In this review, the authors describe the technical approaches that led to these findings. Imprinting in chicks might be a good model of infant and child learning in humans, because infants and chicks share a specific time-window called the "critical period," during which the infant is sensitive to certain experiences. Further study using these techniques might enable a more detailed characterization of this "critical period."

Chapter V - The transmission of sensory information from the peripheral to the central nervous system is modified at each relay station (e.g. spinal cord, thalamus and sensory cortex) by a local circuit and descending and ascending pathways. Peripheral tissue injury/inflammation can also alter the properties of somatic sensory pathways, causing behavioral hypersensitivity and resulting in increased responses to pain caused by noxious stimulation (hyperalgesia) and normally innocuous stimulation (allodynia). Although no synaptic transmission has been found in the primary sensory ganglia, it has been discovered that the activity of neighboring neurons elicits a functional cross-excitation in the somata of affected sensory neurons under normal conditions, indicating that non-synaptically released diffusible chemical messengers modify the neuronal excitability of the sensory ganglia. Non-synaptically released chemical mediators (substance P, ATP and cytokine) were derived from both the neurons and the satellite glia (e.g. neuron-neuron and neuron-glia interactions). More recent studies have suggested that modulation of neuronal excitability within sensory ganglia may trigger chronic pain (hyperalgesia/allodynia) via the autocrine/paracrine mechanism, and this augmented excitability of the primary afferent neurons may also cause the development of changes in the central pain-signaling neurons (central sensitization). Therefore, the present review focuses on the modulation of the neuronal signal by cross-talk in the sensory ganglia, particularly with regard to its contribution to inflammatory and neuropathic pain, and discusses the potential therapeutic target for the prevention of hyperalgesia/allodynia

Chapter VI - Brain white matter, especially in the periventricular region, is vulnerable to damage in hypoxic-ischemic and inflammatory conditions in the developing and adult brain. In the developing brain, damage to the axons and death of immature oligodendrocytes along with myelination disturbances has been widely reported. Release of excess glutamate and enhanced production of nitric oxide, free radicals and inflammatory cytokines have been implicated in this damage. Although microglial cells, the immune effector cells in the central nervous system, have been detected at sites of lesion in the periventricular white matter, their role in the development or amelioration of such lesions is not known. Another important factor which has remained elusive is the vascularity of this region. The blood vessels are thin-walled and are not surrounded completely by astrocyte end feet. In hypoxic-ischemic conditions, they dilate and their permeability is increased due to enhanced production of vascular endothelial growth factor (VEGF) which is known to cause fenestration of endothelial cells. Studies from the authors laboratory have shown increased production of VEGF and vacuolation of endothelial cells in the periventricular white matter in neonatal rats subjected to hypoxia.

White matter damage in the adult brain has been reported in many pathological conditions such as multiple sclerosis, Alzheimer's disease, Parkinson's disease, experimental allergic encephalomyelitis, stroke, traumatic brain injury and due to aging. Factors similar to those observed in the developing brain have been implicated in this damage. The disruption of blood-brain barrier (BBB) may be an important factor underlying production of damage to the white matter. Although many strategies for ameliorating the damage are being tested in animal models or in clinical trials, further research is needed to discover therapeutic agents to combat or prevent the damage. Methods of discovering successful drug delivery through the BBB may be timely, exciting and helpful in mitigating the damage.

Chapter VII - White matter neuronal fiber occupies most of the brain and is one of the bases of our cognitive functions. Recently, diffusion imaging has been applied to the study of neuronal fiber pathways, using tractography algorithms. With this technique, anatomical connections subserving higher cognitive functions in humans can be studied in a whole brain *in vivo*, in a relatively short amount of time. However, there are still a number of concerns that need to be addressed concerning the accuracy of current tractography techniques. Because of the limitations of today's methods, we need to be cautious in how we design tractography studies. Recent experimental data demonstrates that there are large-scale anatomical networks that serve as a basis of episodic memory in humans. In particular, there are two anatomical pathways between frontal and temporal lobes. The existence of segregated lateral and medial parieto-temporal pathways in successful memory retrieval was also suggested. Here, the author reviews a selection of designs and models currently being used in tractography to study cognitive functions. The author also discuss future applications of tractography.

Chapter VIII - Neuroplasticity (brain plasticity or cortical plasticity) refers to the modifications that occur in the organization of the brain, and in particular changes that occur to the location of specific information processing functions, as a result of acquired brain injury. Neuroplasticity allows the neurons to compensate for injury and disease and to adjust their activities in response to new situations or to changes in their enviroment.

Although many plastic changes in the brain and in the receptor organs occur only in developing organisms as a continuation of embryological development, i.e., after birth, neuroplasticity has been also found in the adult.

An increasing interest in neuroplasticity and nerve regeneration within the auditory receptor and pathway has developed in recent years. In humans, the activation of the auditory system starts in the 26th-28th gestational week and continues for several years. Nerve regeneration and neuroplasticity have been demonstrated in the adult auditory receptors as well as throughout the auditory pathway. Many aspects of plasticity in the auditory system, both in the developmental period and during adulthood, are similar to the plasticity in other sensory systems and share similar mechanisms: the auditory system, like other sensory systems, is plastic, and as such is able to reorganize its structure and function after partial or total loss of receptor function.

The receptors of the auditory system are very vulnerable and several pathogenetic mechanisms may significantly alter their function, leading first to temporary threshold shifts and, in the case of prolonged or very intensive exposure, to permanent threshold shifts. Ultimately, the receptors die out, triggering a sequence of changes in the whole auditory system starting with the spiral ganglion cells and ending in the auditory cortex. Regardless of whether the loss of receptors (hearing loss) is based on mechanical lesions, ablations, chemical lesions by ototoxic drugs, noise-induced damage of hair cells, or programmed loss connected with aging, the sequence of basic changes in the central auditory system is similar and results in the reorganization of the cortical projection areas and in some cases the subcortical areas.

A better understanding of the neuroplasticity strategies regarding cortical reorganization and neuroplasticity could potentially be helpful in the designing of effective treatment and management strategies (e.g.,: cochlear implants or stem cells), but also in really understanding the pathogenesis of auditory or language diseases (e.g.: deafness, tinnitus, dyslexia, etc.).

Chapter IX - Brain-derived neurotrophic factor (BDNF) plays a pivotal role in both central nervous system (CNS) development and neural plasticity throughout the lifespan. Indeed, abnormal expression and function of BDNF has been implicated in the pathogenesis of cognitive and mental disorders. Despite the known importance of BDNF function and gene expression in normal neural processes and CNS disorders, there has been little investigation into the molecular mechanisms responsible for complex BDNF transcriptional readout in the brain. In this commentary, the authors will shed light on a relatively new transcriptional regulatory mechanism, epigenetics, which may serve to regulate *bdnf* gene expression in the brain. They will highlight the epigenetic markings of the *bdnf* gene in adult learning and memory formation of both spatial and non-spatial tasks, by social experiences in early development, and in adult cognitive dysfunction following seizures. Such research offers novel concepts for understanding the transcriptional mechanisms subserving adult cognition and mental health, and furthermore promises avenues for therapeutic treatment of adult cognitive and mental dysfunction.

Chapter X - Some patients with a lesion to their primary visual cortex (V1) show a phenomenon called 'affective blindsight': they can correctly guess the emotional expression of a face presented to their blind field. This ability has been attributed to an evolutionary old subcortical route for processing of emotional information. Neuroimaging studies show that this route is also functioning in normal observers: emotional faces, made invisible using backward masking, still elicit responses of the amygdala via a subcortical pathway. However, contrary to blindsight patients, normal observers cannot overtly identify the expression of unseen emotional faces. In this chapter, the author introduces and discusses a model describing the relation between conscious and unconscious processing that explains this discrepancy.

Chapter XI - The classic view of neostriatum placed it among the structures that modulate motor functions. Under that view, research in the field of pathways shed light on the established connections with other nuclei of basal ganglia and motor cortex. However, converging data of the last decades emphasize more and more the role of the neostriatum in non-motor functions, such as behavior and memory. This evidence came from studies on animals with experimental lesions on the neostriatum and clinical studies on the non-motor symptoms of the diseases affecting primarily the neostriatum (Huntington's disease, Parkinson's disease, neuroacanthocytosis). Furthermore, recent data on psychiatric disorders underline the central role of the striatal pathology in the appearance of mood and behavioral manifestations. The anatomical substrate of these neostriatal functions is not yet very clear. The striatum is known to be closely connected to the cerebral cortex, especially the frontal cortex. Besides, the ventral striatum is connected with the limbic system and there are also indications about direct and indirect connections of the neostriatum with the hippocampus, thalamus and amygdala. Future research will possibly reveal the role of these connections in the modulation of behavior and in memory processing.

In: Neural Pathways Research
Editor: F. L. Pichler, pp. 1-30
ISBN: 978-1-60456-214-9

Chapter I

Metabotropic Glutamate Receptors and Neural Processing in the Olfactory System

***Thomas Heinbockel**[*1] **and Matthew Ennis**[2]*
[1]Department of Anatomy, Howard University College of Medicine,
Washington, DC 20059, USA
[2]Department of Anatomy and Neurobiology,
University of Tennessee Health Science Center,
Memphis, TN 38163, USA

Abstract

The main olfactory bulb (MOB) is the first relay station in the CNS for processing sensory information that comes from olfactory receptor neurons in the nasal epithelium. Neural signaling from olfactory nerve terminals to the principal MOB projection neurons, mitral and tufted cells (M/TCs), is in part mediated by glutamate acting at AMPA and NMDA ionotropic glutamate receptor (iGluR) subtypes. However, neuroanatomical studies demonstrate that M/TCs as well as inhibitory interneurons such as GABAergic granule cells (GCs) in the MOB express high levels of metabotropic glutamate receptors (mGluRs), suggesting that these receptors also play a role in the function of the MOB network. Recent experiments that employed a combination of patch-clamp electrophysiology, voltage-sensitive dye imaging, microsurgery, and use of mGluR gene knockout mice have illuminated the role of mGluRs in the MOB and point toward novel and potent regulatory roles of these receptors in shaping olfactory output from the MOB to higher olfactory centers in the brain.

* Correspondence: Thomas Heinbockel, Dept. of Anatomy, Howard University College of Medicine, 520 W Street, NW, Washington, DC 20059. tel.: (202) 806-9873. fax: (202) 265-7055. E-mail: theinbockel@howard.edu

Introduction

The principal role of the olfactory system is the detection of odors. As such the olfactory system plays an important role for food selection through interplay of olfactory and gustatory signals. Olfaction also serves in reproductive and neuroendocrine regulation and is known for its relevance in memory, aggression, emotion, social organization, and recognition of prey and predators (Shipley and Ennis, 1996). Odor molecules in the air that we breathe bind to odorant receptor (OR) proteins on the dendritic surfaces of olfactory receptor neurons (ORNs) and thereby activate ORNs in the nasal epithelium. The OR proteins form a large gene family (1000 genes in rodents, 350 in humans; Buck and Axel, 1991; Young et al., 2002). In the mouse, the nasal epithelium contains about 2000 populations of ORNs (Mombaerts 2006). Each of these ORN populations expresses only one allele of one of the 1000 intact OR genes. The unbranched axon of each ORN projects through the cribiform plate of the ethmoid bone to the ipsilateral MOB (Figure 1). The axons of the ORNs coalesce to form the olfactory nerve (ON) and olfactory nerve layer (ONL) of the MOB.

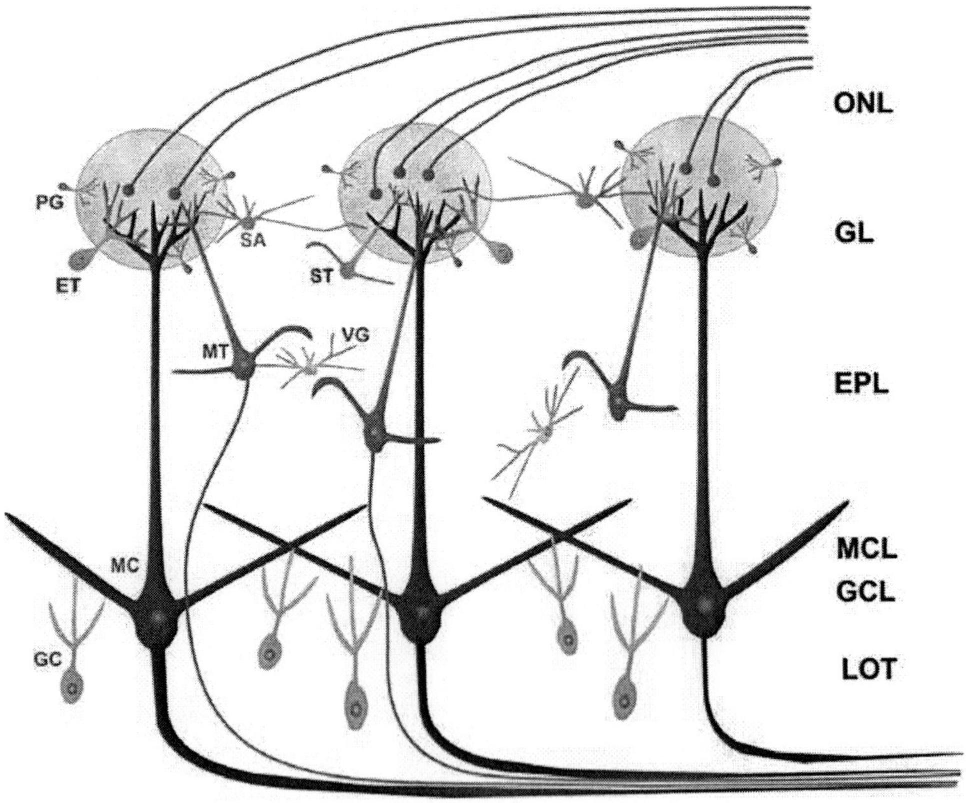

Figure 1. Circuitry diagram of the MOB. ONL – olfactory nerve layer; GL – glomerular layer; EPL – external plexiform layer; MCL – mitral cell layer; GCL – granule cell layer; LOT – lateral olfactory tract; PG – periglomerular cell; ET – external tufted cell; ST – superficial tufted cell; SA – short axon cell; MT – middle tufted cell; VG – Van Gehuchten cell; MC – mitral cell; GC – granule cell. From Ennis et al., 2007 with permission from Springer Science and Business Media..

Immediately deep to the ONL lies the glomerular layer (GL), which houses an array of 1600 to 1800 glomeruli, spheroidal neuropil structures (Klenoff and Greer 1998). Each glomerulus is 80 to 160 μm in diameter and surrounded by neurons and glia (Shipley and Ennis, 1996). The axon terminals of each ORN form synapses with MOB neurons in a single glomerulus. Axons of ORNs expressing the same OR converge onto two or very few topographically fixed glomeruli in the MOB, typically one glomerulus per half-bulb, and thus four glomeruli per mouse (Vassar et al., 1994; Mombaerts, 1999; Mombaerts et al., 1996, Mombaerts, 2006). ORs have two functional roles; they determine the odorant response profile of the ORN and they guide the projection of its axon to a specific glomerulus. In turn, the glomeruli form a map that mirrors ORN activity. Functional imaging and 2-DG data suggest that a given odor activates a distinct pattern of glomeruli across the MOB, and that different odors differentially activate different glomeruli (Astic and Saucier, 1983; Coppersmith et al., 1986; Kauer and Cinelli, 1993; Johnson et al., 1998; Mori et al., 1999). The olfactory system, therefore, must decipher patterns of glomerular activity for odor discrimination and recognition to take place.

Glomeruli are the sites of synaptic interactions among a diverse array of neuronal elements, e.g., between the axon terminals of olfactory afferents and MOB neurons and between MOB neurons themselves. Local interneurons in the GL, collectively referred to as juxtaglomerular cells (JGCs), send dendrites into the glomerular neuropil. Several JG cell types are distinguished based on their morphology, e.g., external tufted (ET) cells, short axon cells, and periglomerular (PG) interneurons. M/TCs are the principal output neurons of the MOB. Each M/TC sends a single apical dendrite into one glomerulus. ORNs make direct synaptic contact with M/TCs and with JGCs (Pinching and Powell, 1971b). JGCs in turn synapse on other glomerular neurons. A large subset of PG cells contains dopamine (Halasz et al., 1981, Davis and Macrides, 1983, McLean and Shipley, 1988) and/or GABA (Ribak et al., 1977). PG cells presynaptically inhibit ORNs through GABAergic and dopaminergic transmission (Hsia et al., 1999; Ennis et al., 2001). PG cells form reciprocal dendrodendritic synapses with M/TCs (Pinching and Powell, 1971b; Shipley and Ennis, 1996). Centrifugal fibers originating from cells in higher olfactory centers project their axons into MOB glomeruli and provide modulatory feedback in different layers of the MOB, including the GL.

The MOB is the initial site that performs neural computations to decipher patterns of ORN activity. The relatively recent advent of mammalian slice preparations has begun to reveal the intrinsic neuronal properties and synaptic mechanisms that govern information processing in the MOB network. The MOB slice preparation was found to provide an adequately preserved intrinsic MOB circuitry including functional synaptic connections (Nickell et al., 1994, Liu and Shipley, 1994; Aroniadou-Anderjaska et al., 1997; Keller et al., 1998; Aroniadou-Anderjaska et al., 1999a). The MOB slice preparation has therefore greatly facilitated the functional exploration and understanding of the mammalian olfactory system. We now know that iGluRs mediate transmission at the synapse between ORNs and M/TCs (Ennis et al., 1996; Aroniadou-Anderjaska et al., 1997), but intrinsic membrane properties of MCs regulate whether such synaptic input to the MC apical dendrite generates somatic spikes that are relayed to higher order olfactory structures (Chen et al., 1997; Bischofberger and Jonas, 1997; Heyward et al., 2001).

The cell sparse external plexiform layer (EPL) is located immediately deep to the GL. Both MCs and TCs are excitatory and glutamatergic, and they release glutamate at their apical dendrites in the GL, their lateral dendrites in the EPL, and at axon terminals. MCs can readily be identified with NIR-DIC microscopy in the MOB slice, by their large cell bodies located in the distinct narrow band of the MC layer (MCL), deep to the EPL (Figure 2). With electrophysiological methods they can be identified by intrinsic membrane properties, such as membrane potential bistability (Heyward et al. 2001) and input resistance (~200 MΩ). Location and bistability distinguish MCs from TCs in the adjacent EPL, and from the small, high resistance (typically >1GΩ) GCs present in the MCL (Heyward et al. 2001; Schoppa et al., 1998, Heinbockel et al., 2007a, b).

The GCs in the MOB are the largest population of GABAergic inhibitory interneurons which modulate and shape the output of the MOB to higher-order olfactory structures (Shepherd et al., 2004). The cell bodies of the deep GCs are densely packed within the GC layer (GCL); superficial GCs (sGCs) occur more superficially, interspersed with MC somata within the MCL. The population of sGCs is substantial. The rat MCL contains ~100,000 of these cells (Frazier and Brunjes, 1988), but only ~40,000 MCs (Meisami, 1989). The apical dendrites of both GC subtypes extend radially into the EPL, where they form dendrodendritic synapses with M/TCs. The M/TC-GC dendrodendritic synapses have been well studied. Sensory transmission from ORN terminals to the M/TCs is mediated by glutamate acting at AMPA and NMDA ionotropic glutamate receptor subtypes (Bardoni et al., 1996; Ennis et al., 1996, 2001; Aroniadou-Anderjaska et al., 1997; Chen and Shepherd, 1997; Keller et al., 1998).

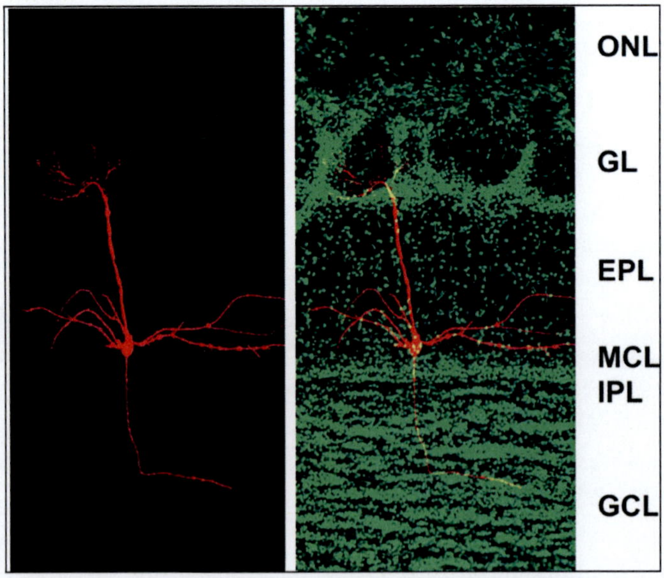

Figure 2. Confocal micrographs of a section of the adult mouse main olfactory bulb with a single mitral cell intracellularly filled with biocytin (red) and nuclei stained with Sytox Green (green). The mitral cell soma is located in the mitral cell layer (MCL). One apical dendrite of the mitral cell reaches into one glomerulus and several lateral dendrites span the main olfactory bulb. ONL – olfactory nerve layer; GL – glomerular layer; EPL – external plexiform layer; MCL – mitral cell layer; IPL – internal plexiform layer; GCL – granule cell layer.

Glutamate released from M/TCs, acting at both ionotropic receptor subtypes, mediates dendrodendritic transmission at synapses with JGCs and GCs (Bardoni et al., 1996; Isaacson and Strowbridge, 1998; Schoppa et al., 1998; Aroniadou-Anderjaska et al., 1999a). Glutamate released by M/TC lateral dendrites activates AMPA and NMDA receptors on GCs, which in turn, drives GABAergic feedback/feedforward inhibition of M/TCs (Aroniadou-Anderjaska et al., 1999a; Chen et al., 2000; Isaacson and Strowbridge, 1988; Schoppa et al., 1998). However, the properties of intrinsic K^+ conductances bias the kinetics of GC responses such that GABA release is thought to be driven entirely by the NMDA receptors (Schoppa and Westbrook, 1999). Recent electrophysiological studies suggest that iGluRs mediate recurrent excitatory interactions among apical and lateral dendrites of MCs/TCs (Isaacson, 1999; Aroniadou-Anderjaska et al., 1999b; Carlson et al., 2000; Salin et al., 2001; Schoppa and Westbrook, 2001).

Although our understanding of the roles of iGluRs in the MOB has significantly improved over the last few years, very little is known about the functions of metabotropic glutamate receptors (mGluRs), the other class of glutamate receptors that are expressed at unusually high density in the MOB. Until recently, this class of receptors was poorly understood, in general due to lack of tools to distinguish the functions of multiple mGluR subtypes. This situation has dramatically changed; there are now specific mGluR pharmacological reagents, as well as mGluR-specific knockout mice, which can be used to investigate these receptors. To date, a wide variety of physiological responses and ion channels have been reported to be modulated by mGluRs (Conn and Pin, 1997; Schoepp et al., 1999).

mGluRs: Structure and Signal Transduction

mGluRs are G protein-linked receptors that are coupled to a variety of second messenger systems (Suzdak et al., 1994). mGluR responses can directly modulate ion-channel permeability via direct G-protein-coupled membrane mechanisms or indirectly via G-protein-activated cytoplasmic second messenger pathways. Eight mGluR subtypes have been cloned thus far and are subdivided into three groups: Grp I (mGluR1 and mGluR5), Grp II (mGluR2 and mGluR3) and Grp III (mGluR4, mGluR6, mGluR7 and mGluR8). These divisions are based on the second messenger linkage, sequence homology and pharmacological sensitivity of the receptors.

The three groups of mGluRs utilize distinct signal transduction mechanisms. Grp I mGluRs (mGluR1, mGluR5) are linked to the G-protein G_q and are, therefore, pertussis toxin (PTX)-insensitive. Activation of Grp I mGluRs stimulates phospholipase C (PLC) and phosphoinositide (PI) hydrolysis, and it subsequently leads to the release of Ca^{2+} from intracellular stores (Murphy and Miller, 1988). mGluR1 has also been shown to stimulate cAMP formation and arachidonic acid release, while mGluR5 does not. Grp II (mGluR2, mGluR3) and Grp III (mGluR4, mGluR6, mGluR7 and mGluR8) mGluRs are negatively coupled to the cAMP cascade and inhibit forskolin-induced cAMP formation (Schoepp et al., 1992). This effect is inhibited by PTX, which indicates a linkage to G_i/G_o proteins.

mGluRs in the MOB

Of the 8 cloned metabotropic glutamate receptors, 6 have been found in the MOB (Figure 3). Although mGluRs are densely expressed by MOB neurons, it is important to point out that mGluRs have not been shown to be present on ON terminals.

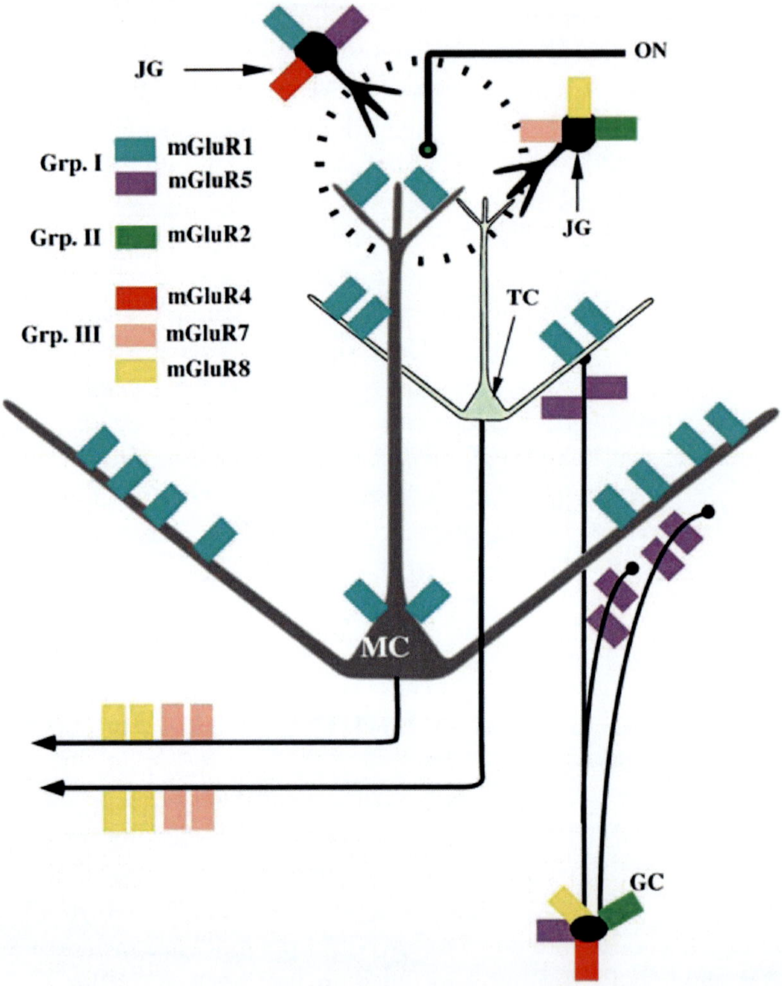

Figure 3. Expression patterns of mGluRs on the cell types of the MOB. JG – juxtaglomerular cells; ON – olfactory nerve; TC – tufted/external tufted cell; MC – mitral cell; GC – granule cell; Grp. I – Group I mGluRs. Adapted from Ennis et al., 2007 with permission from Springer Science and Business Media.

Glomerular layer (GL) neurons: GL neurons arecollectively termed juxtaglomerular cells (JGCs), and they express several types of mGluRs. Deciphering mGluR expression in the GL from the literature is difficult. There are numerous morphological JG cell types, which include ET cells, short axon cells, and PG cells; mGluR expression has not been systematically examined in these cell types. Thus, it is not yet possible to determine mGluR distribution in the GL from existing anatomical studies. However, physiological studies suggest that ETs have the same mGluR makeup as MCs, and TCs in the EPL; i.e., they express high levels of mGluR1, and lower levels of mGluR7 and mGluR8 (Ohishi et al., 1993,1998; Kinzie et al., 1995; Saugstad et al., 1997; Heinbockel et al., 2004). Other, as of yet unidentified JGCs have been reported to express low-to-moderate levels of Grp I (mGluR5) Grp II (mGluR2) and Grp III (mGluR4/7/8) receptors (Romano et al., 1995; Onishi et al., 1998, 1993; Saugstad et al., 1997).

Mitral/Tufted Cells: M/TCs express high levels of mGluR1 (Grp. I; Masu et al., 1991; Martin et al., 1992; Shigemoto et al., 1992; van den Pol, 1995; Heinbockel et al., 2004). Electron microscopy studies demonstrate that mGluR1 is present on the somata, apical dendrite and lateral dendrites of M/TCs (van den Pol, 1995). This expression pattern suggests that mGluR1 could mediate M/TC responses to glutamatergic inputs from ON terminals and/or could function as auto- or heteroreceptors for glutamate released from the apical or lateral dendrites of M/TCs during glutamate-mediated recurrent excitation (i.e., spillover) among MC apical dendrites (Aroniadou-Anderjaska et al., 1999b; Isaacson, 1999; Carlson et al., 2000; Friedman and Strowbridge, 2000; Schoppa and Westbrook, 2001; Salin et al., 2001). M/TCs do not express Grp. II receptors (mGluR2/3), which contrasts markedly with the expression of mGluR2 in MCs in the AOB (Ohishi et al., 1993, 1998). MOB M/TCs express mRNA for mGluR7 and mGluR8 (Kinzie et al., 1995; Saugstad et al., 1997), although immunocytochemical studies indicate that these receptors are present selectively on the axon terminals of M/TCs in piriform cortex (Kinoshita et al., 1998; Wada et al., 1998).

Granule Cells: GCs express the highest levels of mGluR5 (Grp I) in the brain (Romano et al., 1995). At the cellular level, mGluR5 staining is strongest within the deep GCL (Sahara et al., 2001). Electron microscopy studies have shown that mGluR5 is localized on portions of GC dendrites in the external plexiform layer (EPL) apposed to presynaptic glutamatergic synapses from M/TC lateral dendrites (van den Pol, 1995). This suggests that mGluR5 may mediate, at least in part, responses of GCs to glutamatergic inputs from M/TCs. GCs in the GCL do not appear to express mGluR1 (van den Pol, 1995; Heinbockel et al., 2004). The GCL stains only weakly for mGluR1 immunocytochemistry (Martin et al. 1992; Sahara et al. 2001; van den Pol 1995), and faint mGluR1 expression has been observed in this layer with in situ hybridization (Shigemoto et al. 1992). GCs may express low to moderate levels of mGluR2 (Grp. II, Onishi et al., 1998, 1993), and low levels of mGluR4 and mGluR7 (Grp. III; Kinzie et al., 1995; Onishi et al., 1995; Saugstad, 1997; Wada et al., 1998). The precise cellular localization of these Grp. II/III mGluRs on GCs are not known. The expression of mGluRs by other cell types in the GC layer is unknown.

mGluR Actions in Cultured MOB Neurons

Until recently, information available on the actions of mGluRs in MOB came from a few studies in dissociated cultures of frog or neonatal rat neurons. Activation of Grp I mGluRs with (±)-1-aminocyclopentane-trans-1,3-dicarboxylic acid (ACPD) or quisqualate was reported to increase Ca^{2+} release from internal stores in rat and frog M/TCs and unidentified interneurons (Geiling and Schild, 1996; Carlson et al., 1997). The Ca^{2+} increase was thought to originate in the dendrites of M/TCs, suggesting it could play a role in presynaptic facilitation of glutamate release. Consistent with this possibility, Schoppa and Westbrook

(1997) reported that activation of Grp I mGluRs presynaptically increased glutamate release from M/TCs. Grp I mGluRs were additionally found to depolarize M/TCs by closing K^+ channels (Schoppa and Westbrook, 1997). Activation of Grp II (mGluR2/3) receptors was also reported to inhibit high voltage-activated Ca^{2+} currents in cultured frog olfactory bulb interneurons, and to decrease spontaneous GABAergic inhibitory postsynaptic currents (IPSCs) in mitral/tufted cells (Bischofberger and Schild, 1996).

Activation of Grp II and III mGluRs was reported presynaptically to inhibit glutamate release from putative M/TCs in culture (Trombley and Westbrook, 1992; Schoppa and Westbrook, 1997). Conflicting reports attributed the Grp III-evoked inhibition of glutamate release to Ca^{2+}-dependence (Trombley and Westbrook, 1992) or Ca^{2+}-independence (Schoppa and Westbrook, 1997). However, adult rat M/TCs do not express Grp II mGluRs (Ohishi et al., 1993; 1998), and the inhibition of glutamate release evoked by Grp II agonists could not be blocked by specific Grp II receptor antagonists (Schoppa and Westbrook, 1997). Anatomical studies demonstrate that Grp III mGluRs are present only on the axon terminals of M/TCs (Kinoshita et al., 1998; Wada et al., 1998).

mGluR Actions in GL Neurons

The role of mGluRs in the GL remains poorly understood. Specific studies on mGluR actions on M/TC apical dendrites in the GL and on ON-evoked synaptic responses in these cells are discussed below. ET cells in vitro exhibit a distinctive rhythmic bursting activity pattern that is generated by intrinsic membrane properties (Hayar et al., 2004a,b, 2005). Preliminary studies show that the strength of this rhythmic busting is enhanced by Group I/II mGluR agonists and dampened by mGluR antagonists. mGluR1 agonists enhance a TTX-sensitive persistent sodium current necessary for ET cell rhythmic bursting, and also induces a non-selective cation current in these cells (Dong et al., 2007b). Both of these currents appear to be engaged by activation of mGluR1 and mGluR5. These preliminary pharmacological studies indicate that ET cells may express a greater variety of mGluRs than M/TCs because they exhibit excitatory responses to Grp I and II mGluR agonists. Indirect evidence suggests that inhibitory PG cells in the GL are excited by mGluR1 agonists. DHPG applied focally into the GL increases GABAergic IPSCs in ET cells and MCs (Dong et al., 2007a). The increased IPSC activity is likely due to DHPG-evoked excitation of PG cells.

mGluR Effects on Mitral Cells

mGluR Effects on Intrinsic Mitral Cell Properties

Until recently, limited information existed about the physiological actions and function of Group I mGluRs on either MCs or TCs. The goal of our studies, therefore, was to investigate the actions of Group I mGluRs on MC excitability and responsiveness to sensory input from ORN terminals. To address these issues, we studied the consequences of activation or blockade of mGluRs on MCs. Our results demonstrated an important role of mGluR1 in modulating the excitability of MCs (Heinbockel et al., 2004). Broad spectrum (i.e., ACPD) and selective Grp I (i.e., (RS)-3,5-dihydroxyphenylglycine [DHPG]) mGluR agonists potently and dose dependently depolarized (14 mV) and increased the firing rate of MCs. Neither the Grp II agonist (2S,3S,4S)-CCG/(2S,1'S,2'S)-2-(carboxycyclopropyl)glycine (L-CCG-I) nor the Grp III agonist L(+)-2-amino-4-phosphonobutyric acid (AP4) depolarized MCs or increased their firing rates. Similarly, neither L-CCG-I nor AP4 elicited significant shifts of the holding current in voltage clamp recordings. The effects of ACPD and DHPG were due to direct activation of mGluR1 on MCs as they persisted undiminished in the presence of blockers of synaptic transmission (NMDA receptor blocker [DL-2-amino-5-phosphonopentanoic acid (APV)], AMPA receptor blocker [6-cyano-7-nitroquinoxaline-2-3-dione (CNQX)], and $GABA_A$ receptor blocker [gabazine]) and/or TTX. Bath application of (RS)-α-methyl-4-carboxyphenylglycine (MCPG), a nonselective Grp I/II mGluR antagonist, blocked the effects of ACPD. LY341495, a blocker of all mGluR subtypes at micromolar concentrations (Kingston et al., 1998), prevented or reversed the effect of ACPD. The effects of DHPG were also blocked by a preferential mGluR1 antagonist, LY367385, which has negligible actions on Grp II and III mGluRs, and antagonizes mGluR5 only at concentrations in excess of 100 μM (Clark et al., 1997; Salt et al., 1999).

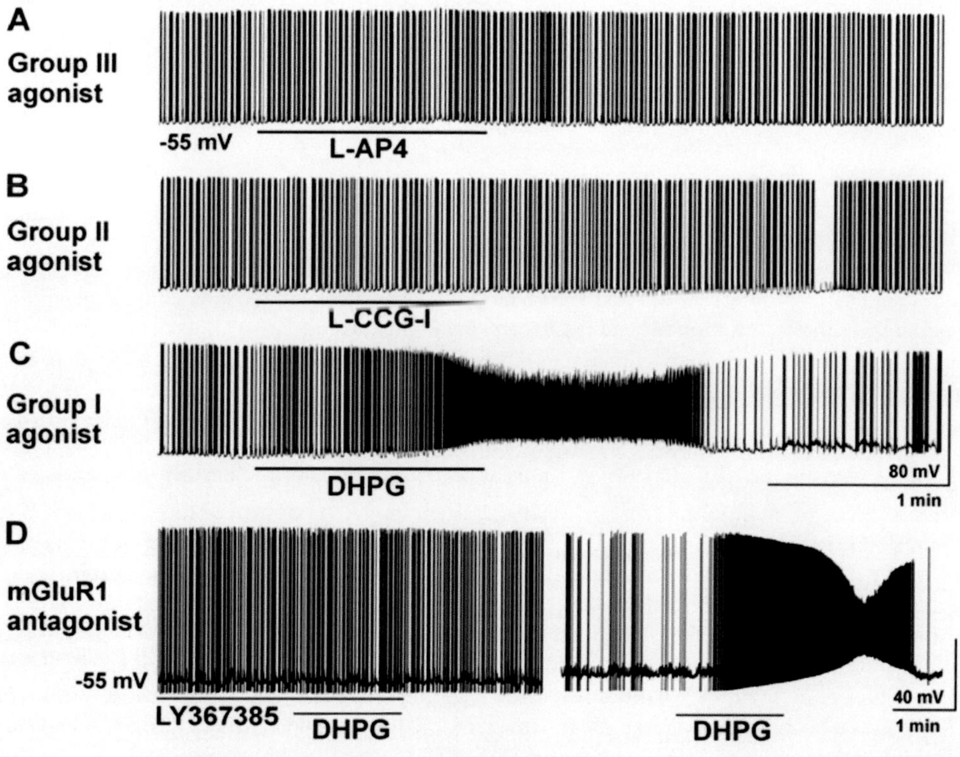

Figure 4. Group I mGluR agonist DHPG activates MCs and the selective mGluR1 antagonist LY367385 blocks the effect of DHPG. Modified from Heinbockel et al., 2004 with permission of The American Physiological Society.

As described above, Grp I mGluRs include both the mGluR1 and mGluR5 subtypes. MCs express high levels of mGluR1. However, mGluR5 appears to be absent or expressed at very low levels in MCs. This suggests that the effects of DHPG are mediated via activation of mGluR1 on MCs, as supported by our pharmacological data. To rule out a role of mGluR5 in MC excitability, we investigated the effects of the preferential mGluR5 agonist (RS)-2-chloro-5-hydroxyphenylglycine (CHPG), because selective mGluR1 agonists are not available. The mGluR5 agonist CHPG was without excitatory effect on MCs, suggesting that the effects of ACPD and DHPG were indeed mediated through the mGluR1 receptor subtype.

Thus our pharmacological results suggested that the effect of DHPG on MCs was mediated by activation of mGluR1. To confirm this, we also used mice with targeted deletion of the mGluR1 receptor gene (Figure 5), and our results support the anatomical results of others (Conquet et al. 1994; Chiamulera et al. 2001). Physiologically, the activating effects of mGluR agonists on MCs were absent in transgenic animals with targeted deletion of the mGluR1 gene and persisted in transgenic animals with targeted deletion of the mGluR5 gene (Heinbockel et al., 2004).

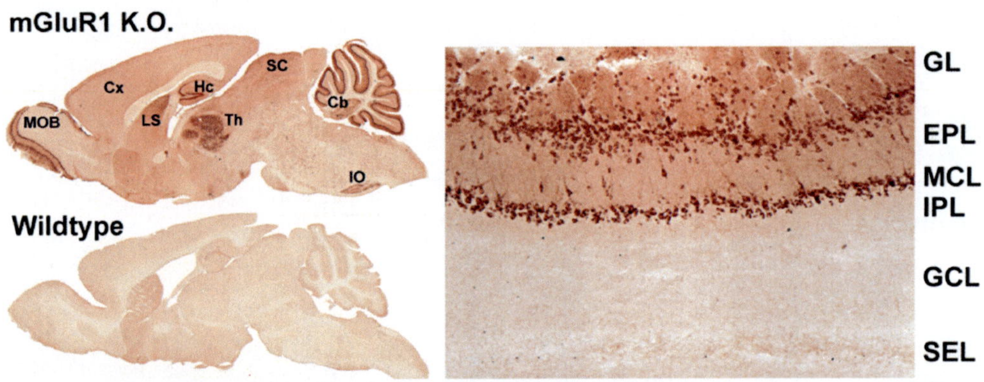

Figure 5. mGluR1 distribution in KO and WT mice. Conventional immunohistochemical techniques were used to map the distribution of neurons expressing β-galactosidase (β-gal) in tissue sections harvested from mGluR1 transgenic animals. In this transgenic strain, the mGluR1 gene sequence has been replaced with that for β-galactosidase (β-gal) by homologous recombination; i.e., in KO mice β-gal is present only in cells that would normally express mGluR1. The distribution of β-gal+ neurons in the MOB and other brain structures is consistent with previous reports on the distribution of mGluR1 (Martin et al., 1992; Shigemoto et al., 1992; van den Pol, 1995; Petralia et al., 1997; Sahara et al., 2001).) In sections of mGluR1 mutant (-/-) mice, β-gal staining is prominent in the cerebellum (Cb), hippocampus (Hc), thalamus (Th), lateral septum (LS), superior colliculus (SC), inferior olive (IO), main olfactory bulb (MOB), and throughout the cortex (Cx). In the MOB, strong and uniform staining was present in MCs. Tufted-like cells in the external plexiform layer, juxtaglomerular cells in the glomerular layer proper and at the glomerular layer-external plexiform layer border were also robustly stained. By contrast, no staining was observed in the internal plexiform or granule cell layers. β-gal staining was absent in mGluR1 WT littermates. GL – glomerular layer; EPL – external plexiform layer; MCL – mitral cell layer; IPL – internal plexiform layer; GCL – granule cell layer; SEL – subependymal layer. From Heinbockel et al., 2004; with permission of The American Physiological Society.

mGluR1-induced depolarizations and inward currents in CNS neurons, including MCs in culture, have been ascribed to the inhibition of K^+ currents (Charpak et al., 1990; Guérineau et al., 1994; Schoppa et al., 1997; Anwyl, 1999). In our experiments, both ACPD and DHPG elicited an inward current in voltage clamp recordings from both rat and mouse MCs (current amplitude: -110 pA) (Heinbockel et al., 2004). The EC_{50} of the DHPG-evoked current, 27.4 μM, was comparable to that reported elsewhere (Schoepp et al., 1999). This inward current, similar to DHPG currents in neurons in several brain areas (see Anwyl 1999 for review) had multiple components. One component of the inward current was mediated by a K^+ conductance, as it was abolished by TEA and Cs^{2+} (in the presence of BAPTA). This result agreed in part with previous observations by Schoppa and Westbrook (1997) that intracellular Cs^{2+} reduces inward currents evoked by DHPG in cultured MCs. A second component required a rise in intracellular Ca^{2+} as it was abolished by BAPTA-AM. The identity of this BAPTA-sensitive current remains to be determined but may involve Ca^{2+}-dependent nonselective cation conductances (Guerineau et al., 1995, Congar et al., 1997, Crepel et al., 1994; Raggenbass et al., 1997) or an electrogenic Na^+/Ca^{2+} exchanger (Keele et al., 1997; Lee and Boden, 1997; Staub et al., 1992). A third component, present at membrane potentials from –60 to –40 mV in the presence of BAPTA, TEA and Cs^{2+}, appeared to involve voltage gated Ca^{2+} channels, as it was eliminated by Cd^{2+} and Ni^{2+}.

mGluR Actions on Mitral Cell Spontaneous and Olfactory Nerve Evoked Discharge

As reported previously (Heyward et al., 2001), MCs *in vitro* exhibit intrinsic membrane potential bistability, spontaneously alternating between two discrete membrane potentials differing by ~10 mV: a more depolarized "upstate", perithreshold for spike generation, and a more hyperpolarized "downstate" devoid of spiking. Membrane bistability significantly influences MC excitability and sensory responses (Heyward et al., 2001). Although bistability is generated by MC intrinsic membrane properties (Heyward et al., 2001) it can be modulated by extrinsic input. Application of mGluR antagonists dramatically altered bistability in rat MCs (Heinbockel et al., 2004) MCPG, LY367385, or LY341495 significantly prolonged the duration of the up- and downstates, specifically, with a shift toward more time spent in the down-state. A hyperpolarizing shift of the upstate membrane potential also occurred. The antagonists significantly reduced the spontaneous firing rate. These results suggest that bistability and MC excitability are tonically modulated, at least in part, by endogenously released glutamate acting at mGlu1 receptors.

Bistability is an important determinant of MC responses to ON input. In the perithreshold upstate, MCs readily launch short latency spikes in response to ON input (Heyward et al., 2001; Hayar et al., 2001). By contrast, in the hyperpolarized downstate, ON input often fails to elicit spikes in MCs or it evokes spikes at long latencies. The increase in latency corresponds to the time to depolarize from the downstate to ramp or upstate potentials close to spike threshold (Heyward et al., 2001). Thus, events that prolong the downstate and/or produce a hyperpolarizing shift of the upstate membrane potentials should render MCs less likely to launch spikes and/or increase the latency of them, in response to ON input. Both of these effects were observed during inactivation of mGluR1 (Heinbockel et al., 2004; Figure 6): ON-evoked spiking in MCs was significantly reduced and the latency of evoked spikes increased. For these experiments, the ONL was focally stimulated with a bipolar electrode, thereby mimicking ORN input to MCs in the MOB. Consistent with *in vitro* findings in rat sensorimotor cortex (Bandrowski et al. 2003), these results indicate that ambient glutamate levels tonically modulate MC excitability via activation of mGluR1.

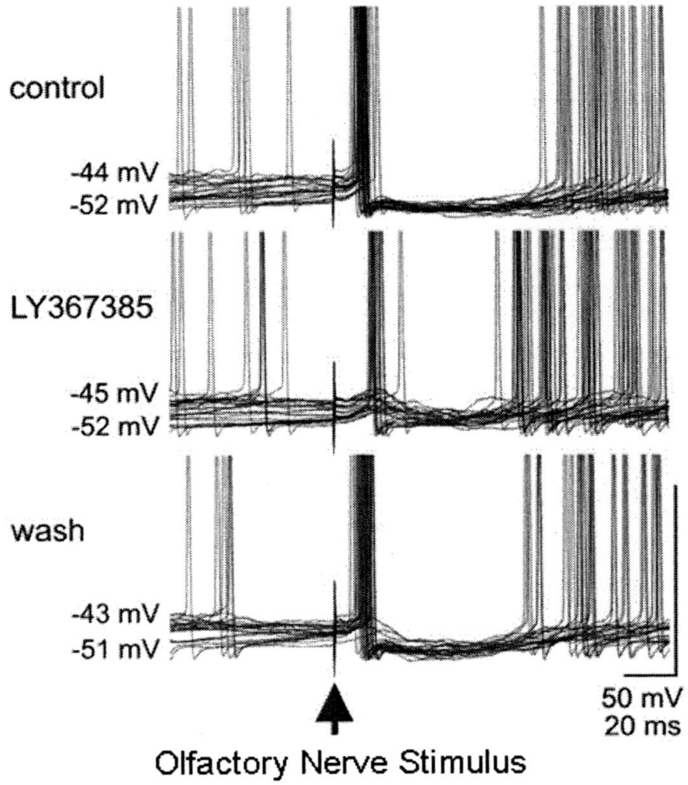

Figure 6. mGluR antagonist attenuates olfactory nerve-evoked excitation of mitral cells. Current clamp recordings show that single olfactory nerve stimuli (stimulus artifact marked by arrow in third trace) evoked short latency spikes in a mitral cell. In the presence of the mGluR antagonist LY367385 (middle trace), the probability of olfactory nerve-evoked short-latency spikes was reduced while the latency to the first spike was increased. These effects were reversible following washout of the antagonist (bottom trace). Twenty-four trials are overlaid in each trace, and spikes are truncated Modified from Heinbockel et al., 2004 with permission of The American Physiological Society.

Olfactory Nerve-Evoked, mGluR-Mediated Synaptic Responses in Mitral Cells

Subsequent work by several groups investigated mGluR-mediated components of MC synaptic responses to ON input. MCs express high levels of mGluR1 on their apical dendrites postsynaptically at ON synapses (Masu et al., 1991; Van den Pol, 1995; Martin et al., 1992; Shigemoto et al., 1992; Sahara et al., 2001). ON-evoked excitatory postsynaptic potentials elicited by single shocks in a subset of MCs (~20%) exhibited a small mGluR1-mediated component in normal physiological conditions (De Saint Jan and Westbrook, 2005, 2007; Yuan and Knöpfel, 2006a,b; Ennis et al., 2006). When iGluRs are blocked, the small ON-evoked mGluR1 synaptic response was increased by glutamate uptake blockers such as D-threo-*b*-benzyloxyaspartate (TBOA) and D-threo-*b*-hydroxyaspartic acid (THA). Thus, in agreement with studies at other CNS synapses, mGluR1-mediated synaptic responses are tightly regulated by glutamate uptake mechanisms.

Previous studies in the cerebellum and other CNS regions (Batchelor and Garthwaite, 1977; Shen and Johnson, 1997; Tempia et al., 1998, 2001; Reichelt and Knöpfel, 2002; Huang et al., 2004; Karakossian and Otis, 2004) demonstrated that mGluRs are preferentially activated by high frequency synaptic activity. Therefore, we determined the contribution of mGluR1 in MC responses elicited over a range of ON stimulation frequencies comparable to odor-evoked firing frequencies of ORNs *in vivo* (Duchamp-Viret et al., 1999). In voltage clamp recordings, the average EPSC evoked in normal media by brief trains of ON stimulation (6 pulses at 50 Hz) was not affected by the non-selective mGluR antagonist LY341495 or the mGluR1 specific antagonist LY367385 (Ennis et al., 2006). In a subset (36%) of cells however, ON-evoked responses were attenuated, but the incidence of an mGluR1-mediated component was comparable to that elicited by single pulse ON stimulation. Other studies using calcium imaging (Yuan and Knöpfel, 2006b) have reported a higher incidence of high frequency ON-evoked, mGluR1-mediated responses in MC apical dendrites in similar recording conditions (i.e., iGluR blockade). In the presence of iGluR blockers, application of the glutamate uptake inhibitors THA and TBOA revealed large amplitude, long duration responses to high frequency ON stimulation (Ennis et al., 2006; Yuan and Knöpfel, 2006a). In contrast, responses elicited by backpropogating action potentials or antidromic activation of MCs were unaffected (Ennis et al., 2006; Yuan and Knöpfel, 2006a). The magnitude of the ON-evoked responses elicited in the presence of THA-TBOA was dependent on stimulation intensity and frequency, and was maximal during high frequency (50 to 100 Hz) bursts of ON spikes (Ennis et al., 2006; Yuan and Knöpfel, 2006a; De Saint Jan and Westbrook, 2007), which can occur in response to odor stimulation. ON-evoked responses elicited in the presence of THA-TBOA were significantly reduced or completely blocked by LY341495 or LY367385. These results demonstrate that glutamate transporters tightly regulate access of synaptically evoked glutamate from ON terminals to postsynaptic mGluR1s on MC apical dendrites. More recent findings indicate that the ON-evoked, mGluR1-mediated response component is facilitated by disynaptic mechanisms involving intraglomerular glutamate spillover (De Saint Jan and Westbrook, 2007). Additionally, ON stimulation activates glial cells in the GL and these responses are dependent on mono- and or disynaptic activation of mGluR1 (De Saint Jan and Westbrook, 2007; Rieger et al., 2007).

Additional studies have demonstrated that single ON shocks or NMDA receptor agonists elicit slow (~ 2Hz) oscillations in MCs (Schoppa and Westbrook, 2001). These oscillations are synchronous among MCs whose apical dendrites extend into the same glomerulus. ON-evoked slow oscillations in MC apical dendrites are triggered by mGluR1 activation in the presence of iGluR blockers (Yuan and Knöpfel, 2006a). These slow oscillations were dampened or eliminated by mGluR1 antagonists, demonstrating that they are driven, in part, by activation of mGluR1. Taken together with the findings above, these results suggest that mGluR1 plays an important role in shaping slow oscillatory activity in MCs.

Other studies suggest that mGluRs participate in activity-dependent regulation of plasticity at ON-MC synapses. Lesion of ORN produces an initial increase followed by longer term (16 days) decrease in mGluR1 RNA in the MOB, without effects on AMPA or NMDA receptor expression (Ferraris et al., 1997; Casabona et al., 1998). Thus, mGluR1 expression in the MOB exhibits plasticity that is dependent upon ON sensory input. Electrophysiological studies show that certain stimulation patterns can produce long-term depression (LTD) at ON synapses (Muthoh et al., 2005). Such LTD can also be induced by DHPG, suggesting that Grp I mGluRs are involved.

mGluR Effects on Granule Cells

Given the high level of expression of mGluR5s on GCs, we hypothesized that mGluRs would modulate GC excitability and responsiveness to glutamatergic synaptic input from M/TCs and, as a result, modulate GC-mediated GABAergic inhibition of M/TCs. To test this hypothesis, we studied the consequences of activation or blockade of mGluR5 on GCs using whole-cell patch clamp recordings and voltage-sensitive dye imaging in MOB slices from rats, WT mice, and mice with targeted gene deletions of Grp I mGluRs.

mGluR Effects on Intrinsic Granule Cell Properties: Direct and Indirect Effects

Activation or inactivation of mGluR5 directly increased or decreased, respectively, the excitability of GABAergic GCs in the MOB (Heinbockel et al., 2007a). The results obtained from GCs in the GCL (deep GCs) indicated that mGluR5 modulates the strength of GC responses to glutamatergic inputs from M/TCs and GC-mediated dendrodendritic feedback inhibition in the MOB. These findings suggest that activation of mGluR5 by glutamate released from M/TCs amplifies lateral inhibition and therefore may increase contrast in the MOB network.

In normal media, ACPD or DHPG robustly depolarized (~20 mV) and increased the firing rate of GCs (Figure 7). These effects were mediated by direct activation of mGluR5 on GCs and indirect excitation (i.e., increased glutamate release) due to activation of mGluR1 on MCs. In the presence of synaptic blockers, the depolarizing response to the mGluR agonists persisted, although it was reduced in amplitude. The latter result indicates that ACPD and DHPG directly depolarize GCs. The effects of ACPD and DHPG were identical in slices harvested from rats and WT mice.

Several results indicate that the direct actions of ACPD and DHPG on GCs are mediated exclusively via activation of mGluR5. First, the mGluR1 antagonist LY367385 did not block the effect of DHPG on GCs, whereas it completely eliminated DHPG-evoked, mGluR1-mediated excitation of MCs at identical concentrations (Heinbockel et al., 2004). Second, DHPG had no discernible effects on GCs in mGluR5 knockout mice, while it readily activated GCs in mGluR1 knockout mice. Third, neither Grp II nor Grp III mGluR agonists elicited currents in GCs.

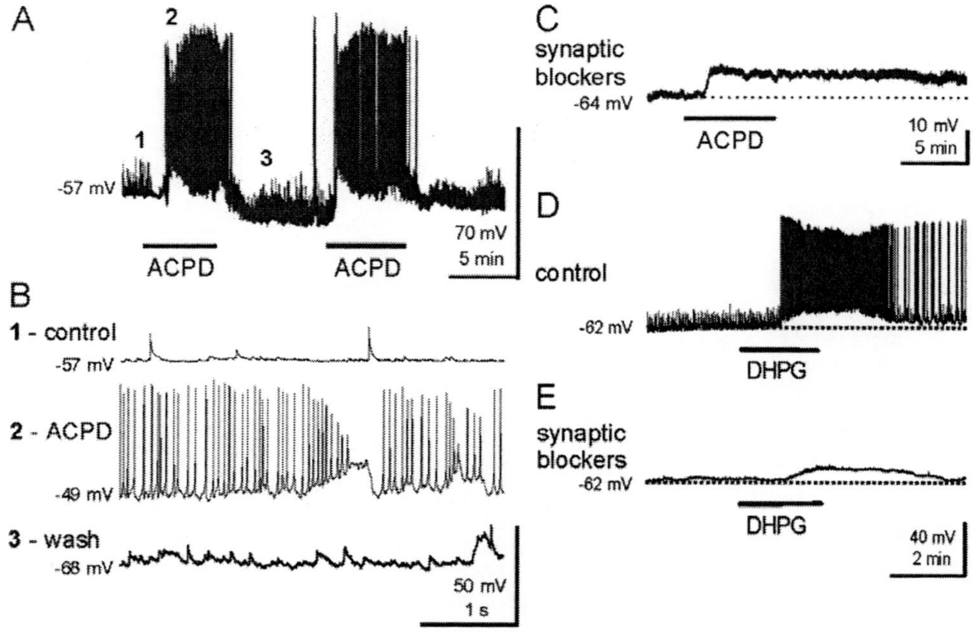

Figure 7. mGluR agonists depolarized and increased the firing rate of GCs. A, B: Response of a GC to ACPD, shown at extended time scale in (B). Traces 1-3 correspond to timepoints in (A). C: In blockers of fast synaptic transmission the cell in (A) had reduced baseline spontaneous synaptic activity and a smaller response to ACPD. D, E: DHPG activated another GC. Blockers of fast synaptic transmission reduced baseline spontaneous synaptic activity (spontaneous EPSCs) and DHPG evoked a more moderate depolarization. From Heinbockel et al., 2007a with permission of The American Physiological Society.

DHPG elicited an inward current (current amplitude: -8 pA) in GCs (Heinbockel et al., 2007a). The I-V relationship of the DHPG current in GCs was nearly linear between -140 mV and -70 mV and reversed in polarity at approximately –100 mV. The current appeared to decrease at potentials positive to -60 mV, suggesting that it involved multiple components, similar to DHPG currents in several other brain areas (Anwyl, 1999).

The identity of the current present at more positive potentials remains to be determined, but might involve voltage-gated Ca^{2+} channels, Ca^{2+}-dependent nonselective cation conductances (Guerineau et al., 1995, Congar et al., 1997, Crepel et al., 1994; Raggenbass et al., 1997) or electrogenic Na^{+}/Ca^{2+} exchangers (Keele et al., 1997; Lee and Boden, 1997; Staub et al., 1992). A major component of the inward current appears to be mediated by closure of K^{+} channels as it was abolished by K^{+} channel blockers. Consistent with this, DHPG increased the membrane input resistance of GCs in normal ACSF, but it did not alter it in the presence of K^{+} channel blockers. Previous studies demonstrate that GCs have a prominent A-like K^{+} current, I_A (Schoppa and Westbrook, 1999). A reduction of I_A is unlikely to account fully for the DHPG-evoked inward current, because I_A in GCs is sharply voltage dependent, activates at potentials positive to -45 mV, and inactivates rapidly (Schoppa and Westbrook, 1999). By contrast, the DHPG-evoked current varied linearly with voltages negative to -45 mV, and its steady state amplitude was measured at a time point several hundred ms after inactivation of I_A. Given their similar properties (e.g., linear I-V relationship), a reasonable mechanism for the DHPG-evoked current is a decrease in a leak

K^+ current. In agreement with this possibility, activation of Grp I mGluRs has been reported to reduce leak K^+ currents (Glaum and Miller, 1992; McCormick and von Krosigk, 1992; Guerineau et al., 1994; Takeshita et al., 1996).

Differential mGluR Expression in Two Distinct Populations of Granule Cells

In the MOB, several populations of GCs can be distinguished based on the soma location either superficially (sGCs), interspersed with MCs within the MCL, or deeper, within the GCL (dGCs). Little is known about the physiological properties of sGCs vs. dGCs or about the role of mGluRs in regulating the activity of the two GC subtypes. dGCs have been reported to express mGluR5, and the GCL and several other non-olfactory regions express the highest mGluR5 levels in the brain (Romano et al., 1995; Sahara et al., 2001). In agreement with these observations, the studies summarized above demonstrate that activation of mGluR5 directly excites dGCs (Heinbockel et al., 2007a). However, little is known about the expression of mGluRs by sGCs within the MCL. mGluR5 staining of MCs and of the entire MCL is weak or absent (van den Pol, 1995). This suggests that sGCs do not express mGluR5. To address this hypothesis, we compared mGluR1 and mGluR5 responses in dGCs vs. sGCs.

Intrinsic properties and resting spontaneous discharge did not differ significantly between sGCs and dGCs of WT mice. Moreover, bath application of the selective Group I mGluR agonist DHPG depolarized and increased the firing rate of both GC subtypes in normal media and in the presence of iGluR blockers (Heinbockel et al., 2007b). We next took advantage of mice with targeted deletions of the mGluR5 or mGluR1 receptor genes (Conquet et al. 1994; Chiamulera et al. 2001) to investigate the role of these two Grp I receptor subtypes in the DHPG-evoked depolarization of sGCs and dGCs. The two GC subtypes responded differentially to DHPG in mGluR1$^{-/-}$ and mGluR5$^{-/-}$ mice. DHPG depolarized sGCs in slices from mGluR5$^{-/-}$ mice, but it had no effect on sGCs in slices from mGluR1$^{-/-}$ mice. By contrast, DHPG depolarized dGCs in slices from mGluR1$^{-/-}$ mice, but it had no effect on dGCs in slices from mGluR5$^{-/-}$ mice. These findings indicate that the two GC subtypes differentially express mGluR1 and mGluR5. Previous studies have shown that MCs express mGluR1, but not mGluR5. These results, therefore, suggest that sGCs are more similar to MCs than dGCs in terms of mGluR expression. Our finding that dGCs express mGluR5 but not mGluR1 is consistent with observations that GCL somata stain for mGluR5, that this staining is strongest within the deep GCL (Sahara et al., 2001), and that the GCL stains weakly for mGluR1 (Martin et al., 1992; Shigemoto et al., 1992; Sahara et al., 2001; van den Pol, 1995).

Activation of mGluR5 Increases GABAergic Inhibition of Mitral and External Tufted Cells

Excitatory mGluRs on GCs may function, in part, to facilitate feedback and feedforward inhibition of MCs. To address this question we used voltage clamp recordings to determine if

DHPG-evoked activation of GCs alters GABAergic inhibitory input to MCs. In the presence of ionotropic glutamate receptor blockers, DHPG increased the frequency of spontaneous IPSCs (sIPSCs) and miniature IPSCs (mIPSCs) in MCs (Heinbockel et al., 2007a; Dong et al., 2007a).

Use of mGluR antagonists showed that the increase was mediated by both mGluR1 and mGluR5; the increase in IPSCs was reduced in mGluR1$^{-/-}$ mice. mGluR1 and mGluR5 antagonists reduced the basal IPSC frequency in MCs (Heinbockel et al., 2007a; Dong et al., 2007a). The increase in mIPSCs in MCs and ET cells depended on voltage-gated Ca^{2+} channels, but persisted when iGluRs and sodium spikes were blocked.

These findings are consistent with those in the accessory olfactory bulb, where activation of mGluR1 increases IPSCs in MCs via a Ca^{2+}-dependent mechanism (Castro et al., 2007). The DHPG-evoked increase in IPSC could result from actions on GCs and/or PG cells in the GL. To address this question, we tested the effects of DHPG in MOB slices in which the GL layer was removed.

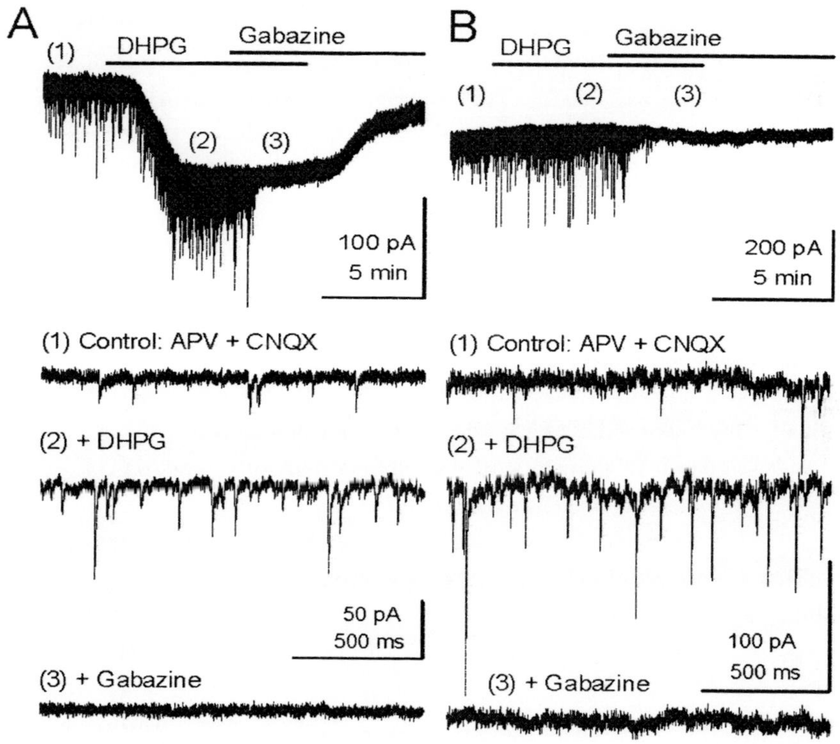

Figure 8. DHPG increased the frequency of GABAergic IPSCs in MCs. Voltage clamp recordings from MCs were made with pipettes with a high chloride concentration during blockade of iGluRs. Under these conditions, chloride-mediated IPSCs are reversed in polarity and appear as downward deflections (inward currents). A: Upper trace - Application of mGluRa agonist DHPG evoked an inward current in MCs recorded in slices from wildtype mice. Timepoints 1-3 are shown at faster timescale in the three lower tracers, respectively. Note the increase in frequency of fast IPSPs (trace 2), which are completely blocked by the $GABA_A$ receptor antagonist gabazine (trace 3). B: In an MC from an mGluR1$^{-/-}$ mouse, the DHPG-evoked inward current was absent, but DHPG substantially increased the frequency of IPSCs. Recording conditions and labeling as in (A). From Heinbockel et al., 2007a with permission of The American Physiological Society.

In such slices, DHPG increased spontaneous, but not miniature, IPSCs in MCs. Additionally, focal DHPG puffs into the GL, but not the GCL, increased mIPSCs in MCs and ET cells (Dong et al., 2007a). These findings indicate that DHPG-evoked GABA release from GCs requires action potentials, whereas in the GL DHPG facilitates PG cell GABA release via both spike-dependent and spike-independent presynaptic mechanisms. Functionally, these results suggest that mGluRs amplify spike-driven lateral inhibition through the MC-to-GC cell circuit, while GL mGluRs may play a more important role in amplifying intraglomerular inhibition following subthreshold input.

Blockade of mGluRs Reduces Mitral Cell-Evoked Excitation of Granule Cells and Lateral Inhibition in the MOB

The preceding findings demonstrate that pharmacological activation of mGluRs increases GABAergic inhibition in the olfactory bulb. Functionally, it is important to determine if endogenous activation of mGluRs plays a role in GC responses to glutamatergic synaptic input from MCs, and if these receptors are involved in GC-mediated feedback inhibition of MCs. To address these issues, we first assessed the effects of the mGluR antagonist LY341495 on GC responses to MC input using optical imaging of a voltage sensitive dye (RH-414) (Heinbockel et al., 2007a). This imaging was used to investigate activity patterns evoked by stimulation of the lateral olfactory tract (LOT), the M/TC output tract to higher olfactory centers. The LOT can be electrically stimulated to activate MCs directly, in turn, activating MC to GC glutamatergic synapses (Schoppa et al., 1998; Schoppa and Westbrook, 1999; Aroniadou-Anderjaska et al., 2000; Halabisky et al., 2000; Egger et al., 2005). As recently reported (Laaris et al., 2007), single shocks applied to the LOT elicited optical responses that were first observed in the EPL. LOT-evoked responses then spread sequentially into the superficial GCL. LOT-evoked responses were abolished by removal of Ca^{2+} from the ACSF, by application of TTX or CNQX-APV, indicating that they are postsynaptically mediated (Laaris et al., 2007). Application of the mGluR antagonist LY341495 reduced the peak amplitude of the LOT-evoked voltage-sensitive dye signal in the EPL by ~50% (Heinbockel et al., 2007a; Figure 9).

Previous studies have shown that such signals primarily reflect glutamatergic excitation of GC dendrites (Laaris et al., 2007). These findings suggest that mGluR blockade reduces GC excitability to M/TC synaptic input. Pharmacological inactivation of mGluRs decreases GABA release from GCs and this reduction in GABA release takes place in response to synaptic input from M/TCs.

To further investigate this issue, current-clamp recordings were used to asses the effects of mGluR blockade on GC responses to M/TC cell input. Single shocks applied to the GL directly activate MCs, and, in turn, elicit spikes in GCs, as previously reported (Schoppa et al., 1998). LY341495 reversibly reduced the mean response probability to M/TC input, i.e., the occurrence of a spike, from ~80% to 50% (Heinbockel et al., 2007a). A cell-by-cell analysis revealed that the reduction of evoked-spiking by LY341495 varied with the degree of hyperpolarization. The response probability to M/TC input decreased with increasing hyperpolarization. Additionally, LY341495 significantly hyperpolarized those cells that exhibited the largest reduction in evoked spiking. The spike threshold was not significantly affected by LY341495 in these cells. LY341495 produced a small, but non-significant reduction in the amplitude of evoked EPSPs, and there was no consistent relationship between the EPSP amplitude and the degree of hyperpolarization.

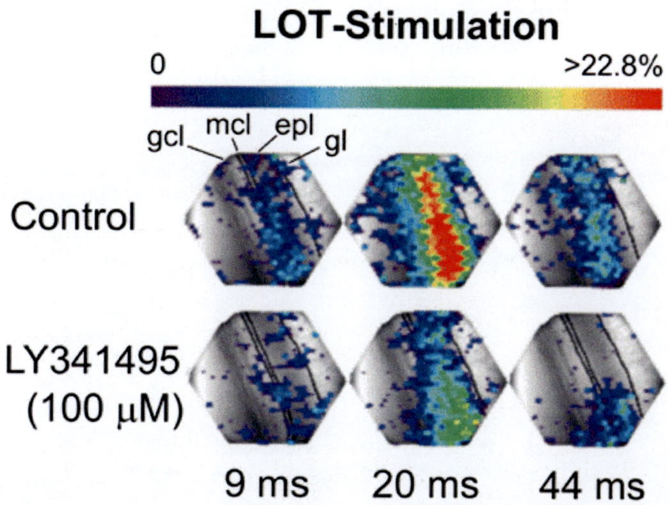

Figure 9. mGluR antagonist LY341495 decreased temporal and spatial spread of evoked activity as revealed by voltage-sensitive dye imaging. Each panel shows the amplitude of optical responses, recorded by each of 464 photodiodes. Signal amplitudes, expressed as a percentage above mean baseline values, are color-coded; color scale at top applies to all images. Time intervals beneath the images are relative to the onset of LOT stimulation. Application of LY341495 reduced the amplitude of LOT-evoked optical signals. gl – glomerular layer; epl – external plexiform layer; mcl – mitral cell layer; gcl – granule cell layer; LOT – lateral olfactory tract. From Heinbockel et al., 2007 with permission of The American Physiological Society.

In parallel voltage-clamp recordings, LY341495 did not significantly affect the amplitude, onset latency, kinetics or integral of the single shock, GL-evoked EPSC in GCs. Taken together, these findings suggest that the reduction of evoked spiking in GCs by LY341495 was not due to presynaptic modulation of glutamate release from M/TC dendrites, but might have been due to the degree of GC hyperpolarization. Although we cannot ascribe these effects solely to blockade of mGluR5, these result suggests that activation of mGluRs by endogenously released glutamate tonically modulates the excitability of GCs.

Blockade of mGluRs Reduces Feedback Inhibition of Mitral Cells

The preceding experiments indicate that inactivation of mGluRs significantly attenuates MC-evoked excitation of GCs. Does blockade of mGluRs decrease GC-mediated inhibition of MCs? To address this question, we determined if LY341495 alters feedback IPSPs elicited by direct depolarization of individual MCs. As previously reported (Isaacson and Strowbridge, 1998), depolarizing current steps applied to MCs in the presence of TTX and TEA were followed by a feedback IPSP. Bath application of LY341495 reduced the peak amplitude and integral of the feedback IPSP (Heinbockel et al., 2007a; Figure 10). These results demonstrated that blockade of mGluRs reduces GC-mediated feedback inhibition of MCs. It is noteworthy that blockade of mGluR1 also reduced feedback inhibition of MCs in the accessory olfactory bulb (Castro et al., 2007).

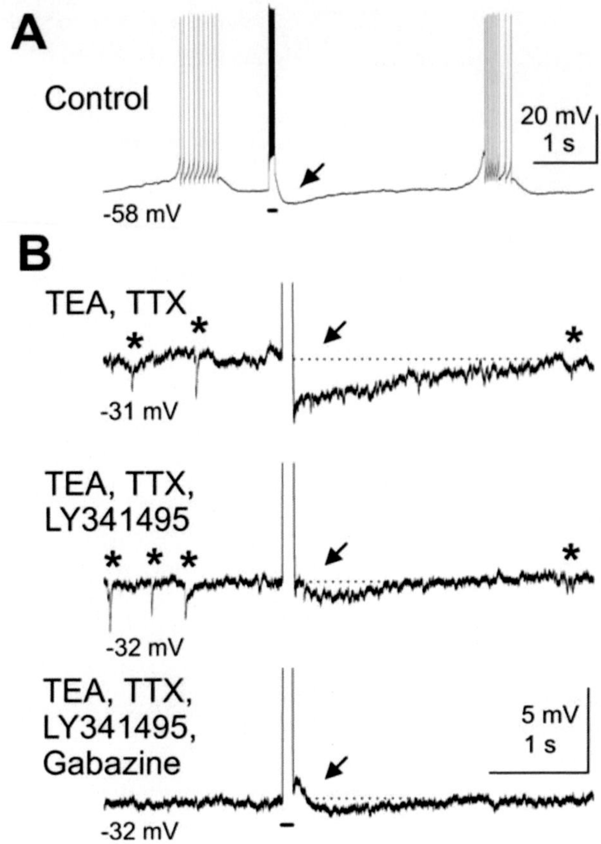

Figure 10. Blockade of mGluRs reduces GC-mediated feedback inhibition of MCs. Depolarizing current pulses injected into an MC evoked GC-mediated feedback IPSPs. A: A depolarizing current pulse (100 ms, 300 pA, indicated by bar beneath the trace) evoked a burst of action potentials followed by an IPSP (at arrow). B: After wash-in of TEA and TTX and current clamp of the MC at ca. –30 mV, spontaneous IPSPs are visible (marked by asterisks above the trace) and current pulses revealed a feedback IPSP (first trace). The mGluR antagonist LY341495 reduced the IPSP in amplitude (second trace). Addition of the $GABA_A$ receptor blocker gabazine completely abolished any remaining feedback IPSP as well as spontaneous IPSPs (third trace). Modified from Heinbockel et al., 2007a with permission of The American Physiological Society.

mGluRs and Network Regulation

A variety of pharmacological effects show that activation of Group I mGluRs potently regulates the intrinsic and synaptic properties of multiple classes of MOB neurons and glia. Use of mGluR KO mice confirmed the pharmacological results.

Functionally, the mGluR-mediated effects have important consequences for two key features of MC activity: mGluR antagonists prolong the periodicity of membrane potential bistability and dramatically attenuate MC responses to sensory input. We conclude that endogenous glutamate tonically activates mGluR1 on MCs and, hence, modulates MC excitability. Activation of mGluR1 dynamically regulates the MOB neural network by magnifying MC synaptic output. By contrast, activation of mGluRs on GCs increases cell excitability through both a direct and an indirect effect and increases GABA release from GCs.

Functional Considerations of mGluR Activation

Tonic activation of mGluR1 by baseline glutamate levels may serve to facilitate or amplify sensory responses of MCs. mGluR1-mediated excitation is typically slow in time-course and preferentially engaged by high frequency activity (Anwyl, 1999). Thus, mGluR1 responses in MCs would tend to be preferentially activated over successive sniff cycles during odor inhalation (Harrison and Scott, 1986; Hamilton and Kauer, 1988, 1989). MC responses to odors are characterized by specific temporal firing patterns, including oscillations and synchronous activity (Kauer, 1998; Kay and Laurent, 1999; Friedrich and Laurent, 2001; Luo and Katz, 2001; Spors and Grinvald, 2002). In this regard, it is noteworthy that mGluR antagonism (i.e., with MCPG) was found to significantly reduce ON-evoked rhythmic oscillations in MCs (Schoppa and Westbrook, 2001). It is reasonable to speculate that activation of mGluR1s on M/TC apical dendrites plays an important role in odor-evoked oscillations and coding of olfactory stimuli.

mGluR1s are also present on portions of MC lateral dendrites presynaptic to GC dendrites (van den Pol, 1995). This suggests that mGluR1 may function, in part, to increase the degree or duration of depolarization of the lateral dendrite following glutamate release from neighboring M/TCs. Spillover-mediated activation of mGluRs among MCs strongly activated by a particular odorant may presynaptically enhance glutamate release onto GCs, increasing GABA release from GCs (Kinzie et al., 1998). mGluR5 is localized on portions of GC dendrites in the EPL directly opposed to presynaptic glutamatergic synapses from M/TC lateral dendrites (van den Pol, 1995). Our results suggest that activation of mGluR5 increases M/TC-evoked excitation of GCs, and in turn, facilitates dendrodendritic GABA release onto M/TCs.

What is the functional consequence of increased GABA release from GCs? mGluRs may participate in feedforward and/or feedback inhibition at dendrodendritic synapses between MCs and GCs such that GC-mediated lateral inhibition can be driven by mGluRs. From functional imaging studies it is known that a given odorant activates a specific glomerulus or set of glomeruli in the olfactory bulb (Wachowiak and Shipley, 2006). A specific odorant activates a set of ORNs in the nasal epithelium and their axons terminate in a given glomerulus in the olfactory bulb such that the glomerulus serves as the address for odorant characteristics (Shepherd et al., 2004). Subsequently, MCs that innervate this glomerulus are activated as well. An MC that was activated by ON input releases glutamate at its long secondary dendrites and excites GCs at its dendrodendritic synapses. The GCs in turn release GABA, which inhibits the same MC (feedback-inhibition) and MCs that innervate other glomeruli. The result is lateral inhibition or feedforward inhibition of neighboring glomeruli. Increased GABA release from GCs would be expected to increase lateral inhibition of neighboring MCs. Lateral inhibition from GCs to MCs is generally considered to provide contrast enhancement by preferentially suppressing populations of MCs that are weakly activated by an odorant or not activated at all (van den Pol, 1995; Kinzie et al., 1998; Urban,

2002, Shepherd et al., 2004). Therefore, one potential functional correlate of mGluR activation in the MOB could be to enhance lateral inhibition, thereby suppressing the activity of weakly odorant-activated M/TCs. To test if mGluRs increase lateral inhibition we mapped neural activity in the MOB using optical imaging of voltage-sensitive dyes (Heinbockel et al., 2007a). The results were very clear and indicated that activation of mGluRs by endogenously released glutamate increases the magnitude, time-course and spatial extent of lateral inhibition. This suggests that MCs that are weakly activated by an odorant would be progressively inhibited over successive sniffs due to the time-course of mGluR components of lateral inhibition. GC-mediated lateral inhibition plays an important role in odor contrast enhancement in the MOB (Yokoi et al., 1995). Activation of mGluR5 on GCs may function to enhance the temporal envelope of lateral inhibition beyond that mediated by ionotropic glutamate receptors or to amplify lateral inhibition during repetitive sensory input associated with high frequency investigative sniffing.

A Novel Hypothesis for mGluR Function

The density of mGluR1 in M/TCs and mGluR5 in GCs are higher than most other regions of the brain. This suggests that mGluRs play important role(s) in the operations of the MOB network. In agreement with this, new findings presented above demonstrate that the excitability of MCs and GCs is directly and potently modulated via activation of mGluRs by endogenously released glutamate in MOB slices. Our studies indicate that specific mGluRs increase the excitability of MCs and GCs via modulation of distinct ionic conductances in these cells. More importantly our findings indicate that the net effect of these mGluR actions is to increase M/TC excitatory drive on GCs, and in turn, to increase GABA release from GCs. Patch clamp recording and voltage-sensitive imaging experiments suggest that a major functional consequence of mGluR activation is to increase both the spatial spread and temporal duration of lateral inhibition in the MOB. We hypothesize, therefore, that mGluRs in the MOB function at the network level to increase contrast for odorant-induced activity patterns.

In view of their abundant expression in MOB neurons and in light of the limited knowledge of mGluR function, a pressing issue in the organization and operation of the olfactory system is the analysis of mGluR actions on adult MOB neurons and upon MOB network function. The MOB is an ideal platform for investigating how mGluRs modulate neural network dynamics to enhance sensory processing. The MOB receives direct sensory input, but this input is strongly modified by intrinsic inhibitory interneurons that provide two discrete tiers for lateral interactions. The MOB also receives vast cortical and subcortical centrifugal inputs that modulate local circuit processing at multiple levels. This organization makes the MOB preparation significantly different from hippocampus, amygdala, neocortex, and cerebellum for addressing functional questions of mGluR modulation in the brain.

A number of studies have demonstrated that mGluRs in these and other brain regions are involved in diseases such as stroke, epilepsy, Alzheimer's, Parkinson's and Huntington's disease, as well as in pain, schizophrenia, and anxiety (Lea and Faden, 2003; Schoepp and Conn, 1993). Therefore, learning about the mechanisms through which mGluR activation shapes synaptic information in a model neural circuit such as the MOB helps to fill major gaps in our understanding of the functioning of both the olfactory system and other brain regions.

Acknowledgements

We gratefully acknowledge the contributions of our many colleagues to the findings presented in this chapter. We acknowledge Adam Puche for his help with confocal microscopy and image processing. The work described here was supported in part by the Whitehall Foundation, the Howard University New Faculty Research Program, and United States Public Health Service grants S06GM08016 (MBRS-SCORE, NIGMS/NIH), U54NS039407 (SNRP, NINDS/NIH) and DC03195.

References

Anwyl, R. (1999). Metabotropic glutamate receptors: electrophysiological properties and role in plasticity. *Brain Res. Rev., 29,* 83-120.

Aroniadou-Anderjaska, V.A., Ennis, M., and Shipley, M.T. (1997). Glomerular synaptic responses to olfactory nerve input in rat olfactory bulb slices. *Neurosci., 79,* 425-434.

Aroniadou-Anderjaska, V., Ennis, M., and Shipley, M.T. (1999a). Current-source density analysis in the rat olfactory bulb: laminar distribution of kainate/AMPA and NMDA receptor-mediated currents. *J. Neurophysiol., 81,* 15-28.

Aroniadou-Anderjaska, V., Ennis, M., and Shipley, M.T. (1999b). Dendrodendritic recurrent excitation in mitral cells of the rat olfactory bulb. *J. Neurophysiol.* 82:489-494, 1999b.

Aroniadou-Anderjaska, V., Zhou, F.-M., Priest, C.A., Ennis, M., and Shipley, M.T. (2000). GABAB receptor-mediated presynaptic inhibition of sensory input to the olfactory bulb. *J. Neurophysiol., 84,* 1194-1203.

Astic, L., and Saucier, D. (1982). Ontogenesis of the functional activity of rat olfactory bulb: Autoradiographic study with the 2-deoxyglucose method. *Dev. Brain Res., 2,* 243-256.

Bandrowski, A.E., Huguenard, J.R., and Prince, D.A. (2003). Baseline glutamate levels affect group I and II mGluRs in layer V pyramidal neurons of rat sensorimotor cortex. *J. Neurophysiol., 89,* 1308-1316.

Bardoni, R., Puopolo, M., Magherini, P.C., and Belluzzi, O. (1996). Potassium currents in periglomerular cells of frog olfactory bulb in vitro. *Neurosci. Lett., 210,* 95-98.

Batchelor, A.M. and Garthwaite, J. (1997). Frequency detection and temporally dispersed synaptic signal association through a metabotropic glutamate receptor pathway. *Nature, 385,* 74-77.

Bischofberger, J., and Jonas, P. (1997) Action potential propagation into the presynaptic dendrites of rat mitral cells. *J. Physiol., 504,* 359-65.

Bischofberger, J., and Schild, D. (1996). Glutamate and N-acetylaspartylglutamate block HVA calcium currents in frog olfactory bulb interneurons via an mGluR2/3-like receptor. *J. Neurophysiol., 76,* 2089-2092.

Carlson, G.C., Shipley, M.T., and Keller, A. (2000). Long-lasting depolarizations in mitral cells of the rat olfactory bulb. *J. Neurosci., 20,* 2011-2021.

Carlson, G.C., Slawecki, M.L., Lancaster, E., and Keller, A. (1997). The distribution and activation of intracellular Ca^{2+} stores in cultured olfactory bulb neurons. *J. Neurophysiol., 78,* 2176-2185.

Casabona, G., Catania, M.V., Storto, M., Ferraris, N., Perroteau, I., Fasolo, A., Nicoletti, F., and Bovolin, P. (1998). Deafferentation up-regulates the expression of the mGlu1a metabotropic glutamate receptor protein in the olfactory bulb. *Eur. J. Neurosci., 10,* 771-776.

Castro, J.B., Hovis, K.R., and Urban, N.N. (2007). Recurrent dendrodendritic inhibition of accessory olfactory bulb mitral cells requires activation of Group I metabotropic glutamate receptors. *J. Neurosci., 27,* 5646-5671.

Charpak, S., Gähwiler, B.H., Do, K.Q., and Knöpfel, T. (1990). Potassium conductances in hippocampal neurons blocked by excitatory amino-acid transmitters. *Nature, 347,* 765-767.

Chen, W.R., and Shepherd, G.M. (1997). Membrane and synaptic properties of mitral cells in slices of rat olfactory bulb. *Brain Res., 745,* 189-196.

Chen, W.R., Midtgaard, J., and Shepherd, G.M. (1997). Forward and backward propagation of dendritic impulses and their synaptic control in mitral cells. *Science, 278,* 463-467.

Chen, W.R., Xiong, W., and Shepherd, G.M. (2000). Analysis of relations between NMDA receptors and GABA release at olfactory bulb reciprocal synapses. *Neuron, 25,* 625-633.

Chiamulera, C., Epping-Jordan, M.P., Zocchi, A., Marcon, C., Cottiny, C., Tacconi, S., Corsi, M., Orzi, F., and Conquet, F. (2001). Reinforcing and locomotor stimulant effects of cocaine are absent in mGluR5 null mutant mice. *Nat. Neurosci., 4,* 873-874.

Clark, B.P., Baker, S.R., Goldsworthy, J., Harris, J.R., and Kingston, A.E. (1997). 2-Methyl-4-carboxyphenylglycine (LY367385) selectively antagonizes metabotropic glutamate mGluR1 receptors. *Bioorg. Med. Chem. Lett.*, 7, 2777-2870.

Congar, P., Leinekugel, X., Ben-Ari, Y., and Crepel, V. (1997) A long-lasting calcium-activated nonselective cationic current is generated by synaptic stimulation or exogenous activation of group I metabotropic glutamate receptors in CA1 pyramidal neurons. *J. Neurosci., 17,* 5366-5379.

Conn, P.J., and Pin, J.P. (1997). Pharmacology and functions of metabotropic glutamate receptors. *Annu. Rev. Pharmacol. Toxicol., 37,* 205-237.

Conquet, F., Bashir, Z.I., Davies, C.H., Daniel, H., Ferraguti, F., Bordi, F., Franz-Bacon, K., Reggiani, A., Matarese, V., Condé, F., Collingridge, G.L., and Crépel, F. (1994). Motor deficit and impairment of synaptic plasticity in mice lacking mGluR1. *Nature, 372,* 237-243.

Crépel, V., Aniksztejn, L., Ben-Ari, Y., and Hammond, C. (1994). Glutamate metabotropic receptors increase a Ca^{2+}-activated nonspecific cationic current in CA1 hippocampal neurons. *J. Neurophysiol., 72,* 1561-1569.

De Saint Jan, D., and Westbrook, G.L. (2005). Detecting activity in olfactory bulb glomeruli with astrocyte recording. *J. Neurosci., 25,* 2917–2924.

De Saint Jan, D., and Westbrook, G.L. (2007). Disynaptic amplification of metabotropic glutamate receptor 1 responses in the olfactory bulb. *J. Neurosci., 27,* 132-140.

Dong, H.W., Hayar, A., and Ennis, M. (2007a). Activation of group I metabotropic glutamate receptors on main olfactory bulb granule cells and periglomerular cells enhances synaptic inhibition of mitral cells. *J. Neurosci., 27,* 5654-63.

Dong, H.W., Hayar, A., and Ennis, M. (2007b). mGluR1 activation enhances nonselective cation currents and rhythmic bursting in external tufted (ET) cells. *Chem. Senses, 32,* A52.

Duchamp-Viret, P., Chaput, M.A., and Duchamp, A. (1999). Odor response properties of rat olfactory receptor neurons. *Science, 284,* 2171-2174.

Egger, V., Svoboda, K., and Mainen, Z.F. (2005). Dendrodendritic synaptic signals in olfactory bulb granule cells: local spine boost and global low-threshold spike. *J. Neurosci., 25,* 3521-3530.

Ennis, M., Hamilton, K.A., and Hayar, A. (2007). Neurochemistry of the main olfactory system. In A. Lajtha (Ed. in Chief) *Handbook of Neurochemistry and Molecular Neurobiology* (3rd edition, vol. 20, pp. 137-204); Sensory Neurochemistry (Johnson, D.A., Volume Editor). Springer: Heidelberg.

Ennis, M., Zimmer, L.A., and Shipley, M.T. (1996). Olfactory nerve stimulation activates rat mitral cells via NMDA and non-NMDA receptors *in vitro. NeuroReport, 7,* 989-992.

Ennis, M., Zhou, F.M., Ciombor, K.J., Aroniadou-Anderjaska, V., Hayar, A., Borrelli, E., Zimmer, L.A., Margolis, F., and Shipley, M.T. (2001). Dopamine D2 receptor-mediated presynaptic inhibition of olfactory nerve terminals. *J. Neurophysiol., 86,* 2986-2997.

Ennis, M., Zhu, M., Heinbockel, T., and Hayar, A. (2006). Olfactory nerve-evoked, metabotropic glutamate receptor-mediated synaptic responses in rat olfactory bulb mitral cells. *J. Neurophysiol., 95,* 2233-2241.

Ferraris, N., Perroteau, I., De Marchis, S., Fasolo, A., and Bovolin, P. (1997). Glutamatergic deafferentation of olfactory bulb modulates the expression of mGluR1a mRNA. *Neuroreport. 8,* 1949-1953.

Frazier, L.L. and Brunjes, P.C. (1988). Unilateral odor deprivation: early postnatal changes in olfactory bulb cell density and number. *J. Comp. Neurol., 269,* 355-370.

Friedman, D., and Strowbridge, B.W. (2000). Functional role of NMDA autoreceptors in olfactory mitral cells. *J. Neurophysiol., 84,* 39-50.

Friedrich, R., and Laurent, G. (2001). Dynamic optimization of odor representations by slow temporal patterning of mitra cell activity. *Science, 291,* 889-894.

Geiling, H., and Schild, D. (1996). Glutamate-mediated release of Ca^{2+} in mitral cells of the olfactory bulb. *J. Neurophysiol., 76,* 563-570.

Glaum, S.R., and Miller, R.J. (1992). Metabotropic glutamate receptors mediate excitatory transmission in the nucleus of the solitary tract. *J. Neurosci., 12,* 2251-2258.

Guerineau, N.C., Bossu, J.L., Gähwiler, B.H., and Gerber, U. (1995). Activation of a nonselective cationic conductance by metabotropic glutamatergic and muscarinic agonist in CA3 pyramidal neurons of the rat hippocampus. *J. Neurosci., 15,* 4395-4407.

Guerineau, N.C., Gähwiler, B.H., and Gerber, U. (1994). Reduction of resting K^+ current by metabotropic glutamate and muscarinic receptors in rat CA3 cells: mediation by G-proteins. *J. Physiol. (Lond), 474,* 27-33.

Halabisky, B., Friedman, D., Radojicic, M., and Strowbridge, B. (2000). Calcium influx through NMDA receptors directly evokes GABA release in olfactory bulb granule cells. *J. Neurosci., 20,* 5124-5134.

Hamilton, K.A., and Kauer, J.S. (1988). Intracellular potentials of salamander mitral/tufted neurons in response to odor stimulation. *Brain Res., 338,* 181-185.

Hamilton, K.A., and Kauer, J.S. (1989). Patterns of intracellular potentials in salamander mitral/tufted cells in response to odor stimulation. *J. Neurophysiol., 62,* 609-625.

Harrison, T.A., and Scott, J.W. (1986). Olfactory bulb responses to odor stimulation: analysis of response pattern and intensity relationships. *J. Neurophysiol., 56,* 1571-1589.

Hayar, A., Karnup, S., Shipley, M.T., and Ennis, M. (2004a). Olfactory bulb glomeruli: external tufted cells intrinsically burst at theta frequency and are entrained by patterned olfactory input. *J. Neurosci., 24,* 1190-1199.

Hayar, A., Karnup, S., Ennis, M., and Shipley, M.T. (2004b). External tufted cells: a major excitatory element that coordinates glomerular activity. *J. Neurosci., 24,* 6676-6685.

Hayar, A., Shipley, M.T., and Ennis, M. (2005). Olfactory bulb external tufted cells are synchronized by multiple intraglomerular mechanisms. *J. Neurosci., 25,* 8197-8208.

Heinbockel, T., Heyward, P., Conquet, F., and Ennis, M. (2004). Regulation of main olfactory bulb mitral cell excitability by metabotropic glutamate receptor mGluR1. *J. Neurophysiol.,* 92, 3085-3096.

Heinbockel, T., Laaris, N., and Ennis, M. (2007a). Metabotropic glutamate receptors in the main olfactory bulb drive granule cell-mediated inhibition. *J. Neurophysiol., 97,* 858-870.

Heinbockel, T., Hamilton, K.A., and Ennis, M. (2007b). Group I metabotropic glutamate receptors are differentially expressed by two populations of olfactory bulb granule cells. *J. Neurophysiol., 97,* 3136-3141.

Heyward, P., Ennis, M., Keller, A., and Shipley, M.T. (2001). Membrane bistability in olfactory bulb mitral cells. *J. Neurosci., 21,* 5311-5320.

Huang, Y.H., Sinha, S.R., Tanaka, K., Rothstein, J.D., and Bergles, D.E. (2004). Astrocyte glutamate transporters regulate metabotropic glutamate receptor-mediated excitation of hippocampal interneurons. *J. Neurosci., 24,* 4551-4559.

Isaacson, J.S. (1999). Glutamate spillover mediates excitatory transmission in the rat olfactory bulb. *Neuron, 23,* 377-384.

Isaacson, J.S., and Strowbridge, B.W. (1998). Olfactory reciprocal synapses: dendritic signaling in the CNS. *Neuron, 20,* 749-761.

Johnson, B.A., Woo, C.C., and Leon, M. (1998). Spatial coding of odorant features in the glomerular layer of the rat olfactory bulb. *J. Comp. Neurol. 393,* 457-471.

Karakossian, M.H., and Otis, T.S. (2004). Excitation of cerebellar interneurons by group I metabotropic glutamate receptors. *J. Neurophysiol., 92,* 1558-1565.

Kauer, J.S. (1998). Olfactory processing: a time and place for everything. *Curr. Biol., 8,* R282-283.

Kauer, J.S., and Cinelli, A.R. (1993). Are there structural and functional modules in the vertebrate olfactory bulb? *Microsc. Res. Tech., 24,* 157-176.

Kay, L.M., and Laurent, G. (1999). Odor- and context-dependent modulation of mitral cell activity in behaving rats. *Nat. Neurosci., 2,* 1003-1009.

Keele, N.B., Arvanov, V.L., and Shinnick-Gallagher, P. (1997). Quisqualate-preferring metabotropic glutamate receptor activates Na^{+}-Ca^{2+} exchange in rat basolateral amygdala neurones. *J. Physiol., 499,* 87-104.

Keller, A., Yagodin, S., Aroniadou-Anderjaska, A., Zimmer, L.A., Ennis, M., Sheppard, N.F., and Shipley, M.T. (1998). Functional organization of rat olfactory bulb glomeruli revealed by optical imaging. *J. Neurosci., 18,* 2602-2612.

Kingston, A.E., Ornstein, P.L., Wright, R.A., Johnson, B.G., Mayne, N.G., Burnett, J.P., Belagaje, R., Wu, S., and Schoepp, D.D. (1998). LY341495 is a nanomolar potent and selective antagonist of group II metabotropic glutamate receptors. *Neuropharmacol., 37,* 1-12.

Kinoshita, A., Shigemoto, R., Ohishi, H., van der Putten, H., and Mizuno, N. (1998). Immunohistochemical localization of metabotropic glutamate receptors, mGluR7a and mGluR7b, in the central nervous system of the adult rat and mouse: a light and electron microscopic study. *J. Comp. Neurol., 393,* 332-52.

Kinzie, J.M., Saugstad, J.A., Westbrook, G.L., and Segerson, T.P. (1995). Distribution of metabotropic glutamate receptor 7 messenger RNA in the developing and adult rat brain. *Neurosci., 69,* 167-76.

Kinzie, J.M., Westbrook, G.L., and Segerson, T.P. (1998). Metabotropic glutamate receptors in the olfactory bulb. In F. Moroni, F. Nicoletti, and D.E. Pellegrini-Giampietro (Eds.), *Metabotropic Glutamate Receptors and Brain Function* (pp. 59-69). London: Portland Press.

Klenoff, J.R., and Greer, C.A. (1998). Postnatal development of olfactory receptor cell axonal arbors. *J. Comp. Neurol., 390,* 256–67.

Laaris, N., Puche, A., and Ennis, M. (2007). Complementary postsynaptic activity patterns elicited in olfactory bulb by stimulation of mitral/tufted and centrifugal fiber inputs to granule cells. *J. Neurophysiol., 97,* 296-306.

Lea, P.M., and Faden, A.I. (2003) Modulation of metabotropic glutamate receptors as potential treatment for acute and chronic neurodegenerative disorders. *Drug News Perspect., 16,* 513-22

Lee, K., and Boden, P.R. (1997). Characterization of the inward current induced by metabotropic glutamate receptor stimulation in rat ventromedial hypothalamic neurones. *J. Physiol., 504.3,* 649-663.

Liu, W.-L., and Shipley, M.T. (1994) Cholecystokinin (CCK) immunocytochemical characteristics and postsynaptic targets of the intrabulbar associational system in the rat olfactory bulb. *J. Comp. Neurol., 346,* 541-558.

Luo, M., and Katz, L.C. (2001). Response correlation maps of neurons in the mammalian olfactory bulb. *Neuron, 32,* 1165-1179.

Martin, L.J., Blackstone, C.D., Huganir, R.L., and Price, D.L. (1992). Cellular localization of a metabotropic glutamate receptor in rat brain. *Neuron, 9,* 259-270.

Masu, M., Tanabe, Y., Tsuchida, K., Shigemoto, R., and Nakanishi, S. (1991). Sequence and expression of a metabotropic glutamate receptor. *Nature, 349,* 760-765.

McCormick, D.A., and von Krosigk, M. (1992). Corticothalamic activation modulates thalamic firing through glutamate "metabotropic" receptors. *Proc. Natl. Acad. Sci. USA., 89,* 2774-2778.

Meisami, E. (1989) A proposed relationship between increases in the number of olfactory receptor neurons, convergence ratio and sensitivity in the developing rat. *Brain Res. Dev. Brain Res., 46,* 9-19.

Mombaerts, P. (1999) Molecular biology of odorant receptors in vertebrates. *Annu. Rev. Neurosci., 22,* 487-509.

Mombaerts, P. (2006) Axonal wiring in the mouse olfactory system. *Annu. Rev. Cell. Dev. Biol., 22,* 713-37.

Mombaerts, P., Wang, F., Dulac, C., Chao, S.K., Nemes, A., Mendelsohn, M., Edmondson, J., and Axel, R. (1996). Visualizing an olfactory sensory map. *Cell, 87,* 675-86.

Mori, K., Nagao, H., and Yoshohara, Y. (1999). The olfactory bulb: coding and processing of odor molecule information. *Science, 286,* 711-715.

Murphy, S.N., and Miller, R.J. (1988). A glutamate receptor regulates Ca2+ mobilization in hippocampal neurons. *Proc. Natl. Acad. Sci. USA., 85,* 8737-8741.

Mutoh, H., Yuan, Q., and Knöpfel, T. (2005). Long-term depression at olfactory nerve synapses. *J. Neurosci., 25,* 4252-4259.

Nickell, W.T., Behbehani, M.M., and Shipley, M.T. (1994). Evidence for GABAB-mediated inhibition of transmission from the olfactory nerve to mitral cells in the rat olfactory bulb. *Brain Res. Bull., 35,* 119-123.

Ohishi, H., Shigemoto, R., Nakanishi, S., and Mizuno, N. (1993). Distribution of the mRNA for a metabotropic glutamate receptor, mGluR2, in the central nervous system of the rat. *Neurosci., 53,* 1009-1018.

Ohishi, H., Neki, A., and Mizuno, N. (1998). Distribution of a metabotropic glutamate receptor, mGluR2, in the central nervous system of the rat and mouse: an immunohistochemical study with a monoclonal antibody. *Neurosci. Res., 30,* 65-82.

Petralia, R.S., Wang, Y.X., Singh, S., Wu, C., Shi, L., Wei, J., and Wenthold, R.J. (1997). A monoclonal antibody shows discrete cellular and subcellular localizations of mGluR1 alpha metabotropic glutamate receptors. *J. Chem. Neuroanat., 13,* 77-93.

Raggenbass, M., Pierson, P., Metzger, D., and Alberi, S. (1997). Action of a metabotropic glutamate receptor agonist in rat lateral septum: induction of a sodium-dependent inward aftercurrent. *Brain Res., 776,* 75-87.

Reichelt, W., and Knöpfel, T. (2002) Glutamate uptake controls expression of a slow postsynaptic current mediated by mGluRs in cerebellar Purkinje cells. *J. Neurophysiol., 87,* 1974-1980.

Ribak, C.E. (1977). The GABAergic neurons of the olfactory bulb in the rat. In J.A. Ferrendelli and G. Guritch (Eds.) *Aspects of Behavioral Neurobiology, Soc. Neurosci. Symposia. III.* (pp. 189-203). Bestheda: Soc. Neurosci.

Rieger, A., Deitmer, J.W., and Lohr, C. (2007). Axon-glia communication evokes calcium signaling in olfactory ensheathing cell of the developing olfactory bulb. *Glia, 55,* 352-359.

Romano, C., Sesma, M.A., McDonald, C.T., O'Malley, K., van den Pol, A.N., and Olney, J.W. (1995). Distribution of metabotropic glutamate receptor mGluR5 immunoreactivity in rat brain. *J. Comp. Neurol., 355,* 455-469.

Sahara, Y., Kubota, T., and Ichikawa, M. (2001). Cellular localization of metabotropic glutamate receptors mGluR1, 2/3, 5 and 7 in the main and accessory olfactory bulb of the rat. *Neurosci. Lett., 312,* 59-62.

Salin, P.A., Lledo, P.M., Vincent, J.D., and Charpak, S. (2001). Dendritic glutamate autoreceptors modulate signal processing in rat mitral cells. *J. Neurophysiol., 85,* 1275-1282.

Salt, T.E., Turner, J.P., and Kingston, A.E. (1999). Evaluation of agonists and antagonists acting at group I metabotropic glutamate receptors in the thalamus in vivo. *Neuropharmacol., 38,* 1505-1510.

Saugstad, J.A., Kinzie, J.M., Shinohara, M.M., Segerson, T.P., and Westbrook, G.L. (1997). Cloning and expression of rat metabotropic glutamate receptor 8 reveals a distinct pharmacological profile. *Mol. Pharmacol., 51,* 119-125.

Schoepp, D.D., and Conn, P.J. (1993). Metabotropic glutamate receptors in brain function and pathology. *Trends Pharmacol. Sci., 14,* 13-20.

Schoepp, D.D., Johnson, B.G., and Monn, J.A. (1992). Inhibition of cyclic AMP formation by a selective metabotropic glutamate agonist. *J. Neurochem., 58,* 1184-1186.

Schoepp, D.D., Jane, D.E., and Monn, J.A. (1999). Pharmacological agents acting at subtypes of metabotropic glutamate receptors. *Neuropharmacol., 38,* 1431-1476.

Schoppa, N.E., and Westbrook, G.L. (1997). Modulation of mEPSCs in olfactory bulb mitral cells by metabotropic glutamate receptors. *J. Neurophysiol., 78,* 1468-1475.

Schoppa, N.E., and Westbrook, G.L. (1999). Regulation of synaptic timing in the olfactory bulb by an A-type potassium current. *Nat. Neurosci., 2,* 1106-1113.

Schoppa, N.E., and Westbrook, G.L. (2001). Glomerulus-specific synchronization of mitral cells in the olfactory bulb. *Neuron, 31,* 639-651.

Schoppa, N.E., Kinzie, J.M., Sahara, Y., Segerson, T.P., and Westbrook, G.L. (1998). Dendrodendritic inhibition in the olfactory bulb is driven by NMDA receptors. *J. Neurosci., 18,* 6790-6802.

Shen, K.Z., and Johnson, S.W. (1997) A slow excitatory postsynaptic current mediated by G-protein-coupled metabotropic glutamate receptors in rat ventral tegmental dopamine neurons. *Eur. J. Neurosci., 9,* 48-54.

Shepherd, G.W., Chen, W.R., and Greer, C.A. (2004). Olfactory bulb. In G. M. Shepherd (Ed.), *The Synaptic Organization of the Brain* (pp.165-216). New York, NY: Oxford.

Shigemoto, R., Nakanishi, S., and Mizuno, N. (1992). Distribution of the mRNA for a metabotropic glutamate receptor (mGluR1) in the central nervous system: an in situ hybridization study in adult and developing rat. *J. Comp. Neurol., 322,* 121-135.

Shipley, M.T., and Ennis, M. (1996). Functional organization of olfactory system. *J. Neurobiol., 30,* 123-176.

Spors, H., and Grinvald, A. (2002). Spatio-temporal dynamics of odor representations in the mammalian olfactory bulb. *Neuron, 34,* 301-315.

Staub, C., Vranesic, I., and Knöpfel, T. (1992). Responses to metabotropic glutamate receptor activation in cerebellar purkinje cells: induction of an inward current. *Eur. J. Neurosci., 4,* 832-839.

Suzdak, P.D., Thomsen, C., Mulvihill, E., and Kristensen, P. (1994). Molecular cloning, expression, and characterization of metabotropic glutamate receptor subtypes. In P.J.

Conn and J. Patel (Eds.) *The metabotropic glutamate receptors* (pp. 1-30). Totowa, NJ: Humana.

Takeshita, Y., Harata, N., and Akaike, N. (1996). Suppression of K^+ conductance by metabotropic glutamate receptor in acutely dissociated large cholinergic neurons of rat caudate putamen. *J. Neurophysiol., 76,* 1545-1558.

Tempia, F., Alojado, M.E., Strata, P., and Knöpfel, T. (2001). Characterization of the mGluR(1)-mediated electrical and calcium signaling in Purkinje cells of mouse cerebellar slices. *J. Neurophysiol., 86,* 1389-1397.

Tempia, F., Miniaci, M.C., Anchisi, D., and Strata, P. (1998). Postsynaptic current mediated by metabotropic glutamate receptors in cerebellar Purkinje cells. *J. Neurophysiol., 80,* 520-528.

Trombley, P.Q., and Westbrook, G.L. (1992). L-AP4 inhibits calcium currents and synaptic transmission via a G-protein-coupled glutamate receptor. *J. Neurosci. 12,* 2043-2050.

Urban, N.N. (2002). Lateral inhibition in the olfactory bulb and in olfaction. *Physiol. Behav., 77,* 607-612.

van den Pol, A. (1995). Presynaptic metabotropic glutamate receptors in adult and developing neurons: autoexcitation in the olfactory bulb. *J. Comp. Neurol., 359,* 253-271.

Vassar, R., Chao, S.K., Sitcheran, R., Nunez, J.M., Vosshall, L.B., and Axel, R. (1994). Topographic organization of sensory projections to the olfactory bulb. *Cell, 79,* 981-991.

Wachowiak, M., and Shipley, M.T. (2006). Coding and synaptic processing of sensory information in the glomerular layer of the olfactory bulb. Sem. *Cell Dev. Biol., 17,* 411-423.

Wada, E., Shigemoto, R., Kinoshita, A., Ohishi, H., and Mizuno, N. (1998). Metabotropic glutamate receptor subtypes in axon terminals of projection fibers from the main and accessory olfactory bulbs: a light and electron microscopic immunohistochemical study in the rat. *J. Comp. Neurol., 393,* 493-504.

Yokoi, M., Mori, K., and Nakanishi, S. (1995). Refinement of odor molecule tuning by dendrodendritic synaptic inhibition in the olfactory bulb. *Proc. Natl. Acad. Sci., USA., 92,* 3371-3375.

Young, J.M., Friedman, C., Williams E.M., Ross, J.A., Tonnes-Priddy, L., and Trask, B.J. (2002). Different evolutionary processes shaped the mouse and human olfactory receptor gene families. *Hum. Mol. Genet., 5,* 535-46.

Yuan, Q., and Knöpfel, T., (2006a). Olfactory nerve stimulation-evoked mGluR1 slow potentials, oscillations and calcium signaling in mouse olfactory bulb mitral cells. *J. Neurophysiol., 95,* 3097-3104.

Yuan, Q., and Knöpfel, T. (2006b). Olfactory nerve stimulation-induced calcium signaling in the mitral cell distal dendritic tuft. *J. Neurophysiol., 95,* 2417-2426.

Reviewed by Dr. Kathryn A. Hamilton, Dept. of Cellular Biology and Anatomy, LSU Health Sciences Center, Shreveport, LA 71130-3932

In: Neural Pathways Research
Editor: F. L. Pichler, pp. 31-71
ISBN: 978-1-60456-214-9

Chapter II

DiI Tracing Is a Useful Tool for Studies of the Hypothalamic Connections during Perinatal Development

Irina G. Makarenko *
Optical Research Group
N. K. Koltzov Institute of Developmental Biology
Russian Academy of Sciences
Vavilov Str. 26, Moscow 119334, Russia

Abstract

Carbocyanine dye tracing was introduced into the neuroanatomical studies in the late 1980s (Godement et al.1987, Honig and Hume, 1989) and has been applied later for numerous central and peripheral axonal systems. The main feature of the fluorescent lipophylic carbocyanine dyes as perinatal tracers is that the markers can be applied on the prefixed brain tissue and thus avoid in vivo stereotaxic insertions specific for most of the axonal brain markers. Even though there are three known dyes the 1,1'-dioctadecyl-3,3,3',3'-tetramethylindocarbocyanine perchlorate (DiI) is widely used as a retrograde and sometimes anterograde tracer.

This review reevaluates advantages and disadvantages of the DiI tracing on the basis of our own experience in using this method in the studies of the development of afferent and efferent hypothalamic connections and literature data. We have studied the time schedule of the formation of the several hypothalamic axonal systems: hypothalamic connections with the posterior and intermediate pituitary lobes and median eminence, mammillary body tracts and reciprocal

* PHD, Head of Optical Research Group, Tel:+7-499-135-82-83; Fax:+7-499-135-80-12;
E-mail: imakarenk@mail.ru , or irina-makarenko@mail.ru

septohypothalamic connections. Some of the hypothalamic fiber systems develop unexpectedly early in embryogenesis but other forms only postnatally.

Principal facts and questions raised in these studies will be discussed: 1) Characteristics of the DiI distribution in the place of application; 2) Specificity of DiI labeling; 3) Comparison of the DiI and DiA; 4) Advantages of the confocal microscopic analysis of the material with carbocianine tracers; 5) What cells of the neuroepithelium are labeled with DiI; 6) Controls in the DiI experiments; 7) Advantages and limitations of the method; 8) Suggestions for new studies.

We are sure that even though DiI tracing needs long exposition time and extremely fine hand manipulations with the fetal brain material, it should be used as a very important tool for studying the early steps of the formation of brain connections. Such data will provide the basic knowledge concerning the main principles of brain network development.

Introduction

Tracing of the Brain Connections (Historic Overview)

Tracing of the brain connections is one of the major directions in the studies of the nervous system providing basic knowledge of brain organization and functioning. Numerous methods were developed for this purpose since the end of the nineteenth century. At the beginning most progressive ones were first neurodegenerative methods (Nauta, Gygax, 1951; Nauta, Ebesson, 1970; Fink and Heimer 1967) and their modifications (Wiitanen, 1969; Eager, 1970). They were based on the surgical damage or extirpation of the brain structures that resulted in the anterograde degeneration of the axons with terminal arborizations and can be revealed using special methods of silver impregnation. Many axonal projection systems were described with these methods in adult mammals and formed the basic knowledge of the brain connections including the hypothalamus (Raisman, 1966; Lammers and Lohman, 1974; etc.). Simultaneously fine details of terminal degeneration were revealed using electron microscopy after similar surgery (Gray and Hemlin, 1962; etc.).

The progress in the studies of brain connections started with the development of the methods based on the axonal transport of horseradish peroxidase (HRP) in vivo (Kristensson, Olsson, 1971; LaVail, LaVail, 1972; Watanabe, Kawana, 1980; Wayner et al., 1983) and later autoradiographyc visualization of the transported labeled amino acids (Cowan et al., 1972; Otellin, Sergeeva, 1982 etc.). Then a series of fluorescent dyes and other tracers such as Fast Blue, Diamidino Yellow, Phaseolus vulgaris leucoagglutinin, biotin-conjugated dextran amine were combined in different ways to provide multicolor labeling of the brain connections. Main principles of these methods, advantages and disadvantages of different tracers were carefully analyzed in the review by Vercelly and coauthors (2000). All these methods and their combination with immunocytochemistry (Bentivoglio, Chen, 1993; Arluison et al., 1994) were successfully used on the adult brain but could not be applied to the fetuses because these markers should be inserted in vivo.

The first axogenesis and tract formation in the central nervous system are obviously important events, as they begin the complex task of forming the network on which neural function depends (Easter et al., 1993). But formation of axonal connections and the time schedule of these processes in different neural systems are the most unknown events in brain development (Keshavan and Murray, 1997). Such studies became possible after advising lipophylic carbocyanine dyes as tracers of axonal connections on the prefixed brain material.

This review will reevaluate advantages and disadvantages of DiI tracing on the basis of our own experience in using this method in the studies of the development of afferent and efferent hypothalamic connections and literature data.

Carbocyanine Dye Tracing Method (General Description)

Carbocyanine dye tracing was introduced into neuroanatomical studies since the late 1980s (Godement et al.1987, Honig and Hume, 1986, 1989). Three lipophilic fluorescent carbocyanine dyes with a different spectrum of the fluorescence were most frequently used: 1,1'-dioctadecyl-3,3,3',3'-tetramethylindocarbocyanine perchlorate (DiI), 4-(4-dihexadecylaminostyryl)-N-methylpyridinium iodide (DiA) also called Di-Asp and 3,3'-dioctadecyloxacarbocyanine perhlorate (DiO). These markers and several of their new modifications are available from Molecular Probes (Eugene, OR). They can be used in the crystal form or as a solution on the fixed and live material on a whole brain, in sections and in the tissue culture. For pressure injections they are usually used as 5% solution in ethanol, 3% in dimethylsulphoxide (Honig, Hume, 1986) or 0.5–10% in dimethylformamide (vonBartheld et al., 1990). It is important that they are not actively transported along the axons by the axoplasmic flow, but diffused in both directions retrogradely and anterogradely via the lipid bilayer of the neuronal membrane and label both cell bodies and processes even in the fixed brain dissected from the scull. Thus they can be used on the fetal and neonatal material that made them very useful for the developmental studies.

DiI is visualized with rhodamine (TRITC) filter set and demonstrates bright orange-red fluorescence but is also visible as pale green in fluorescein (FITC) filter. DiA and DiO have a yellow-green fluorescence in the FITC set, whereas they are hardly seen in light red under the TRITC one (Haugland RP 1996). Thus different markers can be used simultaneously to trace several pathways in the same brain using multiple filter sets and double exposure photos (Mirnics and Koerber, 1995, Gabriele 2006). Such double-labeled material can also be analyzed using confocal microscopy (Bouret et al., 2000; Makarenko 2007). DiI is thought to be the best choice for tracing axonal connections and has been used to trace neural tracts within the peripheral and central nervous systems. Numerous examples of such tracing in the fish, gerbil and mouse with the specific accent on the studies of the human brain were reviewed (Sparks et al., 2000). Our recent data confirm this view and demonstrate that DiI distinctly labels axons for long distances in the fetal and postnatal brain and even visualizes dendrites with all arborizations (Figure 1). DiA labeling demonstrates retrogradely labeled neurons, but fluorescence in the axons and dendrites was always very low and usually it reveals better cell bodies (Figure 2) than fiber tracts. We have no experience in using DiO for neural tracing on the fixed brain but it was thought to be less efficient than DiI and DiA (Haugland, 1996) and was not widely used.

There is only one study aimed to evaluate the influence of different parameters on DiI tracing distances, but it was done on the peripheral nerves and spinal cord of humans and guinea pigs (Lukas et al., 1998) and the conclusions can not be applied for other species and may be for the central nervous system.

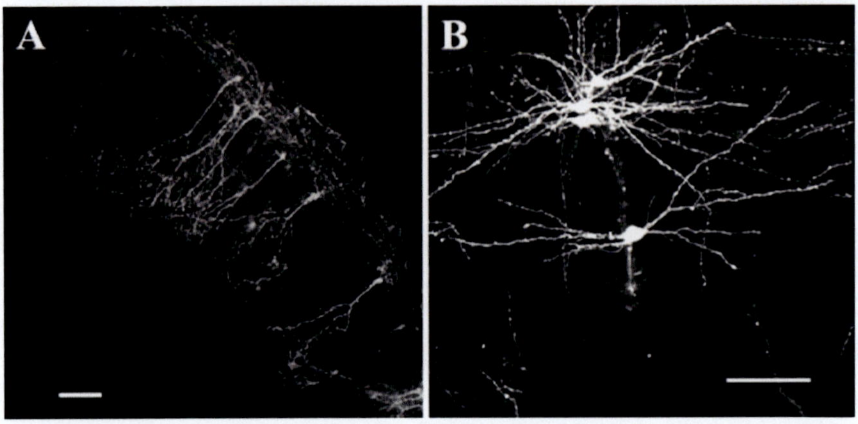

Figure 1. Confocal images demonstrated full dendritic tree of the neurons revealed in the hippocampus on P2 after DiI insertion into dorsolateral part of the lateral septal nucleus (A), and in the pons on E18 (B) after DiI insertion into the preoptic area. Scale bar 100 μm.

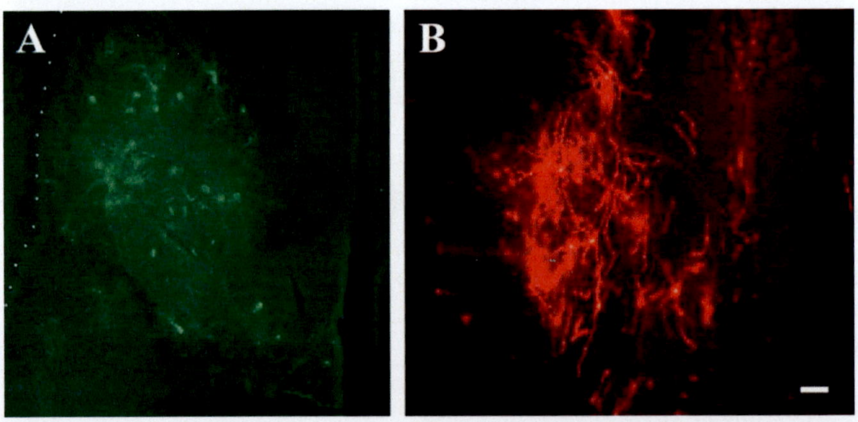

Figure 2. Different quality of DiA (A) and DiI (B) visualization of the lateral septal neurons projecting to the preoptic region on E18. DiI revealed dendrites better than DiA. Scale bar 100 μm.

They estimated that DiI diffuses at the rate 2,5 x 10^{-7}cm/sec. Unfortunately the authors used material storage at 37ºC and thin cryostat sections which can not be regarded as a best methodology for DiI tracing, because it resulted in very fast fading of the fluorescence in the section. Authors came to the conclusion that there are no differences in the tracing rate between adult and newborn animals. Comparison of the diffusion rates for different carbocyanine dyes provided different results. Some authors reported that DiA diffuses faster than DiI or at least at the same rate as DiI (Ozaki, Wahlsten, 1992), but our data and results of Vercelly (2000) were reversed.

Incubation of the labeled brains at 37°C was mentioned to increase the rate of the dye diffusion (Friedmann et al., 1991; Heredia et al., 1991; Lukas et al., 1998) and sometimes was used successfully and provided excellent results (Polston, Simerly, 2006). Unfortunately in some papers where heating was used during material exposure, the authors demonstrated photos with very low magnification (Kouki, Yamanouchi, 2007) or have no labeled dendrites at higher magnification (Lukas et al., 1998). Our own trials with the incubation temperature lead us to an impression, that heating not only produces difficulties in the brain sectioning, but usually resulted in a rather diffuse labeling of fiber tracts in comparison to the material stored at room temperature that is in agreement with the data obtained on a fish brain (Hofmann, Bleckmann, 1999) who additionally visualized transneuronally labeled cells and fibers. Thus we can not recommend such kind of storage for a DiI-labeled brain. Of course tracing time for each system under investigation will vary based on the developmental stage (size of the brain) and will be always longer for the postnatal brain. We have used storage time not less than 3-4 months for the fetal brains and 6 months for the postnatal material. Our data obtained on the different hypothalamic axonal systems strongly confirmed the great stability of the DiI labeling even after long storage (1-2 years). We had an example when a good DiI picture was preserved in the 100 µm vibratome section photographed at first in phosphate buffered saline and stored then in the paraformaldehide more than one year.

Long storage of the labeled brain material was regarded as a main problem of DiI tracing. To avoid it delayed aldehyde fixation and Fast Di-I was used to trace neural circuitry in the human material for distances of at least 20–40 mm over a few days (Sparks et al., 2000). This kind of methodological approach may be useful for the human material because of the inevitable postmortem intervals for human autopsy samples that are often as lengthy as 24–48 hours. Unfortunately, the resulted pictures had very low magnification and did not reveal clear neuronal labeling and fine details and consequently they could be used only for a general overview.

There are different views on the DiI as a tracer of the adult brain connections. Several authors mentioned that it is rarely effective in mature brains *in vitro* (Baltazart et al., 1994; Vercelly et al., 2000) but the other reported that DiI works well in the studies of the human postmortem brain (Zec et al., 1997; Dai et al., 1998). Our negative results in the DiI labeling of the hippocampal projections to the mammillary body and adjacent posterior hypothalamic regions during perinatal period (preliminary data) and difficulties in tracing adult hypothalamic projections to the posterior pituitary lobe support the possibility of the differential ability of DiI for tracing some fiber systems in adult rats. The quality of visualization of the hypothalamo-posterior pituitary projections in adult rats was also lower than in fetuses and pups. It can be supposed that such problems may be due to mielinization of the adult brain fiber systems.

DiI was used in the studies of the graft integration into the host brain. When crystals were placed on the graft inserted into the third ventricle, labeled axons were found in the host median eminence or in the host hypothalamus taking an arching trajectory toward the median eminence. Retrogradely labeled neurons in the host were few in number and largely confined to the host arcuate nucleus most DiI positive cells in the graft contained GnRH (Silverman et al., 1992). Reevaluating the outgrowth of fetal rat SCN/AH grafts implanted in the third ventricle of hamsters by using two methods shows that the use of a donor-specific neurofilament marker reveals more extensive SCN/AH graft outgrowth than DiI (Lehman et al., 1995, 1998).

Besides tracing axonal connections in the whole brain, carbocyanine dyes are also useful in studies of connections in slice cultures (Molnár and Blakemore, 1991), in dissociated cell culture (usually DiI and DiO) and in fixed cultures (DiI and DiA) (Vercelly et al., 2000). Time-lapse videomicroscopy of the DiI labeled living slice cultures raised the possibility of studying the development of axons in vitro (Yamamoto et al., 1997).

Photoconversion of DiI resulted in electron-dense labeling which can be analyzed using an electron microscope that was supposed to be used as a tool for studying of synaptogenesis (Bruce et al., 1997) but this methodology was not widely distributed.

Thus the carbocianine dye tracing method provides new possibilities for tracing of the brain connections and can be applied for solving different questions.

Tracing of the Hypothalamic Connections

The hypothalamus is a relatively small but complex division of the vertebrate brain that plays an especially important role in the integration of various neuroendocrine homeostatic processes, endocrine, autonomic, somatomotor control mechanisms and behavioral responses that guarantee the survival of the organism (Swanson, 2000; Simerly 2004). All these functions are realized using a complex of axonal intrinsic hypothalamic connections and by numerous relations with other brain structures and pituitary, including limbic nervous circuits and well described in adult vertebrates.

Special attention was paid for hypothalamic neurosecretory magnocellular nuclei and their projections known as hypothalamo-pituitary tract to the posterior pituitary lobe (PL) and projections from the parvocellular neurons to the median eminence (ME) (Reyes et al., 2005). Numerous studies have evaluated neuronal hypothalamic connections in adult mammals using in vivo axonal tracing techniques (Wiegand and Price, 1980; Ju et al., 1986; Arai et al., 1990; Larsen et al., 1991; Merchenthaler, 1992; Kalsbeek et al., 1993; Gayrard et al.1995; Gu, Simerly, 1997; Swanson 1986, 1987, 1991). Non-stereotactic approach was used for detection of axonal projections to the neurohemal organs in adults, based on the intravascular injections of axonal tracers (Armstrong, Hatton, 1980; Markakis, Swanson, 1997). Absence of a blood-brain barrier during the perinatal period of ontogenesis allows the tracer to reach not only neurohemal regions but the other brain regions. Several hypothalamic connections were described using immunocytochemical visualization of the neuropeptides and neurotransmitters in the neurons, fiber tracts and terminal fields inside the hypothalamus (Swanson, Kuipers 1980; Rhodes et al., 1981; Sofroniev 1982, 1983; Zimmerman, 1984; Whitnall et al., 1985; Hoffman et al., 1986; Watts et al., 1987a, b; Arai et al., 1990; Horvath et al., 1992; Dong, Swanson, 2003, 2004) or by combination retrograde tracing with immunocytochemistry (Watts, Swanson 1987; Kawano, Daikoku 1987 DA ; Mikkelsen JD; Niimi ea 1991, 1992; Morin ea. 1992; Vereczki ea., 2006; Dong, Swanson, 2006).

Recent studies reexamined hypothalamic structural organization and revealed specific neurocircuits of interconnected cell groups referred to as visceromotor or somatomotor (Thompson and Swanson, 2003) or belong to the neural circuits involved in regulation of stress integration (Herman et al., 2002). As a result a general knowledge of hypothalamic structure and connections was developed.

Hypothalamic neurosecretory neurons are the principal source of neurohormones, contributing to the neuroendocrine regulation of the adenohypophysis and some visceral organs (Baulieu, Kelly, 1990). Most neurons send their axons toward the median eminence and the pituitary posterior lobe (PL), lacking the blood–brain barrier. In these so-called neurohemal organs, neurohormones are released to the hypophysial portal circulation or general circulation reaching target cells in the pituitary anterior lobe and some visceral organs (kidney, uterine, and blood vessels) respectively (Ugrumov, 1992). The pituitary intermediate lobe (IL) in mammals is the only nonbrain target receiving hypothalamic afferent fibers (Chronwall et al., 1998). Indeed, the axons of dopaminergic neurons of the hypothalamic periventricular nucleus (Pe) and arcuate nucleus innervate melanotrophs of the IL providing the inhibitory control of the alfa-melanocyte-stimulating hormone secretion (Godement et al., 1987; Kawano, Daikoku, 1987; Luppi, 1986). Most of the neurons of the PV send their axons either to the external zone of the ME, or to other brain regions (Armstrong, Hatton, 1980; Swanson, Kuypers, 1980; Poulain, Jard, 1992).

Extensive reciprocal connections were described between the septum and nucleus of the diagonal band with the hypothalamus (Tavares et al., 2005). Septo-hypothalamic relations are topographically organized (Risold, Swanson, 1997) but do not involve magnocellular hypothalamic nuclei. Detailed tracing of the projections from the subfornical region of the lateral hypothalamus with *Phaseolus vulgaris*-leucoagglutinin revealed massive inputs to the lateral septum (Goto et al., 2005). Thalamus, amygdala, habenula, hippocampus and lower brain structures also have specific connections with the hypothalamus which will not be described here.

DiI was used as an axonal tracer in several studies of the hypothalamic fiber systems. Thus, an unknown pathway from the bed nucleus of stria terminalis to the arcuate hypothalamic nucleus via stria terminalis was visualized in the rat (Magoul et al., 1994). It was shown immunocytochemically that this fiber system also contains fibers immunoreactive for peptide adrenocorticotropin specific for the neurons of arcuate nucleus. DiI was successfully used on the postmortem human brain for visualization of bilateral optic nerve projections to both suprachiasmatic nuclei (Friedman et al., 1991). In amphibians DiI was used to trace tracts from the pituitary to the preoptic nucleus, chiasmatic hypothalamus (Artero et al., 1994). First trials of the combination of the electrophysiology with DiI tracing have shown that PVN-projecting neurons of the dorsomedial hypothalamic nucleus are less excitable than neurons of unknown projection (Bailey et al., 2003). Another electrophysiological study performed on the DiI-labeled PVN neurons provides the evidence that NO suppresses the activity of spinally projecting PVN neurons through potentiation of the GABAergic synaptic input (Li et al., 2002).

The development of hypothalamic connections was not studied systematically and some of the studies were performed only on the postnatal material. It was shown that the number of GnRH neurons projecting to the ME distributed in a diffuse ventromedial continuum from the septum through the anterior hypothalamus decreased in the more rostral regions with sexual maturation (Buchnan and Yellon, 1993). Ingrowing of retinal axons into the suprachiasmatic hypothalamic nucleus was revealed in the hamster after the fifth postnatal day, but suprachiasmatic efferents began to invade the appropriate targets during the second postnatal day and increased gradually in the subsequent days (Müller and Torrealba, 1998).

In mice lacking transcription factor Nkx2.1 that localized in the medial ganglionic eminence and the ventromedial part of the hypothalamus in normal developing animals, DiI demonstrated the abnormal crossing of ascending DA axons directed through the lateral hypothalamus to the striatum (Kawano et al., 2003). Authors regarded this fact as attributed to the elimination of exact chemorepulsive signals in the medial part of the mutant hypothalamus.

Video microscopy registered the migration of DiI-labeled cells in the coronal brain slices through the preoptic region from E15 mice and revealed both media-lateral and significant dorsal-ventral migration of cells in the preoptic region and anterior hypothalamus (Henderson et el., 1999). Authors suggested that this fact support the view that this rostral hypothalamic region develops from both telencephalic and diencephalic compartments.

Recently the main attention was devoted to the hypothalamic circuitry that modulates feeding and energy expenditure as disturbances in their normal formation during critical periods of the perinatal development may have long-term consequences on feeding behavior and body weight management (Grove et al., 2005). Elegant work used DiI tracing inside the hypothalamus for a short distance had revealed that arcuate nucleus projections to paraventricular and dorsomedial hypothalamic nuclei and lateral hypothalamic area are immature at birth (Bouret et al., 2004). Authors even attempted to quantify confocal images of the innervation of the paraventricular nucleus, but such trials are rather questionable, because of the serious difficulties in the standardization of the marker application on the brains at different developmental stages. An additional problem in such experiments is that DiI visualized all terminal arborisations but not the pure nerve terminals. In the later paper from the same laboratory (Polston, Simerly, 2006) authors described a new methodological approach for quantified study of the anteroventral periventricular (AVPV) hypothalamic nucleus projections to the GnRH neurons close to the OVLT region. To shorten the time of the brain storage in the fixative and combination with GnRH immunocytochemistry, they used CM-DiI (Molecular Probes). They carefully analyzed synaptic oppositions labeled with CM-DiI on the dendrites of GnRH neurons and developed a detailed method for their quantification on the volume reconstructions of the confocal images. This hard work did not reveal significant differences at studied ages (E19, P0 and P60). Other projections of AVPV were described without quantification. They have found that caudal projections of AVPV were delayed relative to projections to the OVLT. Labeled AVPV fibers reached the PVH during the first postnatal week, and fibers targeting the BSTp and LSv were not observed until the second and third postnatal weeks, respectively. Our preliminary results are in agreement with the last fact, because we did not find labeled neurons in the AVPV although described them in the ventral part of the medial preoptic area from E18. Developmental profile for each fiber system has spatial and temporal specificity but we know little about how hypothalamic circuitries are constructed during development (Polston, Simerly, 2006). The knowledge of the precise time consequence of the formation of hypothalamic connections may be important in understanding the potential of therapeutics/interventions for the treatment of obesity in humans (Grayson et al., 2006). As the hypothalamus is a famous integrator of several homeostatic processes that are required for the survival of vertebrates, it was predicted, that disruption of its development may be an unsuspected cause of later disorders (Caqueret et al., 2005).

The main purpose of our own serial studies was to perform general and systematic evaluation of the first steps of the hypothalamic connections development during perinatal ontogenesis in rats. Efficacy of carbocyanine dye tracing was proved in our first studies of the formation of hypothalamic projections to the posterior pituitary lobe and median eminence (Makarenko et al., 2000, 2001) and later applied for several afferent and efferent hypothalamic projection systems (Makarenko, 2007; Makarenko et al., 2005; Alpeeva, Makarenko, 2007). The main results of these works and our new findings will be reviewed here.

Experimental Procedures

Wistar rats provided by the Scientific Center of Biomedical Technology Russian Academy of Medical Sciences (Stolbovaya) mated overnight were used in our studies. The day of conception was defined as E0 and the day of birth as P0. Fetuses and postnatal animals were used at different developmental stages: from E13 to E21 and from P0 to P16 respectively. All animal procedures were made under anesthesia with pentobarbital (50 mg/kg body weight). All protocols of the manipulations with the animals have been approved by the animal care committee of Institute of Developmental Biology of Russian Academy of Sciences and experiments were carried out in accordance with European Communities Council Directive of 24 November 1986 (86/609/EEC).

Fixation. On early embryonic stages (E13-E15) fetuses were removed from the uterus, decapitated and dissected brains fixed by the immersion in 4% paraformaldehyde in 0.1 M sodium phosphate buffer (pH 7,2-7,4) for 12-24 hours at room temperature. Older fetuses (E16-E21) and postnatal rats were perfused transcardially, first with saline, and then with 4% paraformaldehyde and postfixed overnight by the immersion in the same solution.

Carbocyanine dye insertion, distribution and visualization in the place of insertion. DiI (DiA in few cases) were inserted into the fixed brain dissected from the scull. Methodology of the carbocyanine dye application can vary significantly depending of the task, studied projection system and age of the animal. Usually crystals of the marker were applied on the surface of the brain section or inserted into the small pine-hole into the brain using insect pine or fine glass electrode. Our insertions of the DiI or DiA crystals were performed under a dissecting microscope. Brain meninges were removed from ventral hypothalamic surface in the place of future insertion, the brain was dried by filter paper, and the tracer crystals were placed into a small incision. Large insertions are useful for general evaluation of the consequent formation of the hypothalamic connections. For this purpose we have used dye insertions which cover the whole part of the hypothalamus or other place of interest. It is important in fetuses when the process of neuronal migration is not finished and this approach gives us the possibility to avoid negative results. Changes of the exact position and volume of the insertion resulted in clear differences in the final labeling. Analysis of these data with subtraction of the results revealed a real source of studied projections, as in the case with tracing of the intermediate pituitary labeling or combination of the intermediate and posterior lobes (Makarenko et al., 2005). We have found that dye distribution in the place of insertion depends on the morphological features of this region and presence of the fiber tracts and capsules may prevent marker distribution to the adjacent structure that will be demonstrated later.

Brains with DiI were stored for 3-24 months in the paraformaldehyde in a dark at room temperature for DiI retrograde and anterograde diffusion along the neuronal membranes. Long storage mentioned previously as the main disadvantage of the DiI tracing method (Lucas, et al., 1998; Vercelly et al., 2000) is very important as it helps avoid negative results and makes sure that the marker had reached neurons even at a large distance from the place of insertion. Additionally it helps to label fine details of the neuronal morphology. After the storage brains were photographed under a dissecting microscope equipped with digital photocamera for documentation of the insertion places (Figure 3).

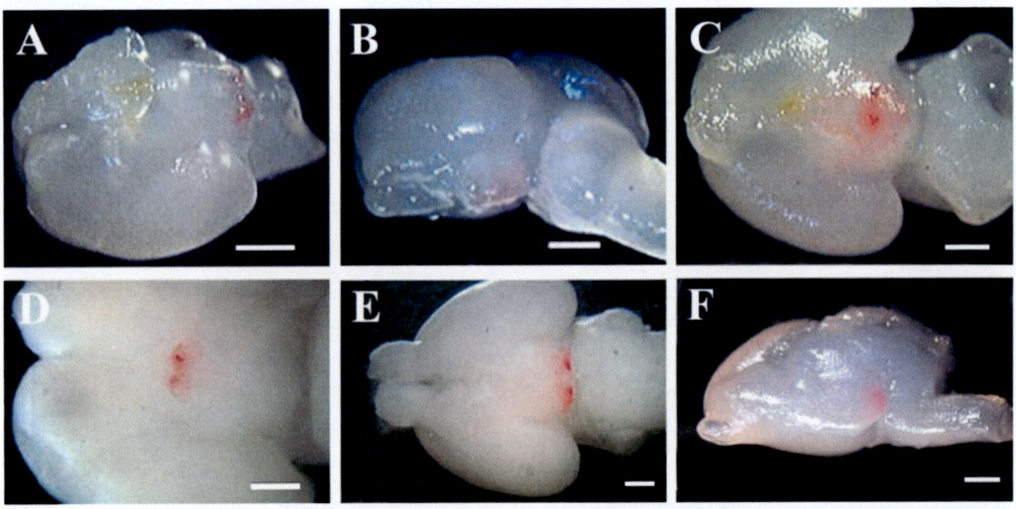

Figure 3. Examples of the rat embryonic brains with different types of carbocyanine dyes insertions (DiA – yellow, DiI – pink). A – preoptic area and posterior hypothalamus; B – mediobasal hypothalamus; C – preoptic area and mediobasal hypothalamus; D – mammillary body on the midsagittal section of the brain; E – mammillary bodies bilaterally; F – Anterior hypothalamus. Scale bar 1 mm.

Transport of DiI from the region with crystals looked as a diffuse or patch-like pinc staining (DiA – yellow). Region including insertion was analyzed first on the sections to identify its position and volume. We have found that an ultraviolet filter is very helpful for identifying the exact position of the marker crystals and for visualizing the real size of the dye distribution around them (Figure 4) in comparison to that visible with a rhodamine filter.

We performed several types of insertions into the perinatal brain material depending on the studied system:

(a) *Posterior pituitary lobe (PL). For tracing magnocellular hypothalamo-pituitary system (Makarenko et al., 2000).* Brain was removed from the scull with attached pituitary. The pituitary stalk was bent rostrally, displacing the pituitary on the median eminence. This manipulation resulted in turning of the PL from an inferior position to a superior one. Small cuts or pin-holes were made in the PL under a dissecting microscope for subsequent insertion of DiI crystals. Pituitary with crystals were fixed by a drop of liquid agar. It was found that the boundary between PL and IL may prevent DiI distribution from one lobe to another. No DiI diffusion to the IL was observed after its insertion into the PL in the cases where insertions did not damage this boundary (Figure 5 A).

(b) *Intermediate pituitary lobe (IL) or combinatory labeling of the PL and IL. For tracing of hypothalamic projections to the IL (Makarenko et al., 2005).* Fetal brains were prepared as for the PL insertions. Then, the DiI crystals were placed into several pinholes, running through the PL and penetrating the IL, and on the free edge of the IL around the PL. This was followed by DiI diffusion through both PL and along the tracs into IL (Figure 5B).

In young rats (P1-P20) perfused only with saline dissection of the pituitary was made. Anterior lobe was mechanically detached from the IL and bent rostrally on the median eminence, but the PL, covered by IL, remained in its original position without interruption of the pituitary stalk. The brains with attached splinted pituitaries were fixed overnight by immersion in 4% paraformaldehyde. Then, DiI crystals were placed on the surface of the IL and attached by a drop of 3% agar. Usually such applications resulted in the labeling of both IL and PL. In a few cases with flat DiI application on the surface of IL after pituitary dissection it resulted in single IL labeling, without penetration into the PL.

(c) *Median eminence (ME). For tracing parvicellular hypothalamic projections to the portal vessels into the median eminence (Makarenko et al., 2001).* DiI was inserted into the incision in the ME rostrally to the pituitary stalk.

(d) *Different rostrocaudal levels of the hypothalamus. For retrograde tracing of the septal projections to different rostrocaudal levels of the hypothalamus (Makarenko, 2007).* DiI was inserted in the pine holes made from the ventral surface into the preoptic area, mediobasal hypothalamus and posterior hypothalamus. Most anterior insertions were made rostrally or caudally to the optic chiasm, located predominantly in the medial or lateral parts of the preoptic area or covering both. Mediobasal hypothalamic insertions were performed rostrally to the pituitary stalk at the level of the median eminence laterally to it. Posterior hypothalamic DiI applications were made caudally from the pituitary stalk into the mammillary bodies or adjacent regions. Several cases on E18–E21 received applications of two markers: DiI was inserted in the mediobasal hypothalamus and DiA into the preoptic area and anterior hypothalamus.

(e) *Mammillary body (MB). For retrograde and anterograde tracing of the mammillary body tracts (partly in Alpeeva and Makarenko, 2007).* DiI crystals were placed into a small incision made in the MB ventral surface on the whole brain or applied on the MB on the brain dissected sagittaly along the midline. At early embryonic stages (E14-E16), MB had no capsule of nerve fibers, characteristic for adult rats. Application was proved by visualization the thick principal mammillary tract extending dorsally from the site of DiI insertion caudally to the pituitary stalk (Figure 4C-E; Figure 6D). Displacement of the insertion place rostrally from the MB to the area of the mediobasal hypothalamus did not reveal this tract and was used only as a control. If the place of insertion was localized just rostrally to MB without damaging of its fiber capsule the marker spreads in the mediobasal hypothalamus and occupies premammillary, supramammillary and posterior hypothalamic regions along the MB capsule, but do not penetrate the proper MB (Figure 6B). Development of the fiber capsule of the mammillary body from E18 onwards restricted DiI spreading outside the MB margins in the cases when incision with crystals was located precisely inside MB (Figure 6C, D). Penetration of the MB capsule led to the DiI diffusion from the proper MB to adjacent nuclei of the posterior hypothalamus.

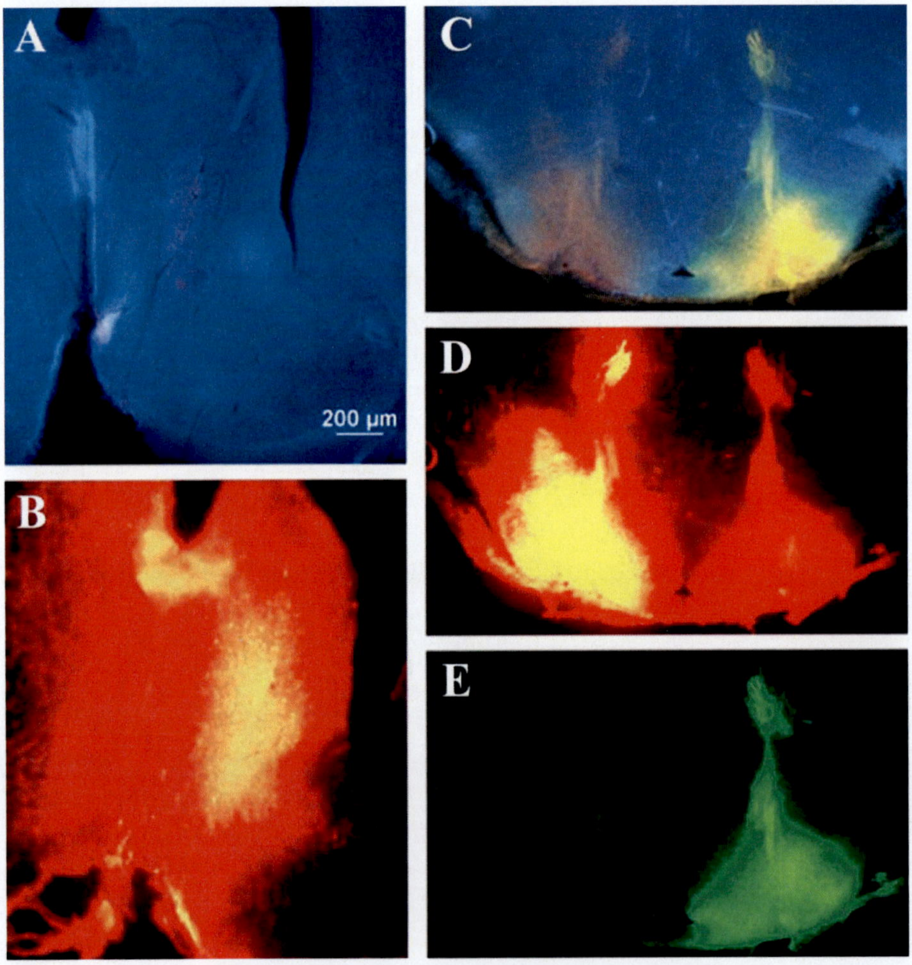

Figure 4. Places of DiA and DiI insertion in the lateral septum (A, B) and mammillary bodies (C-E) visualized with different filter sets: ultraviolet (A, C); rhodamine (B, D) and fluoresceine (E). LS - lateral septal nucleus; LV – lateral ventricle; MB – mammillary body. Arrow – principal mammillary tract. Scale bar 200 μm for all.

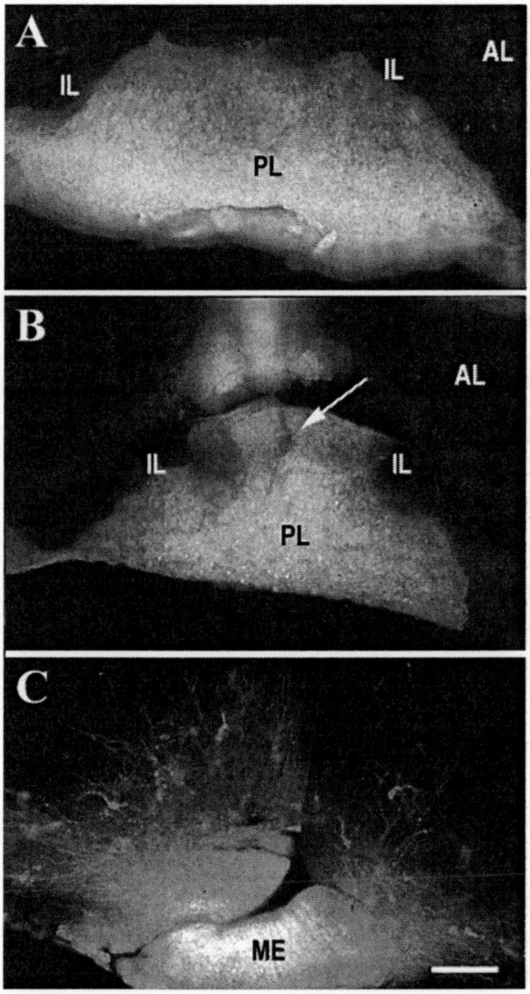

Figure 5. DiI distribution after insertion into the posterior pituitary (A, B) and median eminence (ME). Labeling occupies posterior (PL) and part of the intermediate (IL) pituitary lobes. Scale bar 100 μm.

(f) *Lateral septal nucleus. For tracing hypothalamic projections to the lateral septum (preliminary data).* For this purpose rostral part of the brain was cut in the coronal plane at the level of the rostral septum and DiI crystals were inserted in the small incision in the lateral septal nucleus.

(g) *Control insertions which did not provide neuronal labeling in the brain tissue.*

- DiI application on the surface of non-perforated meninges on the ventral surface of the hypothalamus or posterior pituitary.
- Interruption of the hypophysial stalk following the DiI insertion into PL
- DiI administration to the adenohypophysis

(h) *Control insertions to the regions adjacent to the place of main insertion to confirm the stability of the results.*

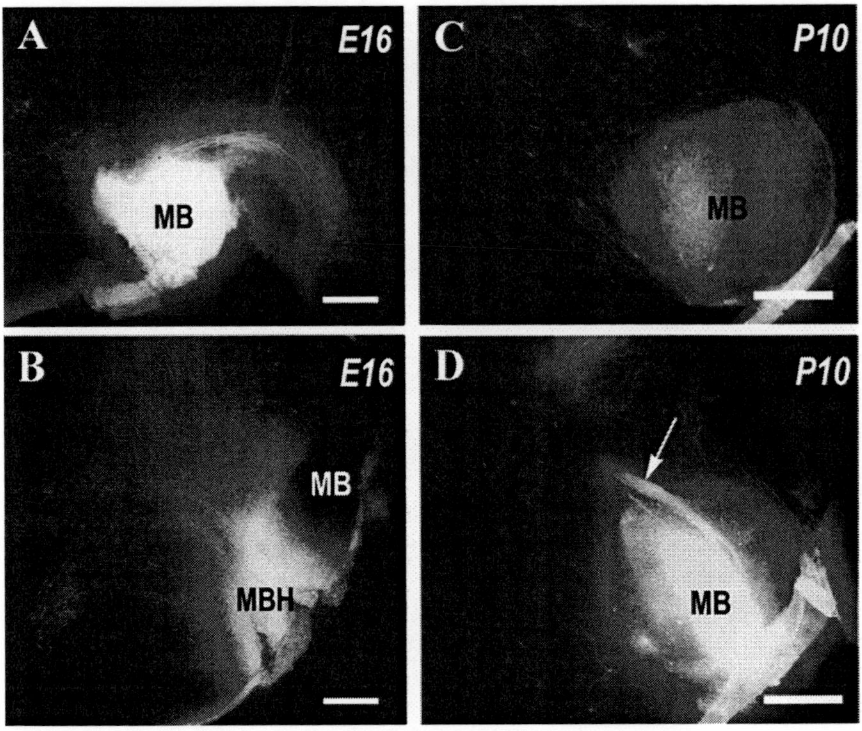

Figure 6. DiI inserted in the mammillary body (MB) without the damage of its fiber capsule localizes precisely inside MB (A, C, D), but after insertion into the mediobasal hypothalamus (MBH) it does not label MB (B). Arrow – principal mammillary tract, arising from MB. Scale bar 200μm.

Sectioning and Microscopy

Serial vibratome brain sections, 80-100 μm-thick, were made in the coronal or sagittal plane, collected in PBS, mounted on the gelatinized or Super Frost/Plus slides. Different embedding media were used for coverslipping: PBS, PBS-glycerol, Vectashield (Vector, USA) and moviol (Calbiochem, Germany). Slides were stored at 4-8°C until examination. The last one is the best for serial sections, because it became tightened and thus prevents the movement of the cover glass.

Sections were examined under different types of fluorescent microscopes equipped with a rhodamine, and fluorescein filter sets for viewing the orange-red DiI fluorescence, yellow-green fluorescein one and ultraviolet: Olympus AH-3, (Japan), Diaplan (Leitz, Wetzlar, Germany), Leica DMRXA2 (Germany) and photographed using negative photo film Ilford-400, Kodak Ectachrom-400 or with a digital photo camera.

Visualisation of DiI in the place of application or insertion deals with several problems. It is often very difficult to justify the exact size of the place because of high brightness of the labeling, as it spreads along the processes in both directions and labels numerous afferent and efferent fibers around the real place of the crystals insertion. Viewing and photographing of such section with rhodamine filter gives different pictures depending of the brightness adjustment (Figure 4, 6). Moreover, in the cases with both markers close to each other, DiA place of insertion is visible with rhodamine filter (Figure 4 D right side).

Some sections were visualized additionally using blue ultraviolet filter, to clarify the place of the marker insertion (Figure 4 A, C). This was important because the larger area visualized as 'bright' in the actual channels of DiI and DiA fluorescence on the thick vibratome sections did not represent the valid site of implant because of the dye diffusion and additional fluorescence of the labeled fibers and neurons close to the insertion place. Examination in the ultraviolet channel did not reveal such fibers and thus provided the real position of the crystals and the field of the dye diffusion.

Some sections were scanned in the confocal microscope Leica TCS SP (Germany). Confocal images were processed using the standard Leica LCS software. The brain regions were defined according to the atlases of the developing (Paxinos et al., 1991) and adult rat brain (Paxinos, Watson, 1986).

The fading of the dye labeling in the sections was mentioned as one of the problems (Vercelly, 2000). It can be prevented by anti-fading mounting mediums such as Vectashield. Even more there are recommendations to use a phosphate buffer or even 4% paraformaldehyde as a mounting medium. We have used such methods at first, but it is not easy because one needs to store the slides flat in the moisture chamber. Both solutions dried quickly under the cover glass. We recommend the Moviol (Calbiochem) which can be prepared and stored in aliquots in the freezer and which polimeraized on the slides at room temperature. DiI labeled structures can be stored after coverslipping in moviol for several months.

Specificity of the DiI Staining

We did not find non-specific DiI or DiA staining in any experiments. There were no signs of transsynaptic DiI diffusion in our material independent of the direction of its diffusion. In the cases with DiI insertions into the PL only supraoptic, paraventricular and accessory peptidergic hypothalamic nuclei contained retrogradely labeled magnocellular neurons, although they receive heavy innervation from other hypothalamic nuclei. Insertions into ME resulted in neuronal labeling in the same nuclei, but there are significantly more parvocellular cells projecting to ME. Anterograde diffusion along the axons of the mammillo-thalamic tract resulted in heavy preterminal network in all anterior thalamic nuclei, but not in the neurons. No staining of hypothalamic neurons was observed following: a) DiI implantation into PL with subsequent interruption of the hypophysial stalk, b) DiI insertion into the anterior lobe, c) DiI application on the surface of the meninges without their damage. Nevertheless, in the last case DiI-labeled neuroepithelial cells or tanycytes were observed. This was most probably a result of the DiI diffusion, first, along the meninges and, then, via basal processes of the neuroepithelial cells or tanycytes toward their cell bodies located rather near to the surface of the third ventricle.

Developmental Schedule of the Formation of Hypothalamic Connections (Main Results)

According to our data, DiI insertions in the pituitary and brain structures listed above resulted in staining of neuronal cell bodies and fibers starting from E15. Elier on E13-14 DiI insertions in the preoptic-anterior hypothalamic regions or in the mediobasal hypothalamus did not result in the labeling of the neurons and fibers even close to this place (Figure 7 A). Rhodamine fluorescence was visible only in the blood vessels (Figure 7 B) penetrating into the brain tissue and in some of the neuroepithelial cells.

Hypothalamo-Neurohypophysial Magnocellular System

Bilateral supraoptic (SO) and paraventricular (PV) nuclei, are the main magnocellular neurosecretory centers of the hypothalamus. Their neurons synthesize the peptide neurohormones vasopressin and oxytocin, which are released in the blood flow in the posterior pituitary lobe and portal system of circulation in the median eminence (Hatton, 1990; Falke, 1991). In addition to these centers, there are so-called accessory magnocellular nuclei described in rats (Peterson, 1966) and other mammals (Grinevich and Polenov, 1994). Neurons projecting axons to the posterior pituitary lobe were found in many accessory hypothalamic nuclei in studies on adult rats with the use of various markers (Sherlock et al., 1975; Fisher et al., 1979).

DiI insertion to the PL at E15-16 resulted in staining of large bilateral accumulations of cells in the primordium of the SO extending toward the dorsal diamond-like swelling of the third ventricle (Figure 8 A). These uni- and bipolar cells were characterised by long, beaded, bifurcated processes, suggesting that they are neurons. This location is similar to the distribution of differentiating magnocellular vasopressin and oxytocin neurons known as originated from the neuroepithelium of the diamond-like swelling of the third ventricle, in rats at E12-E14 (Altman and Bayer, 1978a, b, 1986; Buijs, 1992; Trembleau et al., 1995). Later on E17-19, the major DiI-labeled neurons were gathered along the lateral edges of the optic chiasm and tracts, i.e., in the SO (Figure 8 B, C), whereas only a few fluorescent neurons still were seen between the SO and the presumptive PV at the third ventricle. Thus labeled cells described earlier between PV and SO represent migrating magnocellular neurons, although they were identified immunocytochemically only from E16 and their axonal projections were not seen in the ME and the PL until E17 and E18, respectively (Buijs, 1992; Buijs et al., 1980; Whitnall et al., 1985). The possible view, that these cells may belong to radial glia was discussed and declined in the original paper (Makarenko et al., 2000).

In contrast to the SO, only occasional fluorescent neurons were seen in the primordium of the PV at E16 and E17 (Figure 8 A) suggesting that the axons of differentiating neurons of the proper PV reach the PL significantly later than neurons of the SO. These data appear to be consistent with earlier observations of the slower settling and differentiation of magnocellular neurons of the PV when compared to those of the SO (Laurent et al., 1989; Trembleau et al., 1995). A moderate number of the labeled neurons projecting to PL from the magnocellular part of PV in young rats is in agreement with earlier observations of a rather limited number of axonal projections from the PV to the PL even in adults (Arai et al., 1990).

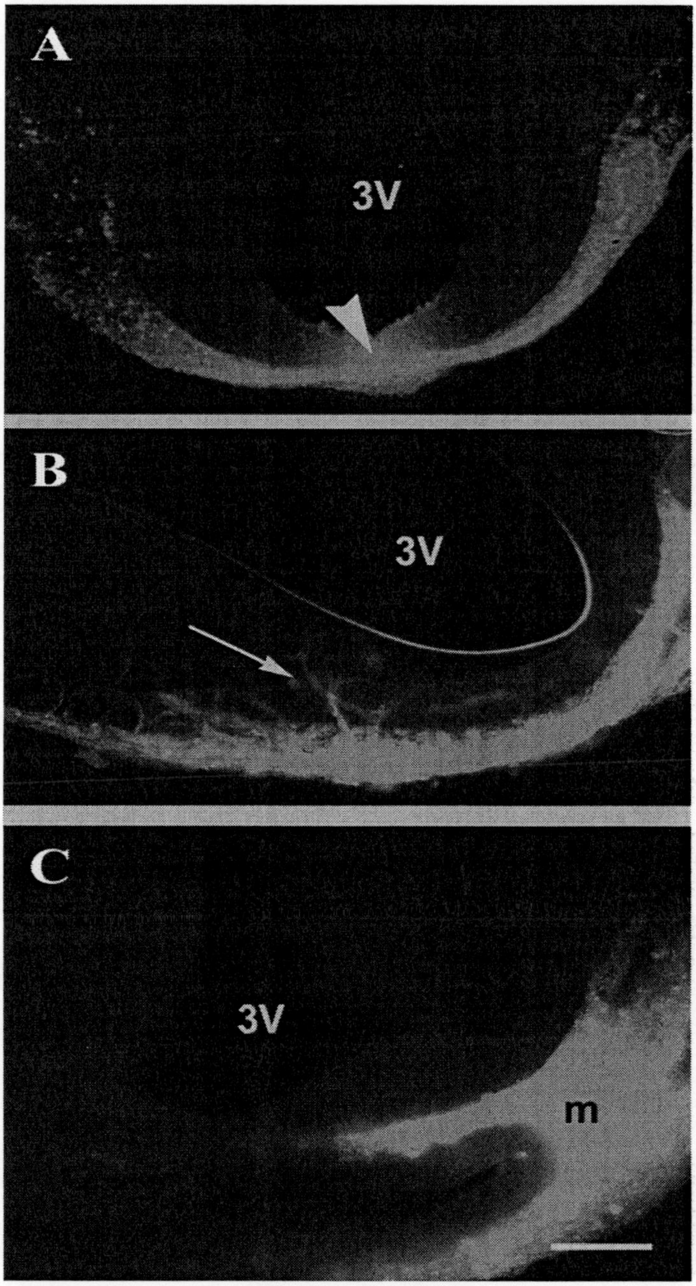

Figure 7. DiI insertion into the ventral hypothalamus on E13 (arrow head) does not reveal any labeled structures in the brain tissue (A), sometimes labeled vessels (arrow) were ingrowing into the anterior hypothalamus (B). DiI labeling of the meninges (m) on E16 and later stages does not produce neuronal labeling in the brain (C). Scale bar 100 μm.

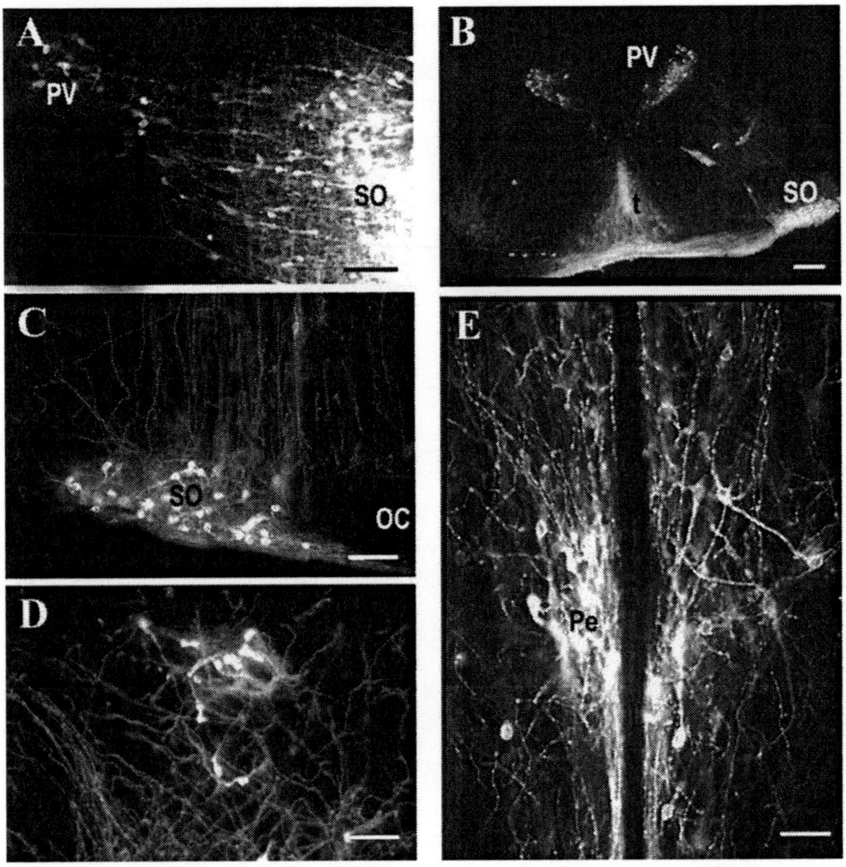

Figure 8. Hypothalamic neurons projecting axons to the posterior (A-D) and intermediate (E) pituitary lobes. A) First neurons occupying presumptive supraoptic (SO) and paraventricular (PV) nuclei; B) Magnocellular neurons organized compact SO and PV nuclei; C) SO nucleus at higher magnification; D) Dorsolateral accessory nucleus revealed postnatally; E) Periventricular nucleus is the main source of intermediate lobe innervation. Scale bar 100 μm.

Accessory peptidergic nuclei also consists of magnocellular neurons projecting axons to the PL (Rhodes et.al., 1981; Sofroniew, 1983; Ju et al., 1986). DiI tracing demonstrated that bilateral retrochiasmatic nucleus sent axons to the PL first on E16-17, that is later than those from the SO but slightly earlier than the axons from the PV (Makarenko et al., 2000). In contrast to the retrochiasmatic nucleus, the axons of neurons of other accessory nuclei (the dorsolateral nucleus (Figure 8 D), the nucleus of the medial forebrain bundle, the circular nucleus and the commissural nucleus) first reached the PL postnatally (Makarenko et al., 2002) although these neurons begin to express specific synthesis before birth (Buijs et al., 1980; Whitnall et al., 1985).

Rare DiI-labeled cells were also observed postnatally in the periventricular region but the labeling in this region may be the result of DiI staining of the tanycytes (Figure 8B). These data are in agreement with previous studies showing that in adults the PL receives major projections from the vasopressin and oxytocin magnocellular neurons and minor projections from the hypothalamic parvocellular neurons, e.g., dopaminergic neurons (Swanson, Kuipers, 1980; Kawano, Daikoku, 1987).

Our results indicate that the initial axons of magnocellular neurons reach the PL unexpectedly early in embryogenesis, namely, shortly after birth of these neurons and, perhaps, before occupation of their final position in the SO. On the other hand, this coincides with the onset of neuropeptide synthesis, but before visible accumulation of vasopressin- and oxytocin-immunoreactive materials in axonal terminals in the PL.

Hypothalamic Projections to the Intermediate Pituitary Lobe (IL)

Although the distribution of the DiI-labeled neuronal cell bodies and fibers in the hypothalamus of rat fetuses following the DiI administration to the IL + PL together resembled that following the DiI staining of the PL (Makarenko et al., 2000), certain features in the distribution of the DiI-labeled structures in the former have been distinguished. The fluorescent neuronal cell bodies were observed in these fetuses not only in magnocellular nuclei (SO, dorsolateral part of the PV, the accessory retrochiasmatic nucleus), but also in the parvocellular nuclei (periventricular (Pe), arcuate nuclei). According to our data, the first few axons arrive from the Pe in the IL as early as at the end of fetal life that coincides with the first appearance of TH-immunoreactive neurons in the Pe (Jaeger, 1986; Ugrumov et al., 1989). From these data taken together it indirectly concludes that the IL is innervated by dopaminergic neurons of the Pe from the end of fetal life. The number of the fluorescent cell bodies in the Pe increased progressively in postnatal rats with age (Figure 8 E), which coincides with the first appearance of the tyrosine hydroxylase-immunoreactive fibers in the IL (Gary et al., 1992). This most likely means that axonal projections to the IL in neonatal rats belong to dopaminergic neurons of the Pe.

In addition to the Pe, the parvocellular part of the PV contained DiI-labeled neuronal cell bodies in neonatal rats following the DiI labeling of a single IL (Makarenko et al., 2005). The neurons located in the magnocellular part of the PV projected their axons to the PL (Makarenko et al., 2000; Swanson, Kuipers, 1980; Swanson, Sawchenko, 1983) were not labeled with DiI in the same animals. Besides, the Pe and the parvocellular PV, the fluorescent neurons were regularly seen in the arcuate nucleus of neonatal rats following the DiI staining of the IL + PL or only in IL. These data show that the IL receives afferents from the arcuate nucleus in rats shortly after the birth.

The number of the hypothalamic fluorescent neuronal cell bodies and fibers increased progressively from P10 to P20. From comparison of their distribution after the DiI administration to the IL + PL or to the PL alone it follows that most fluorescent neurons in the Pe and arcuate nucleus project their axons to the IL. This conclusion is in agreement with previous data obtained by other techniques in adult rats (Goudreau et al., 1992; Kawano, Daykoku, 1987; Vanhatalo et al., 1995). Additional DiI labeled neurons projecting to the IL were revealed in young rats along the dorsolateral border of the ventromedial nucleus.

Thus, the hypothalamic parvocellular neurons, first, of the Pe and, then, of the mediobasal hypothalamus, provide an innervation of the IL in rats from the end of fetal life to prepuberty.

Hypothalamic Projections to the Median Eminence

It was repeatedly demonstrated that the majority of the hypothalamic nuclei consist of so-called parvocellular neurons, producing adenohypophysiotropic neurohormones, peptides and dopamine (Halász et al. 1962; Burgus et al. 1969; Zimmerman et al. 1984; Swanson 1987; Ugrumov 1991). A considerable number of parvocellular neurons project their axons to the median eminence (ME) providing the pathway for the neurohormone transfer via hypophysial portal circulation to the adenohypophysial target cells (Hoffman et al. 1986; Ugrumov 1992). The axons containing adenohypophysiotropic neurohormones or the enzymes of their synthesis were first detected with immunocytochemistry usually not earlier than on E16 (Daikoku et al. 1985; Ishikava et al. 1986; Okamura et al.1991). However, immunocytochemistry defined really the accumulation of neurohormones in the axons or terminals that could be visualized immunocytochemically but not the first arrival of the axons in the ME.

DiI implanted into the median eminence rostrally to pituitary stalk labeled both fibers and neurons in the hypothalamus from first rare cells on E14 with subsequent increase during the next few days (Makarenko et al., 2001). They were widely distributed through the hypothalamus and in the ventromedial region of the more rostral forebrain. Till the 20th embryonic day, the majority of the fluorescent neurons were concentrated mainly in the paraventricular nucleus (dorsal and medial parvocellular parts), the arcuate nucleus and to a lesser extent in the medial preoptic nucleus, the supraoptic nucleus, the diagonal band, and the retrochiasmatic nucleus. In neonates, DiI-labelled neurons appeared additionally in the accessory dorsolateral nucleus, medial preoptic area lateral to the diagonal band, anterior hypothalamic area, and in the anterior periventricular nucleus. Thus, the axons of differentiating neurons arrive in the median eminence from the 14th embryonic day till the neonatal period, providing the pathway for the neurohormone transfer to the hypophysial portal circulation.

Mammillary Body Connections

Mammillary bodies are paired brain structures surrounded by capsules of nervous fiber located in the ventral part of the posterior hypothalamus. Their axonal connections are organized as rather compact fiber tracts which represent important pathways that are related functionally to the limbic system that plays a major role in emotional and motivational activity and other basic psychological functions of the brain (Morgane et al., 2005). They were included in the limbic "Papez circuit" were information is transferred from the hippocampus to MB and then through the anterior thalamic nuclei to the cingular gyrus. MB tracts were demonstrated in adult mammals using the silver impregnation method and vital tracers: most neurons medial and lateral MB nuclei send their axons to the anterior thalamus via mammillothalamic tract (Watanabe and Kawana, 1980; Seki and Zyo, 1984) and to the midbrain via mammillotegmental tract (Petrovicky, 1973; Cruce, 1977; Allen and Hopkins, 1990). Afferent projections to MB from the ventral and dorsal tegmental nuclei are realized via the fiber system called mammillary peduncle, rather than via the mammillotegmental tract (Petrovicky, 1973; Allen and Hopkins, 1990).

We have first described successively the development of the main projection systems connecting MB with the midbrain tegmental area, and the formation of the midbrain ventral and dorsal tegmental nuclei using carbocyanine tracers (Alpeeva, Makarenko, 2007). Axons of MB neurons start to grow and form the mammillo-tegmental tract curved along the mesencephalic flexura even on E14 before the MB is formed as an independent brain structure surrounded by a fibrous capsule (Figure 9 A). Growth cones on the ends of the fibers demonstrated that mammillo-tegmental tract on E14-E19 consisted of the growing MB efferents. On E19-E21, when DiI was distributed precisely inside MB, a distinct group of labeled neurons was visualized in the midbrain, which is supposed to be the primordium of the tegmental nuclei (Fig, 9 B). This was confirmed by the fact that DiI stained also the mammillary peduncle on these stages that consists of axons from the tegmental neurons and terminates in the MB. Prominent differentiation of the tegmental nuclei took place postnatally. From P2 to P10 single tegmental group divides in two (dorsal and ventral tegmental nuclei) with specific neuronal localization (Alpeeva, Makarenko, 2007). DiI application onto the midbrain tegmentum on E19-E21 revealed many DiI-positive nerve fibers in the mammillotegmental tract with retrogradely labeled cells in the MB; fasciculus retroflexus with the neurons in the habenula and even in the growing mammillothalamic tract. The last feature confirms the view, that mammillothalamic tract is formed by the collaterals of the fibers of the mammillotegmental tract (Van der Kooy et al., 1978; Hayakawa and Zyo, 1989).

Our recent results have revealed that the mammillothalamic tract starts this dividing from the mammillotegmental tract in the rat on E18 (Figure 9 D).

During next days of the prenatal development it growth through the hypothalamus and thalamus and reach ventral level of anterior thalamic nuclei on E21 (Figure 9 E). We have found an interesting feature of its development: confocal scanning of the growing end of mammillothalamic tract revealed growth cones on all fibers (Figure 9 C). This fact was described for the first time and assumed that in the compact mammillothalamic tract all fibers grow simultaneously. From E21 the first terminal network was visualized in the anteromedial thalamic nucleus. Just after birth this innervation increased gradually and to P5 spread inside all three ipsilateral anterior thalamic nuclei. These projections are mostly unilateral, but contralateral terminal DiI labeling was revealed only in the anterodorsal thalamic nucleus (Fig 10 A, B). Terminal network consisted of fine beaded nervous processes tightly surrounded all neurons inside anterior thalamic nuclei but not in the surrounded brain regions (Figure 10 C). Sometimes single terminal arborizations so called "terminal brush" can be found at the periphery of the nuclei (Figure 10 D). Visualization of numerous terminals labeled with DiI raised the question whether they are functionally active or not. To realize it we performed immunocytochemical visualization of synapsin known as a protein marker of functioning brain synapses (Melloni, De Gennaro, 1992; Orlando, 2004; Castejón, et al., 2004). Because of very long storage of our material in the paraformaldehyde additional brains from the P5 rats fixed by perfusion and postfixation for 2-16 hours were used. Synapsin staining was visualized in all antherior thalamic nuclei as small dots in the neuropil around neurons (Figure 11) and density of their distribution correlated with density of the DiI labeled terminal network in each nucleus.

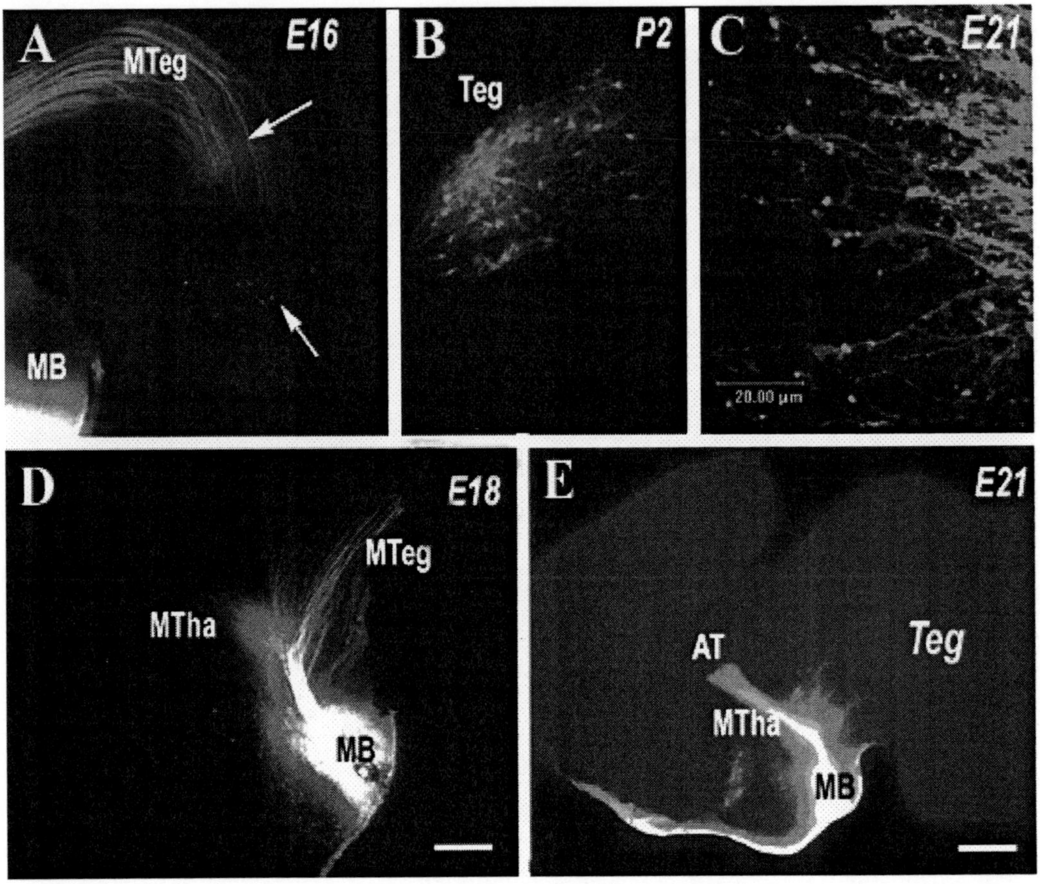

Figure 9. Growing mammillary body tracts on the sagittal brain sections on different stages of perinatal development after DiI insertion into the mammillary body (MB). A) Mammillotegmental tract (MTeg) was visualized in the midbrain from E14; B) Neurons in the tegmentum (Teg); C) Confocal scanning of growth comes on the ends of all fibers of the mammillothlamic tract (MTha); D) First signs of MTha dividing from MTeg; E) MTha reach the ventral level of anterior thalamus (AT); arrows – labeled neurons in the developing midbrain. Scale bar 200 µm for A, B, D, E and 20 µm for C.

The results obtained confirmed existing concepts on early development of the mammillotegmental tract in the brain of mice and rats (Mastick and Easter, 1996; Valverde et al., 2000) and specify the time schedule of the formation of reciprocal mammillotegmental connections. Thus, the axons of MB neurons start to grow and form a tract even before the MB is formed as an independent brain structure surrounded by a fibrous capsule. Formation of mammillothalamic tract starts prenatally from E18 and developed dense and functionally active innervation of anterior thalamic nuclei on P5.

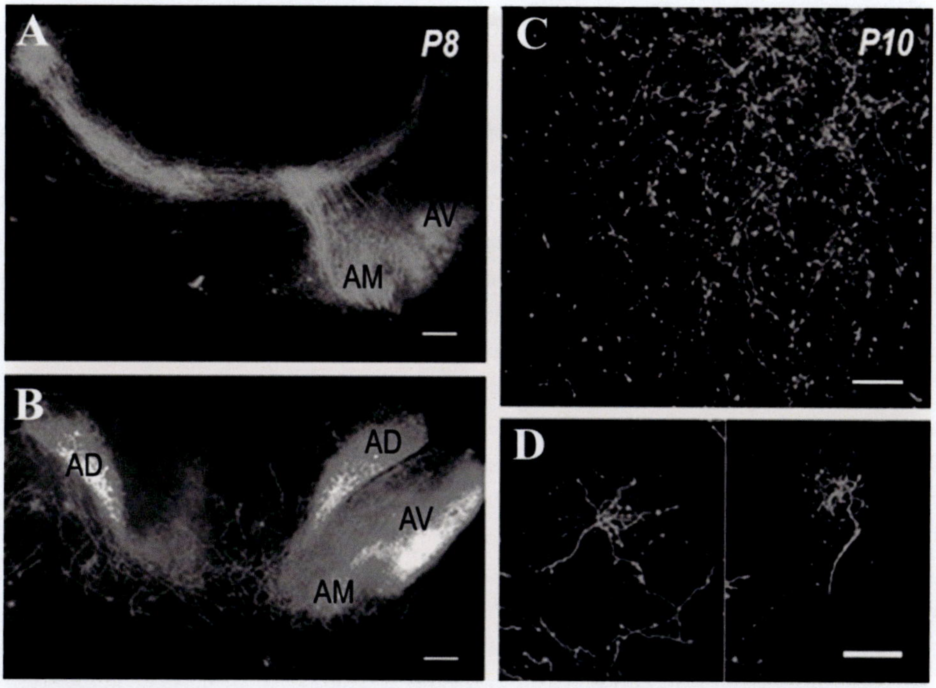

Figure 10. Unilateral DiI insertion into the mammillary body visualized innervation of all three ipsilateral anterior thalamic nuclei (anteroventral = AV; anteromedial = AM; anterodoral=AD) and only one AD on the controlateral side (A, B). Confocal images of the terminal network in the anteromedial thalamic nucleus and single terminal brushes at the periphery of the nucleus (C, D). Scale bar 200 μm.for A and B; 40 for C and D.

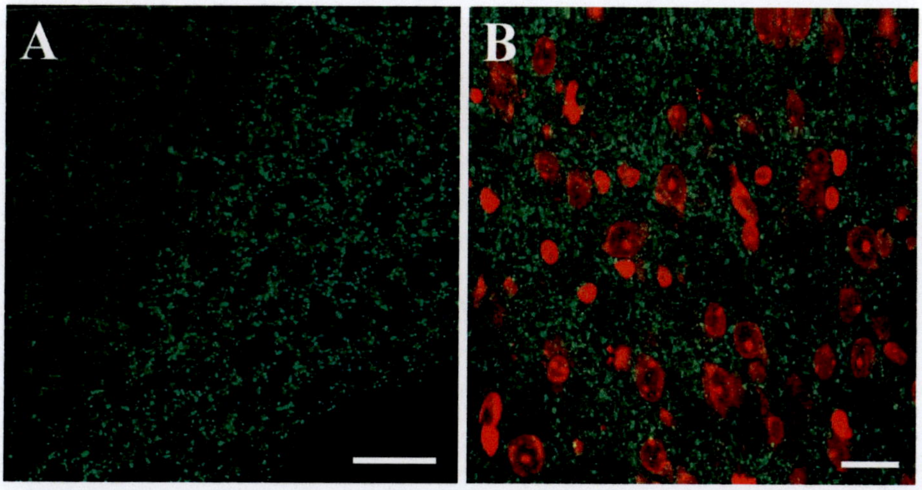

Figure 11. Immunocytochemical visualization of synapsin I in the anteromedial (A) and anteroventral (B) thalamic nuclei on P6. On B cell nuclei were stained with propidium iodide. Scale bar 40 μm for A and 20 μm for B.

Projections of the Lateral Septal Nucleus (LS) to Different Hypothalamic Regions

The septum is an integrative relay station of the limbic system, interconnecting the limbic forebrain with the hypothalamus and brainstem (Risold and Swanson, 1997; Herman et al., 2002). Its extensive connections with the hypothalamus were visualized and described in adult rats using different methods, such as, degeneration techniques, autoradiography (Swanson and Cowan, 1979), horseradish peroxidase and other kinds of tracing (Wayner et al., 1983; Chiba and Murata, 1985; Simerly and Swanson, 1986) and electrophysiology. The organization of the rat lateral septal projections to the hypothalamus were re-examined and described in detail with new phaseolus-vulgaris and fluorogold tracing methods (Risold and Swanson, 1997). It was shown that these connections are topographically organized and belong to specific hypothalamic behavioral systems.

We have first described the prenatal development of septal projections to the hypothalamus in rats, using carbocyanine dyes (DiI and DiA) as retrograde tracers (Makarenko, 2007). It was shown that septo-hypothalamic projections are unilateral and the number of retrograde-labeled neurons revealed in the septum correlates with the region and size of the DiI application in the hypothalamus. The starting point of the development of the septal afferents to the hypothalamus was about E14.5, when first septal neurons send axons to the preoptic area. From E15 numerous labeled neurons with short dendrites were visualized in the septum after DiI insertion into the preoptic region, but there is no clear differentiation of the septal nuclei and the neuroepithelial layer is very thick (Figure 12 A). From E18 septal neurons projected to the preoptic area increased in number and developed spiny dendrites (Figure 12 B) as in adult rats (Risold and Swanson, 1997). LS projections to the mediobasal hypothalamus are not so numerous at all studied ages. This was confirmed with a double labeling study, when DiA was inserted into the preoptic region and DiI into the mediobasal hypothalamus on E18-E21. These insertions resulted in visualization of few DiI-labeled cells and numerous DiA-labeled neurons in the lateral septum (Fig 12 C, D). No septal connections with the posterior hypothalamus specifically, the mammillary bodies were revealed prenatally, although they were described in adult rats (Swanson, Cowan, 1979; Risold, Swanson, 1997b). According to the data that septal projections to the hypothalamus provide regulation of several neuroendocrine functions associated with water and salt intake, food intake, thermoregulation, aggressiveness, and sexually related behaviors (for review, see Herman, et al., 2002; Sheehan, et al., 2004) our results are of interest as providing knowledge that a significant part of septal projections to the preoptic area are developed prenatally.

Hypothalamic Projections to the Lateral Septal Nucleus

Our recent study is aimed to describe development of the hypothalamic projections to LS. Main sources of this innervation in the adult rat are located in the preoptic area and lateral hypothalamus and specifically in the perifornical region (Chiba and Murata 1985; Risold and Swanson, 1997). DiI inserted into the LS revealed numerous retrogradely labeled neurons in all these structures just after birth. Preoptic region contained labeled cells in the ventral parts of the medial and lateral preoptic area (Figure 13 A). In the lateral hypothalamus neurons with long processes were visualized as a group surrounded the fornix and distributed ventrolaterally (Figure 13 B) which can be regarded as subfornical region of the lateral hypothalamus (LHsf) described by Goto and coauthors (2005) as an exact hypothalamic cell group projecting to LS. This LHsf group can be traced from the anterior hypothalamic level where few cells were located around the fornix. At more caudal levels laterally from the ventromedial hypothalamic nucleus DiI-labeled cells formed large group ventrolaterally from the fornix and several multipolar neurons with long dendrites were found ventromedially from the main subfornical group in the tuberal region. Experiments on the prenatal material on E20 and E18 also visualized neurons projecting to the LS in preoptic area (Figure 13 C, D) and lateral hypothalamus. On both prenatal stages, a distinct subfornical group was visualized in the lateral hypothalamus. The number of the neurons in it was not large, but they have well developed dendrites and specific orientation (Figure 13 E, F). Thus hypothalamic innervation of the LS develops prenatally and the starting point of its formation will be revealed in our future experiments.

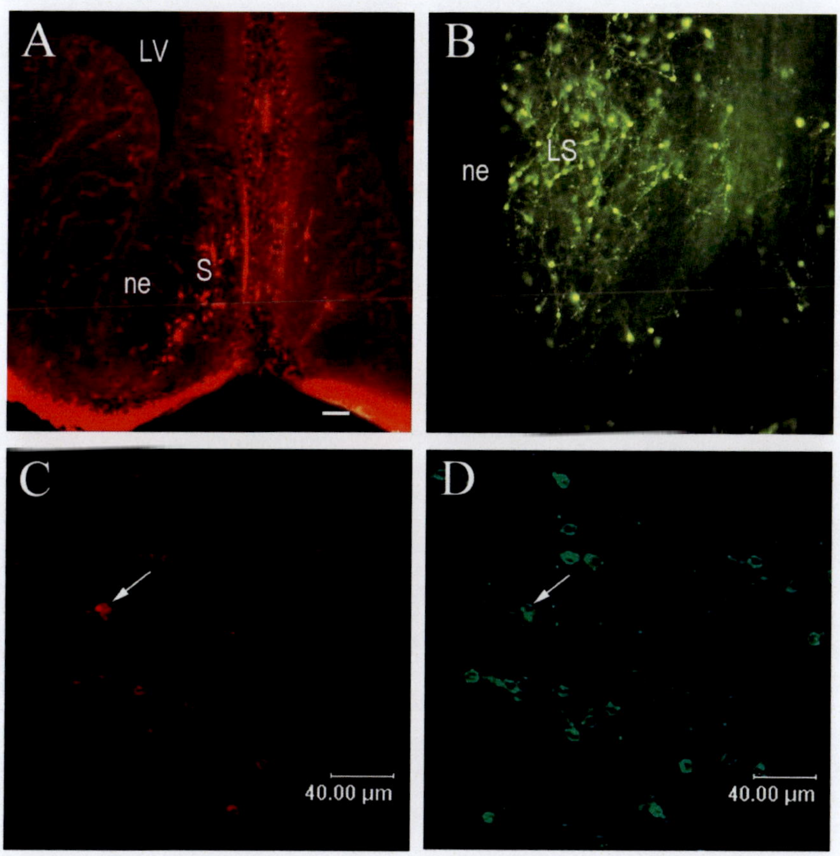

Figure 12. Septal neurons innervating preoptic area on E15 revealed with DiI (A) and on E20 revealed with DiA (B). Confocal visualization of the results of double labeling with DiI and DiA: single septal neuron (arrow) projecting to mediobasal hypothalamus (C) and numerous cells, projecting to the preoptic area (D) on the same brain on E18. Scale bar 100µm for A and B; 40µm for C and D.

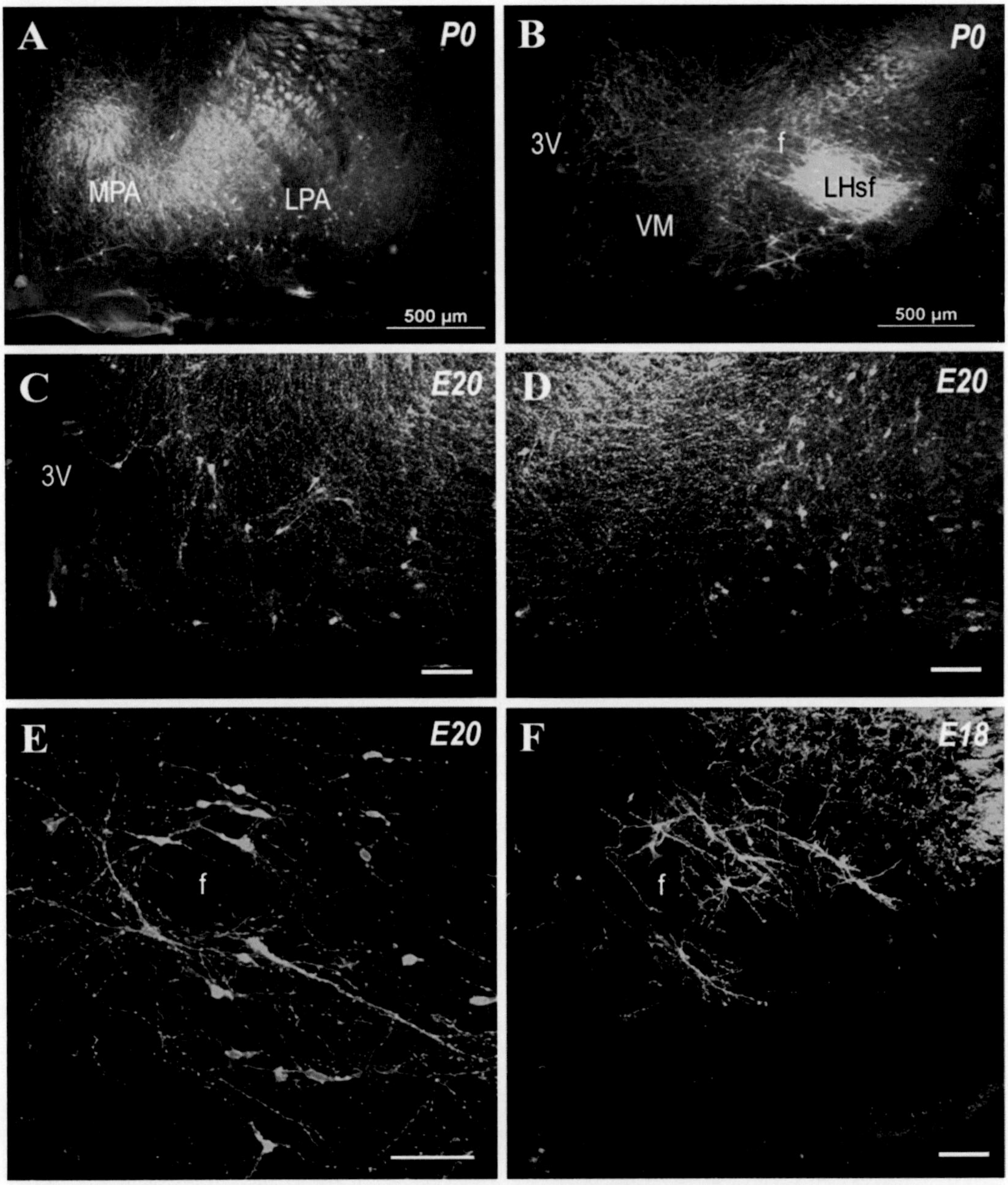

Figure 13. Hypothalamic neurons innervated lateral septal nucleus during perinatal development llocalized in the medial and lateral preoptic area (MPA, LPA) and in the subfornical area of the lateral hypothalamus (LHsf) but not in the ventromedial nucleus (VM).

Non-Neuronal Labeling (Neuroepithelium or Ependima)

We have some preliminary data, which should be fullfilled and analyzed in the future. Our DiI and DiA insertions into different hypothalamic and forebrain regions in addition to neuronal labeling often revealed labeled cells in the neuroepithelial layer on the early fetal stages and in the ependimal layer on late prenatal stages. Localization of these cells correlated with the place of the marker insertion. For example:

1. DiI insertions into the posterior pituitary lobe revealed labeled cells in the wall of the third ventricle at the level of its diamond-like swelling that were regarded as neuroblasts on the first stages of differentiation that later will migrate to the final positioning in the magnocellular hypothalamic nuclei. Sometimes DiI labeled cells located in the ventral part of the ependyma of the third ventricle and have processes directed in arch-like manner to the ventral surface of the hypothalamus (Figure 14 A). These cells resemble tanycytes specific type of the ependymal cells of the third ventricle. It was regarded that their visualization is a result of the DiI diffusion, first, along the meninges and, then, via basal processes of the neuroepithelial cells or tanycytes toward their cell bodies located rather near to the surface of the third ventricle.
2. DiI or DiA insertions in the preoptic region on E15-18 visualized neuroepithelial cells close to the ventral edge of the lateral ventricle on the side of the septum (Figure 14 A). Some labeled cells were found also in the wall of the third ventricle at the level of insertion. Both types of labeled cells can be regarded as neuroblasts, because dual direction of migration of the differentiating neurons in the preoptic area (along a dorsal-ventral orientation and addition to medial-lateral) was also described in mice (Henderson et al., 1999), confirms our finding.
3. Several insertions into the dorsal striatum in the prenatal brains resulted in labeling of the segment of the adjacent neuroepithelium of the lateral ventricle wall just above the crystals (Figure 14 A). These cells are packed very tightly to each other. This region never covered both angles of the ventricle. This observation is in agreement with the notion that during the formation of the striatum the large wave of neuronal cell migration originates from the subventricular zone and supplies the striatum with the striatal projection neurons (Hamasaki et al., 2003).

These data led us to an assumption that carbocyanine dyes can label not only differentiating neurons, but even progenitor cells in the ventricular wall and thus identify the source of the origin of the neurons of specific neuronal projection system. This suggestion was confirmed by the finding that DiI may label neural progenitor cells existing in the ependymal layer of the adult rat third ventricle (Xu, et al., 2005). The authors demonstrated that such cells at least some of the tanycytes may differentiate later into the hypothalamic neurons. Although the transection of the hypothalamo-pituitary axons at the level of ME which stimulated axonal regeneration and growth along tanycytic processes labeled with DiI applicated on the ventricular surface did not viualizet any signs of tanycyte transformation (Chauvet et al., 1995).

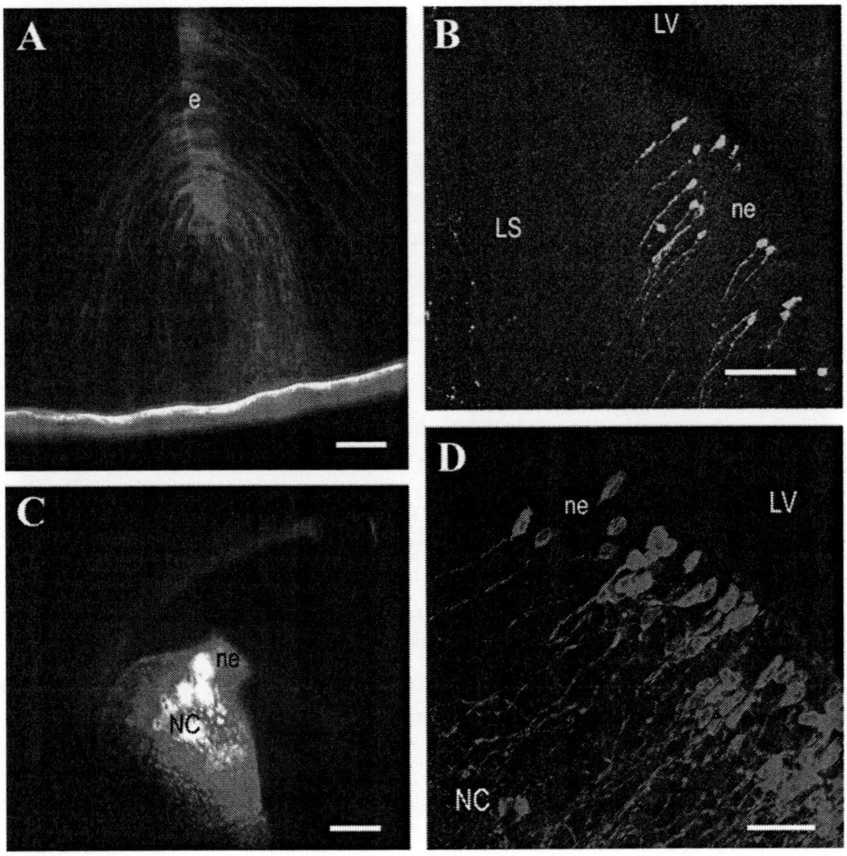

Figure 14. Carbocyanine dye labeling of cells in the ependyma of the third ventricle (A) and in the neuroepithelium of the lateral ventricle (B-D). A) Tanycytes in the ventral part of the third ventrice; B) Neuroepithelial cells in the neuroepithelium adjacent to the lateral septum (LS) after DiA insertion into the preoptic area (confocal image); C) DiI insertion into the nucleus caudatus (NC); D) Confocal image of the neuroepithelium above the NC. Scale bar μm for A, B; 500 μm for C; 40 μm for D.

Summary of DiI Tracing of the Developing Hypothalamic Connections

Taken together our data demonstrate that hypothalamic connections develop with both spatial and temporal specificity, innervating each target with a unique developmental schedule. Studied projection systems can be divided into three groups according to the time of their formation:

(1) Prenatally developed connections (started from E14-E15):

- Hypothalamo-posterior pituitary and median eminence projections, except projections of several accessory magnocellular peptidergic nuclei
- Mammillo-tegmental tract
- Lateral septal projections to the preoptic area – anterior hypothalamic region

- Hypothalamic projections to the lateral septum

(2) Systems developed from the end of intrauterine life and differentiate progressively on early postnatal stages:

- hypothalamic projections to the intermediate pituitary lobe
- mammillo-thalamic tract

(3) Projection systems which were not revealed perinatally:

- septal projections to the posterior hypothalamus
- hippocampal projections to the posterior hypothalamus

These data provide basic knowledge of the early steps of the development of the hypothalamic projection systems, but future careful analysis is necessary for understanding the functional significance of these fiber systems during ontogenesis and how changes of their normal development may influence the health of postnatal animals.

Own experience and analysis of literature data led us to a more detailed summary of the advantages and limitations of the DiI tracing method because this marker is the best of the carbocyanine dyes for tracing of the brain connections during perinatal development.

Advantages of DiI Tracing of Developing Brain Connections

1. DiI can be used on the dissected prefixed brain tissue and thus is applicable for the prenatal and human material.
2. DiI application allows avoiding surgical procedures and stereotaxic insertions in vivo.
3. All central nervous system regions are accessible.
4. DiI labeling is very stable for a long time in the whole brain and in the vibratome sections after mounting in different mounting media.
5. DiI visualized fine details of the neurons and nervous fibers (dendritic spines, growth cones, terminal arborizations).
6. Thick vibratome sections can be easily used and keep neurons with whole dendrites or a large part of the labeled tract inside one section.
7. DiI can be used in combination with DiA for double tracing
8. In some cases DiI can be combined with immunocytochemistry, but not after very long storage (This question and methodology should be adjusted in each case).
9. Confocal analysis of the DiI labeled material provides several advantages: a) Sharp visualization of the fine detail; b) Possibility to make volume reconstruction of the scanned object; c) Possibility to distinguish separately two carbocyanine dyes in the double labeling studies.
10. DiI can be used in combination with electrophysiological studies in vivo.

11. DiI labeled material can be used as a graft or it can be inserted in the host brain in the transplantation experiments.
12. DiI can be used for regeneration study.

Limitations of DiI Tracing

1. Long storage time of the labeled brain material for DiI diffusion.
2. DiI does not equally visualize all adult axonal systems and should be confirmed in each case.
3. High brightness of the DiI fluorescence in the labeled structures makes it visible with the FITC filter that should be taken in mind using fluorescent microscopy for analysis of the double labeling.
4. Frozen sections can not be reccommended for the DiI-labeled material, because of fast diffusion of the marker from the labeled structures in any mounting medium. Frozen sections can be used only dry mounted without coverslipping during a short period after sectioning.
5. If regions under investigation have reciprocal connections, DiI results should be analyzed carefully as it will visualize simultaneously afferent and efferent fibers in the labeled fiber tract.

New Directions in the DiI Tracing Studies

Development of the brain connections during ontogenesis is determined by many factors (Tessier-Lavigne, Goodman, 1996). Some of them have been reported as specific for diencephalic structures (Schmahl et al., 1993; Warren a. Price, 1997). Pax-6 and some other genes are expressed in the regions of the developing brain and implicated in the differentiation and establishment of proper neuronal connections within and across the different prosomeric boundaries. In Pax-6 mutant mice MB and mammillo-tegmental tract have normal development, but thalamus is severely disorganized and mammillo-thalamic tract is absent and radial glia changes were described (Valverde et al., 2000). Nkx-2.2 (Price et al., 1992; Kitamura et al., 1997) and Mash-1 (Lo et al., 1991; Guillemot and Joyner, 1993) were visualized in the zona limitans intrathalamica dividing thalamus on the dorsal and ventral parts. In Fkh5-deficient mice medial mammillary nucleus is absent (Wehr et al., 1997). Absence of orphan nuclear receptor steroidogenic factor 1 (SF-1) resulted in disorganization of the ventromedial hypothalamic nucleus (Ikeda et al., 1995; Shinoda et al., 1995).

Detailed analysis of different mutant models will help to reveal fundamental mechanisms of the development of major fiber tracts and carbocyanine dye tracing will be extremely useful in such studies.

Conclusion

Although there are numerous neuronal tracers lipophilic carbocyanine dyes and specifically DiI are unique and the most suitable choice for the neuroanatomical studies of the development of axonal brain connections on the prenatal end early postnatal stages of ontogenesis. Combination of ordinary fluorescent visualization of labeled structures with the confocal microscopy should be recommended for the analysis of the fine details of the growing neurons, their dendrites and innervation of brain nuclei.

We are sure that even though DiI tracing needs long exposition time and extremely fine hand manipulations with the fetal brain material, it should be used as a very important tool for studying the early steps in the formation of the brain connections. Such data will provide the basic knowledge concerning the main principles of brain network development.

Acknowledgements

The authors would like to thank Elena Alpeeva for participation in tracing of mammillary body connections. This work was supported by research grants of the Russian Foundation for Basic Research 99-0449134; # 00-15-97840, #04-04-48073 and #07-04-00798; INTAS-RFBR 95-IN-RU 1246, PICS-RFBR 98-04-22018.

References

Allen, GV; Hopkins, DA. Topography and Synaptology of Mammillary Body Projections to the Mesencephalon and Pons in the Rat. *J. Comp. Neurol.*, 1990 301, 214-231.

Alpeeva, EV; Makarenko, IG. Perinatal development of mammillotegmental connections in rats, *Russian J. of Developmental Biology*. 2007 38, 58-65. Translated from *Ontogenez*, 2007 38, 86-93.

Altman, J; Bayer, SA. Development of the diencephalon in the rat. Autoradiographic study of the time of origin and settling patterns of neurons of the hypothalamus. *J. Comp. Neurol.*, 1978a 182, 945-972.

Altman, J; Bayer, SA. Development of the diencephalon in the rat. II. Correlation of the embryonic development of the hypothalamus with the time of origin of its neurons. *J. Comp. Neurol.*, 1978b 182, 973-974.

Altman, J; Bayer, SA. The development of the rat hypothalamus. *Adv. Anat. Embryol. Cell Biol.*, 1986 100, 1-173.

Arai, R; Onteniente, B; Trembleau, A; Landry, MA; Calas, A. Hypothalamic galanin-immunoreactive neurons projecting to the posterior lobe of the rat pituitary: A combined retrograde tracing and immunocytochemical study. *J. Comp. Neurol.*, 1990 229, 405-420.

Arluison, M; Brochier, G; Vankova, M; Leviel, V; Villalobos, J; Tramu G. Demonstration of peptidergic afferents to the bed nucleus of the stria terminalis using local injections of colchicine. A combined immunohistochemical and retrograde tracing study. *Brain Res. Bull.*, 1994 34, 319-337.

Armstrong, WE; Hatton, GI. The localization of projection neurons in the rat hypothalamic paraventricular nucleus following vascular and neurohypophysial injection of HRP. *Brain Res. Bull.*, 1980 5, 473–477.

Artero, C; Fasolo, A; Franzoni, MF. Multiple sources of the pituitary pars intermedia innervation in amphibians: a Di-I retrograde tract-tracing study. *Neurosci. Lett.*, 1994 169,163–166.

Bailey, TW; Nicol, GD; Schild, JH; DiMicco, JA. Synaptic and membrane properties of neurons in the dorsomedial hypothalamus. *Brain Res.*, 2003 985, 150-162.

Balthazart, J; Dupiereux, V; Aste, N; Viglietti-Panzica, C; Barrese, M; Panzica, GC. Afferent and efferent connections of the sexually dimorphic medial preoptic nucleus of the male quail revealed by in vitro transport of DiI. *Cell Tissue Res.*, 1994 276, 455-475.

Bentivoglio, M; Chen, S. Retrograde neuronal tracing combined with immunocytochemistry. In: Cuello AC, editor. *Immunohistochemistry* II. New York: Wiley; 1993; 301–328.

Bouret, SG; Draper, SJ; Simerly, RB. Formation of projection pathways from the arcuate nucleus of the hypothalamus to hypothalamic regions implicated in the neural control of feeding behavior in mice. *J. Neurosci.*, 2004 24, 2797-2805.

Bruce, LL; Christensen, MA; Fritzsch, B. Electron microscopic differentiation of directly and transneuronally transported DiI and applications for studies of synaptogenesis. *J. Neurosci. Methods,* 1997 73, 107-112.

Buijs, RM. The development of vasopressin and oxytocin systems in the brain. In: Björklund A, Hökfelt T, editors. *Handbook of Chemical Neuroanatomy*, vol 10: Ontogeny of Transmitters and Peptide in the CNS. Amsterdam: Elsevier; 1992; 547-569.

Buijs, RM; Velis, DN; Swaab, DF. Ontogeny of vasopressin and oxytocin in the fetal rat: early vasopressinergic innervation of the fetal brain. *Peptides* 1980 1, 315-324.

Caqueret, A; Yang, C; Duplan, S; Boucher, F; Michaud, JL. Looking for trouble: A search for developmental defects of the hypothalamus. *Horm. Res.*, 2005 64, 222-230.

Castejón, OJ; Fuller, L; Dailey, ME. Localization of synapsin-I and PSD-95 in developing postnatal rat cerebellar cortex. *Developmental Brain Research*, 2004 151, 25-32.

Chauvet, N; Parmentier, ML; Alonso, G. Transected axons of adult hypothalamo-neurohypophysial neurons regenerate along tanycytic processes. *J. Neurosci. Res.*, 1995 41, 129-144.

Chiba, T; Murata, Y. Afferent and efferent connections of the medial preoptic area in the rat: a WGA-HRP study. *Brain Res. Bull.*, 1985 14, 261–272.

Chronwall, BM; S.A. Sands, SA; Cummings 3rd, KC; Hagler,KE; M. Norberg, M; Morris, SJ; Gary, KA. Differential innervation of individual melanotropes suggests a role for nonsynaptic inhibitory regulation of the developing and adult rat pituitary IL. *Synapse*, 1998 28, 227-243.

Choy, VJ; Watkins, WB. Maturation of the hypothalamoneurohypophysial system I. Localization of neurophysin, oxytocin and vasopressin in the hypothalamus and neural lobe of the developing rat brain. *Cell Tissue Res.*, 1979 197, 325–336.

Cowan, WM; Gottlieb, DI; Hendrickson, AE; Price, JL; Woolsey, TA. The autoradiographic demonstration of axonal connections in the central nervous system. *Brain Res.*, 1972 37, 21–51.

Cruce, JA. An autoradiographic study of the descending connections of the mammillary nuclei of the rat. *J. Comp. Neurol.*, 1977 176, 631-644.

Cvetkovic, V; Brischoux, F; Griffond, B; Bernard, G; Jacquemard, C; Fellmann, D; Risold, PY. Evidence of melanin-concentrating hormone-containing neurons supplying both cortical and neuroendocrine projections. *Neuroscience*, 2003 116, 31-35.

Deurveilher, S; Semba, K. Indirect projections from the suprachiasmatic nucleus to the median preoptic nucleus in rat. *Brain Research*, 2003 987, 100–106.

Dai, J; Swaab, DF; Van Der Vliet, J; Buijs, RM. Postmortem tracing reveals the organization of hypothalamic projections of the suprachiasmatic nucleus in the human brain. *J. Comp. Neurol.*, 1998 400, 87–102.

Davis, AM; Seney, ML; Stallings, NR; Zhao, L; Parker, KL; Tobet, SA. Loss of steroidogenic factor 1 alters cellular topography in the mouse ventromedial nucleus of the hypothalamus. *J. Neurobiol.*, 2004 60, 424-436.

Dong, HW; Swanson, LW. Projections from the rhomboid nucleus of the bed nuclei of the stria terminalis: implications for cerebral hemisphere regulation of ingestive behaviors. *J. Comp. Neurol.*, 2003 463, 434-472.

Dong, HW; Swanson, LW. Organization of axonal projections from the anterolateral area of the bed nuclei of the stria terminalis. *J. Comp. Neurol.*, 2004 468, 277-298.

Dong, HW; Swanson, LW. Projections from bed nuclei of the stria terminalis, dorsomedial nucleus: implications for cerebral hemisphere integration of neuroendocrine, autonomic, and drinking responses. *J. Comp. Neurol.*, 2006 494, 75-107.

Dunlop, SA; Tran, N; Tee, LB; Papadimitriou, J; Beazley, LD. Retinal projections throughout optic nerve regeneration in the ornate dragon lizard, Ctenophorus ornatus. *J. Comp. Neurol.*, 2000 416, 188-200.

Eager, RP. Selective staining of degenerating axons in the central nervous system by a simplified silver method: spinal cord projections to external cuneate and inferior olivary nuclei in the cat, *Brain Res.*, 1970 22, 137-141.

Easter, SSJr; Ross, LS; Frankfurter, A. Initial tract formation in the mouse brain. *J. Neuroscience*, 1993 13, 285-299.

Elmquist, JK. Hypothalamic pathways underlying the endocrine, autonomic, and behavioral effects of leptin. *Physiol. Behav.*, 2001 74, 703–708.

Enquist LW, Card JP. Recent advances in the use of neurotropic viruses for circuit analysis. *Curr. Opin. Neurobiol.*, 2003 13, 603-606.

Fahrbach, SE; Morrell, JI; Pfaff, DW. Studies of ventromedial hypothalamic afferents in the rat using three methods of HRP application. *Exp. Brain Res.*, 1989 77, 221-233.

Falke, N. Modulation of oxytocin and vasopressin release at the level of the neurohypophysis. *Prog. Neurobiol.*, 1991 36, 465-484.

Fink, RP; Heimer, L. Two methods for selective silver impregnation of degenerating axons and their synaptic endings in the central nervous system. *Brain Res.*, 1967 4, 369–374.

Fisher, AW; Price, PO; Lederis, K A. 3-Dimensional Reconstruction of the Hypothalamo-Neurohypophysial Sys tem of the Rat. The Neurons Projecting to the Neuro/Intermediate Lobe and Those Containing Vasopressin and Somatostatin. *Cell Tiss. Res.*, 1979 204, 343–354.

Friedman, DI; Johnson, JK; Chorsky, RL; Stopa, EG. Labeling of human retinohypothalamic tract with the carbocyanine dye, DiI. *Brain Res.*, 1991 560, 297-302.

Gabriele, ML; Smoot, JE; Jiang, H; Stein, BE; McHaffie, JG. Early establishment of adult-like nigrotectal architecture in the neonatal cat: a double-labeling study using carbocyanine dyes. *Neuroscience*, 2006 137, 1309-1319.

Gary, KA; Chronwall, BM. The onset of dopaminergic innervation during ontogeny decreases melanotrope proliferation in the IL of the rat pituitary, *Int. J. Dev. Neurosci.*, 1992 10, 131-142.

Gayrard, V; Thiéry, J-C; Thibault, J; Tillet, Y. Efferent projections from the retrochiasmatic area to the median eminence and to the pars nervosa of the hypophysis with special reference to the A15 dopaminergic cell group in the sheep. *Cell Tissue Res.*, 1995 281, 561–567.

Godement, P; Vanselow, J; Thanos, S; Bonhoeffer, F. A study in developing visual systems with a new method of staining neurons and their processes in fixed tissue. *Development*, 1987 101, 697-713.

Goto, M; Canteras, NS; Burns, GY; Swanson, LW. Projections from the Subfornical Region of the Lateral Hypothalamic Area. *J. Comp. Neurol.*, 2005 493, 412-438.

Goudreau, JL; Lindley, SE; Lookingland, KJ; Moore, KE. Evidence that hypothalamic periventricular dopamine neurons innervate the IL of the rat pituitary. *Neuroendocrinology*, 1992 56, 100-105.

Gray, EG; Hemlin LH. Electron microscopy of experimental degeneration in the optic tectum. J Anat (Lond), 1962 19, 111-182.

Grayson, BE; Allen SE; Billes, SK; Williams, SM; Smith, MS; Grove, KL. Prenatal development of hypothalamic neuropeptide systems in the nonhuman primate. *Neuroscience*, 2006 143, 975-986.

Grinevich, VV; Polenov, AL. Evolution of Nonapeptidergic Neurosecretory Hypothalamic Formations in Vertebrates. *Zh. Evol. Biokhim. Fiziol*, 1994 30, 270–292 (in Russian).

Grove, KL; Allen, S; Grayson, BE; Smith, MS. Postnatal development of the hypothalamic neuropeptide Y system. *Neuroscience*, 2003 116, 393–406.

Grove, KL; Grayson, BE; Glavas, MM; Xiao, XQ; Smith, MS. Development of metabolic systems. *Physiology and Behavior*, 2005 86, 646–660.

Gu, GB; Simerly, RB. Projections of the sexually dimorphic anteroventral periventricular nucleus in the female rat. *J. Comp. Neurol.*, 1997 384, 142-164.

Guillemot, F; Joyner, AL. Dynamic expression of the murine Achaete- Scute homologue Mash-1 in the developing nervous system. *Mech. Dev.*, 1993 42, 171–185.

Hamasaki, T; Goto, S; Nishikawa, S; Ushio, Y. Neuronal cell migration for the developmental formation of the mammalian striatum. *Brain Res. Brain Res. Rev.*, 2003 41, 1-12.

Hatton, GI. Emerging Concepts of Structure-Function Dynamics in Adult Brain: The Hypothalamo-Neurohypophysial System. *Prog. Neurobiol.*, 1990 34, 437–504.

Haugland, RP. Handbook of fluorescent probes and research chemicals.Eugene, OR: *Molecular Probes Inc*; 1996.

Hayakawa, T; Zyo, K. Quantitative and Ultrastructural Study of Ascending Projections to the Medial Mammillary Nucleus in the Rat. *Anat Embryol (Berl.),* 1991 184, 611-622.

Heredia, M; Santacana, M; Valverde, F. A method using DiI to study the connectivity of cortical transplants. *J. Neurosci. Methods*, 1991 36, 17–25.

Herman, JP; Tasker, JG; Ziegler, DR; Cullinan, WE. Local circuit regulation of paraventricular nucleus stress integration: glutamate-GABA connections. *Pharmacol. Biochem. Behav.*, 2002 71, 457-468.

Henderson, RG; Brown, AE; Tobet, SA. Sex differences in cell migration in the preoptic area/anterior hypothalamus of mice. *J. Neurobiol.*, 1999 41, 252-266.

Hoffman, GE; Phelps, CJ; Khachaturian, H; Sladek, JR. Neuroendocrine projections to the median eminence. *Curr. Top Neuroendocrinol.*, 1986 7, 161–317.

Hofmann, MH; Bleckmann, H. Effect of temperature and calcium on transneuronal diffusion of DiI in fixed brain preparations. *J. Neurosci. Methods*, 1999 88, 27-31.

Honig, MG; Hume, RI. Fluorescent carbocyanine dyes allow living neurons of identified origin to be studied in long-term cultures. *J. Cell Biol.*, 1986 103, 171–187.

Honig, MG; Hume, RI. DiI and DiO: versatile fluorescent dyes for neuronal labeling and pathway tracing. *Trends Neurosci.*, 1989 12, 333-341.

Honig MG. DiI labeling. *Neurosci. Protocols*, 1993 50, 1–20.

Horváth, TL; Naftolin, F; Leranth, C. GABAergic and catecolaminergic innervation of mediobasal hypothalamic betaendorphin cells projecting to the medial preoptic area. *Neuroscience*, 1992 51, 391–399.

Hutton, LA; Gu, GB; Simerly, RB. Development of a sexually dimorphic projection from the bed nuclei of the stria terminalis to the anteroventral periventricular nucleus in the rat. *J. Neurosci.*, 1998 18, 3003-3013.

Ikeda, Y; Luo, X; Abbud, R; Nilson, JH; Parker, KL. The nuclear receptor steroidogenic factor 1 is essential for the formation of the ventromedial hypothalamic nucleus. *Mol. Endocrinol.*, 1995 9, 478–486.

Jaeger, CB. Aromatic L-amino acid decarboxylase in the rat brain: I immunocytochemical localization during prenatal development. *Neuroscience*, 1986 8, 121-150.

Ju, G; Liu, S; Tao, J. Projections from the hypothalamus and its adjacent areas to the posterior pituitary in the rat. *Neuroscience,* 1986 19, 803-828.

Kalsbeek, A; Teclemariam-Mesbah, R; Pevet, P. Efferent projections of the suprachiasmatic nucleus in the golden hamster (Mesocricetus auratus). *J. Comp. Neurol.*, 1993 332, 293–314.

Kawano, H; Daikoku, S. Functional topography of the rat hypothalamic dopamine neuron systems: retrograde tracing and immunohistochemical study. *J. Comp. Neurol.*, 1987 265, 242-253.

Kawano, H; Horie, M; Honma, S; Kawamura, K; Takeuchi, K; Kimura, S. Aberrant trajectory of ascending dopaminergic pathway in mice lacking Nkx2.1. *Exp. Neurol.*, 2003 182, 103-112.

Keshavan, MS; Murray, RM. *Neurodevelopment and adult psychopatology*. Cambridge:Cambridge University Press; 1997.

Kitamura, K; Miura, H; Yanazawa, M; Miyashita, T; Kato, K. Expression patterns of Brx1 (Rieg gene), Sonic hedgehog, Nkx2.2, Dlx1 and Arx during zona limitans intrathalamica and embryonic ventral lateral geniculate nuclear formation. *Mech. Dev.*, 1997 67, 83–96.

Kouki, T; Yamanouchi, K. Postnatal development of septal projections to the midbrain central gray in female rats: tract-tracing analysis with DiI. *Neurosci. Lett.*, 2007 411, 37-41.

Kristensson, K; Olsson, Y. Retrograde axonal transport of protein. *Brain Res.*, 1971 29, 363–365.

Lammers, HJ, Lohman, AH. Structure and fiber connections of the hypothalamus in mammals. *Prog. Brain Res.*, 1974 41, 61-78.

Larsen, PJ; Moller, M; Mikkelsen, JD. Efferent projections from the periventricular and medial parvocellular subnuclei of the hypothalamic paraventricular nucleus to circumventricular organs of the rat: A Phaseolus vulgaris-leucoagglutinin (PHAL) tracing study. *J. Comp. Neurol.*, 1991 306, 462–479.

LaVail, JH; LaVail, MM. Retrograde axonal transport in the central nervous system. *Science*, 1972 176,1416 –1417.

Laurent, FM; Hindelang, C; Klein, MJ; Stoeckel, ME; Felix, JM. Expression of the oxytocin and vasopressin genes in the rat hypothalamus during development: an in situ hybridization study. *Dev. Brain Res.*, 1989 46, 145-154.

Lehman, MN; LeSauter J; Kim, C; Berriman, SJ; Tresco, PA; Silver, R. How do fetal grafts of the suprachiasmatic nucleus communicate with the host brain? *Cell Transplant.*, 1995 4, 75-81.

Lehman, MN; Lesauter, J; Silver, R. Fiber outgrowth from anterior hypothalamic and cortical xenografts in the third ventricle. *J. Comp. Neurol.*, 1998 391, 133-145.

Li, DP; Chen, SR; Pan, HL. Nitric oxide inhibits spinally projecting paraventricular neurons through potentiation of presynaptic GABA release. *J. Neurophysiol.*, 2002 88, 2664-2674.

Lo, L-C; Johnson, JE; Wuenschell, CW; Saito, T; Anderson, D. Mammalian achaete-scute homolog 1 is transiently expressed by spatially restricted subsets of early neuroepithelial and neural crest cells. *Genes Dev.*, 1991 5, 1524–1537.

Lukas, JR; Aigner, M; Denk, M; Heinzl, H; Burian, M; Mayr, R. Carbocyanine postmortem neuronal tracing: Influence of different parameters on tracing distance and combination with immunocytochemistry. *J. Histochem. Cytochem.*, 1998 46, 901–910.

Luppi, PH; Sakai, K; Salvert, D; Berod, A; Jouvet, M. Periventricular dopaminergic neurons terminating in the neuro-IL of the cat hypophysis. *J. Comp. Neurol.*, 1986 244, 204-212.

Magoul, R; Ciofi, P; Tramu, G. Visualization of an efferent projection route of the hypothalamic rat arcuate nucleus through the stria terminalis after labeling with carbocyanine dye (DiI) or proopiomelanocortin-immunohistochemistry. *Neurosci. Lett.*, 1994 172, 134-138.

Makarenko, IG; Ugriumov, MV; Kalas, A. The development of connections of the magnocellular hypothalamic nuclei with the posterior hypophyseal lobe in rat prenatal ontogeny. *Ontogenez*, 1999 30, 296-301. (In Russian).

Makarenko, IG; Ugrumov, MV; Derer, P; Calas, A. Projections from the hypothalamus to the posterior lobe in rats during ontogenesis: 1,1'-dioctadecyl-3,3,3', 3'-tetramethylindocarbocyanine perchlorate tracing study. *J. Comp. Neurol.*, 2000 422, 327-337.

Makarenko, IG; Ugrumov, MV; Calas, A. Axonal projections from the hypothalamus to the median eminence in rats during ontogenesis: DiI tracing study. *Anat. Embryol.* (Berl), 2001 204, 239-252.

Makarenko, IG; Ugryumov, MV; Kalas, A. Involvement of accessory neurosecretory nuclei of hypothalamus in the formation of hypothalamohypophyseal system during prenatal and postnatal development in rats. *Russ. J. Devel. Biol.,* 2002 33, 37-42. *Translated from Ontogenez,* 2002 33, 43-49.

Makarenko, IG; Ugrumov, MV; Calas, A. Axonal projections from the hypothalamus to the pituitary intermediate lobe in rats during ontogenesis: DiI tracing study. *Brain Res. Dev. Brain Res.,* 2005 155, 117-126.

Makarenko, IG. Prenatal carbocyanine dye tracing of septo-hypothalamic connections. *Brain Res.*, 2007 1130, 38-47.

Markakis, EA; Swanson, LW. Spatiotemporal patterns of secretomotor neuron generation in the parvocellular neuroendocrine system. *Brain Res. Rev.*, 1997 24, 255–291.

Mastick, GS; Easter, SS. Initial Organization of Neurons and Tracts in the Embryonic Mouse Fore- and Midbrain. *Devel. Biol.*, 1996 173, 79-94.

Melloni, RH Jr; DeGennaro, LJ. Temporal onset of synapsin I gene expression coincides with neuronal differentiation during the development of the nervous system. *J. Comp. Neurol.*, 1994 342, 449-462.

Merchenthaler, I. Enkephalin-immunoreactive neurons in the parvocellular subdivisions of the paraventricular nucleus project to the external zone of the median eminence. *J. Comp. Neurol.*, 1992 326, 112-120.

Mikkelsen, JD. Projections from the lateral geniculate nucleus to the hypothalamus of the Mongolian gerbil (Meriones unguiculatus): an anterograde and retrograde tracing study. *Comp. Neurol.*, 1990 299, 493-508.

Mirnics, K; Koerber, HR. Prenatal development of rat primary afferent fibers: I. Peripheral projections. *J. Comp. Neurol.*, 1995 355, 589–600.

Morin, LP; Blanchard, J; Moore, RY. Intergeniculate leaflet and suprachiasmatic nucleus organization and connections in the golden hamster. *Vis. Neurosci.*, 1992 8, 219-230.

Morgane, PJ; Galler, JR; Mokler, DJ. A review of systems and networks of the limbic forebrain/limbic midbrain. *Prog. Neurobiol.*, 2005 75, 143-160.

Müller, C; Torrealba, F. Postnatal development of neuron number and connections in the suprachiasmatic nucleus of the hamster. *Brain Res. Dev. Brain Res.*, 1998 110, 203-213.

Nauta, WJH; Gygax, PA. Silver impregnation of degenerating axon terminals in the central nervous system: (1) Technique, (2) Chemical notes. *Stain Technol.*, 1951 26, 5–11.

Nauta, WJH; Haymaker, W. Hypothalamic nuclei and fiber connections. In: *The hypothalamus.* Haymaker W; Anderson E; Nauta WJH, editors. Illinois: Thomas, Springfield, 1969, 136-209.

Nauta, WJH; Ebbesson, SOE. *Contemporary research methods in neuroanatomy.* Berlin: Springer-Verlag; 1970.

Niimi, M; Takahara, J; Sato, M; Kawanishi, K. Neurotensin and growth hormone-releasing factor-containing neurons projecting to the median eminence of the rat: a combined retrograde tracing and immunohystochemical study. *Neurosci. Lett.*, 1991 133, 183–186.

Niimi, M; Takahara, J; Kawaniahi, K. Corticotropin releasing factor and galanin-containing neurons projecting to the median eminence of the rat. *Neurosci. Res.*, 1992 14, 295–299.

Otellin, VA; Sergeeva, IG. Autoradiographic study of the connections between field 17 of the visual cortex and the hypothalamus in the cat. *Neuroscience and Behavioral Physiology*, 1982 2, 428-434.

Ozaki, HS; Wahlsten, D. Prenatal formation of the normal mouse corpus callosum: A quantitative study with carbocyanine dyes. *J. Comp. Neurol.*, 1992 323, 81–90.

Paxinos, G; Watson, C. *The rat brain in Stereotaxic coordinates*. II Edition. London: Academic Press, 1986.

Paxinos, G; Tork, I; Tecon, LH; Valentino, KL. *Atlas of the developing rat brain*. San Diego: Academic Press Inc, 1991.

Peterson, RP. Magnocellular Neurosecretory Centers in the Rat Hypothalamus. *J. Comp. Neurol.*, 1966 128, 181–190.

Petrovicky, P. Note on the Connections of Gudden's Tegmental Nuclei. 1. Efferent Ascending Connections in the Mammillary Peduncle. *Acta Anat.* (Basel), 1973 86, 165-190.

Polston, EK; Simerly, RB. Ontogeny of the projections from the anteroventral periventricular nucleus of the hypothalamus in the female rat. *J. Comp. Neurol.*, 2006 495,122-132.

Price, M; Lazzaro, D; Pohl, T; Mattei, MG; Ruther, U; Olivo, JC; Duboule, D; Lauro, DR. Regional expression of the homeobox gene Nkx-2.2 in the developing mammalian forebrain. *Neuron,* 1992 8, 241–255.

Raisman, G. Neural connections of the hypothalamus. *Brit. Med. Bull.*, 1966 22, 197-201.

Reyes, BA; Valentino, RJ; Xu, G; Van Bockstaele, EJ. Hypothalamic projections to locus coeruleus neurons in rat brain. *Eur. J. Neurosci.*, 2005 22, 93-106.

Rhodes, CH; Morrell, JI; Pfaff, DW. Immunohistochemical analysis of magnocellular elements in rat hypothalamus. Distribution and numbers of cells containing neurophysin, oxytocin, and vasopressin. *J. Comp. Neurol.*, 1981 198, 45–64.

Risold, PY; Swanson, LW. Connections of the rat lateral septal complex. *Brain Res. Reviews*, 1997 24, 115–195.

Seki, M; Zyo, K. Anterior Thalamic Afferents from the Mammillary Body and the Limbic Cortex in the Rat. *J. Comp. Neurol.*, 1984 229, 242-256.

Sergeeva, IG. Connections between field 17 of the optic cortex /cerebrum/ and the mammillary nuclein cats. *Neuroscience and Bechavioral Physiology*, 1980 10, 17- 21.

Sheehan, TP; Chambers, RA; Russell, DS. Regulation of affect by the lateral septum: implications for neuropsychiatry. *Brain Research Reviews*, 2004 46, 71–117.

Silverman, RC; Gibson, MJ; Silverman, AJ. Application of a fluorescent dye to study connectivity between third ventricular preoptic area grafts and host hypothalamus. *J. Neurosci. Res.*, 1992 31, 156-165.

Simerly, RB; Swanson, LW. The organization of neural inputs to the medial preoptic nucleus of the rat. *J. Comp. Neurol.*, 1986 246, 312–342.

Simerly, RB. Anatomical substrates of hypothalamic integration. In: Paxinos G, editor. *The rat nervous system*. Amsterdam: Elsevier Academic Press; 2004; 335-368.

Sherlock, DA; Field, PM; Raisman, G. Retrograde transport of horseradish peroxidase in the magnocellular neurosecretory system of the rat. *Brain Res.*, 1975 88, 403–414.

Shinoda, K; Lei, H; Yoshii, H; Nomura, M; Nagano, M; Shiba, H; Sasaki, H; Osawa, Y; Ninomiya, Y; Niwa, O; et al. Developmental defects of the ventromedial hypothalamic nucleus and pituitary gonadotroph in the Ftz-F1 disrupted mice. *Dev. Dyn.*, 1995 204, 22–29.

Schmahl, W; Knoedlseder, M; Favor, J; Davisdon, D. Defects of neuronal migration and the pathogenesis of cortical malformations are associated with Small eye (Sey) in the mouse, a point mutation at the Pax-6-locus. *Acta Neuropathol.*, 1993 86, 126–135.

Sofroniew, MV. Vasopressin and oxytocin in the mammalian brain and spinal cord. *Trends Neurosci.*, 1983 6, 467–472.

Sparks, DL; Lue, L-F; Martin, TA; Rogers, J. Neural tract tracing using Di-I: a review and a new method to make fast Di-I faster in human brain. *J. of Neurosci. Methods*, 2000 103, 3–10.

Swanson, LW. Organization of mammalian neuroendocrine system, In: Bloom FE, editor. *Handbook of Physiology: The Nervous System*, Vol. IV, Baltimore: Waverly Press; 1986; 317–363.

Swanson, LW. Cerebral hemisphere regulation of motivated behavior. *Brain Res.* 2000 886, 113-164.

Swanson, LW; Cowan, WM. The connections of the septal region in the rat. *J. Comp. Neurol.,* 1979 186, 621–656.

Swanson, LW; Kuypers, HGJM. The paraventricular nucleus of the hypothalamus: Cytoarchitectonic subdivisions and organization of projections to the pituitary, dorsal vagal complex, and spinal cord as demonstrated by retrograde fluorescence double-labeling methods. *J. Comp. Neurol.*, 1980 194, 555–570.

Swanson, LW; Sawchenko, PE. Hypothalamic integration: Organization of the paraventricular and supraoptic nuclei, *Ann. Rev. Neurosci.*, 1983 6, 269-324.

Tavares, RF; Fernandes, KB; Pajolla, GP; Nascimento, IA; Correa, FM. Neural connections between prosencephalic structures involved in vasopressin release. *Cell Mol. Neurobiol.*, 2005 25, 663-672.

Thompson, RH; Swanson, LW. Structural characterization of a hypothalamic visceromotor pattern generator network. *Brain Res. Rev.*, 2003 41, 153–202.

Trembleau, A; Ugrumov, M; Roche, D; Calas, A. Vasopressin and oxytocin gene expressions in intact rats and under the catecholamine deficiency during ontogenesis. *Brain Res. Bull.*, 1995 37, 437-448.

Ugrumov, MV. Developing hypothalamus in differentiation of neurosecretory neurons and in establishment of pathways for neurohormone transport. *Int. Rev. Cytol.*, 1991 129, 207–267.

Ugrumov, MV. Development of the median eminence during ontogenesis (morpho-functional aspects). *Prog. Brain Res.*, 1992 91, 349-356.

Ugrumov, MV; Taxi, J; Tixier-Vidal, A; Thibault, J; Mitskevich, MS. Ontogenesis of tyrosine hydroxylase-immunopositive structures in the rat hypothalamus. An atlas of neuronal cell bodies. *Neuroscience*, 1989 29, 135-156.

Valverde, F; Garcia, C; Lopez-Mascaraque, L; De Carlos, JA. Development of the mammillothalamic tract in normal and Pax-6 mutant mice. *J. Comp. Neurol.*, 2000 419, 485-504.

Van der Kooy, D; Kuypers, HG; Catsman-Berrevoets, CE. Single Mammillary Body Cells with Divergent Axon Collaterals. Demonstration by a Simple, Fluorescent Retrograde Double-Labeling Technique in the Rat. *Brain Res*., 1978 158, 189-196.

Vanhatalo, S; Soinila, S; Kaartinen, K; Back, N. Colocalization of dopamine and serotonin in the rat pituitary gland and in the nuclei innervating it. *Brain Res.*, 1995 669, 275-284.

Vercelli, A; Repici, M; Garbossa, D; Grimald, A. Recent techniques for tracing pathways in the central nervous system of developing and adult mammals. *Brain Res. Bull*., 2000 51, 11-28.

Vereczki, V; Koves, K; Csaki, A; Grosz, K; Hoffman, GE; Fiskum, G. Distribution of hypothalamic, hippocampal and other limbic peptidergic neuronal cell bodies giving rise to retinopetal fibers: anterograde and retrograde tracing and neuropeptide immunohistochemical studies. *Neuroscience*, 2006 140, 1089-1100.

von Bartheld, CS; Cunningham, DE; Rubel, EW. Neuronal tracing with DiI: Decalcification, cryosectioning and photoconversion for light and electron microscopic analysis. *J. Histochem. Cytochem.*, 1990 38, 725–733.

Watanabe, K; Kawana, E. A horseradish peroxidase study on the mammillothalamic tract in the rat. *Acta Anat*., 1980 108, 394–401.

Warren, N; Price, DJ. Roles of Pax-6 in murine diencephalic development, *Development,* 1997 124, 1573–1582.

Watts, AG; Swanson, LW; Sanchez-Watts, G. Efferent projections of the suprachiasmatic nucleus: I. Studies using anterograde transport of Phaseolus vulgaris leucoagglutinin in the rat. *J. Comp. Neurol*., 1987a 258, 204–229.

Watts, AG; Swanson, LW. Efferent projections of the suprachiasmatic nucleus: II. Studies using retrograde transport of fluorescent dyes and simultaneous peptide immunocytochemistry in the rat. *J. Comp. Neurol*., 1987b 258, 230–252.

Wayner, MJ; Barone, FC; Scharoun, SL; Guevara-Aguilar, R; Aguilar-Baturoni, HU. Limbic connections to the lateral preoptic area: a horseradish peroxidase study in the rat. *Neurosci. Biobehav. Rev*., 1983 7, 375-384.

Wehr, R; Mansouri, A; de Maeyer, T; Gruss, P. Fkh-5-deficient mice show dysgenesis in the caudal midbrain and hypothalamic mammillary body. *Development*, 1997 124, 4447–4456.

Whitnall, MH; Key, S; Ben-Barak, Y; Ozato, K; Gainer, H. Neurophysin in the hypothalamo-neurohypophysial system. II. Immunocytochemical studies of the ontogeny of oxytocinergic and vasopressinergic neurons. *J. Neurosci*., 1985 5, 98–109.

Wiegand, SJ; Price, JL. Cells of origin of the afferent fibers to the median eminence in the rat. *J. Comp. Neurol*., 1980 192, 1-19.

Wiitanen, JT. Selective silver impregnation of degenerating axons and axon terminals in the central nervous system of the monkey (Macaca mulatta). *Brain Res*., 1969 14, 546-547.

Xu, Y; Tamamaki, N; Noda, T; Kimura, K; Itokazu, Y; Matsumoto, N; Dezawa, M; Ide, C. Neurogenesis in the ependymal layer of the adult rat 3rd ventricle. *Exp. Neurol*., 2005 192, 251-264.

Yamamoto, N; Higashi, S; Toyama, K. Stop and branch behaviors of geniculocortical axons: A time-lapse study in organotypic cocultures. *J. Neurosci*., 1997 17, 3653–3663.

Zec, N; Filiano, JJ; Kinney, HC. Anatomic relationships of the human arcuate nucleus of the medulla: A DiI-labeling study. *J. Neuropathol. Exp. Neurol.*, 1997 56, 509 –522.

Zimmerman, EA; Nilaver, G; Hou-Yu, A; Silverman, AJ. Vasopressinergic and oxytocinergic pathways in the central nervous system. *Fed. Proc.*, 1984 43, 91–96.

In: Neural Pathways Research
Editor: F. L. Pichler, pp. 73-93
ISBN: 978-1-60456-214-9

Chapter III

Involvement of Zinc via Crosstalk with Calcium in Synaptic Plasticity and Neurodegeneration in the Hippocampus

Atsushi Takeda**
Department of Medical Biochemistry,
School of Pharmaceutical Sciences, University of Shizuoka,
52-1 Yada, Suruga-ku, Shizuoka 422-8526, Japan

Abstract

Zinc is released with glutamate from neuron terminals in the hippocampus. Zinc may serve as a negative-feedback factor of presynaptic activity and negatively modulate postsynaptic calcium mobilization. On the other hand, the hippocampus is vulnerable to glutamate excitotoxicity, a final common pathway for numerous pathological processes such as Alzheimer's disease and amyotrophic lateral sclerosis, in addition to stroke/ischemia, temporal lobe epilepsy. The excitotoxicity is linked to the excessive influx of zinc and calcium. The crosstalk between zinc and calcium via calcium channels may play a role in both synaptic plasticity and excitotoxicity. This paper summarizes the involvement of zinc in functional and toxic aspects in the hippocampus focused on the crosstalk. The enhanced excitotoxicity in the hippocampus in zinc deficiency is also summarized.

Keywords: zinc, glutamate, excitotoxicity, hippocampus, calcium, crosstalk, zinc deficiency.

* TEL: 81-54-264-5700; FAX: 81-54-264-5705; E-mail: takedaa@u-shizuoka-ken.ac.jp

1. Introduction

Zinc is the second most abundant trace element in the body and powerfully influences cell division and differentiation [1, 2]. In microorganisms, plants and animals, over 300 enzymes require zinc for their functions. Zinc has three functions in zinc enzymes: catalytic, coactive (or cocatalytic) and structural [3]. In the brain, zinc turnover is strictly regulated via the brain-barrier system [4, 5]. Averaged intracellular zinc concentration is estimated to be approximately 150 μM, judging from zinc concentration in the total brain, while extracellular zinc concentration is estimated to be approximately 0.15−1 μM, judging from zinc concentration in the cerebrospinal fluid and extracellular zinc concentration measured by in vivo microdialysis. Zinc serves as an intracellular and an extracellular signal factor in synaptic neurotransmission; approximately 90 % of the total brain zinc forms zinc metalloproteins. The rest exists in the presynaptic vesicles and is histochemically reactive (as revealed by Timm's sulfide-silver staining method) [6, 7].

Zinc concentration is relatively high in the hippocampus in the brain [8] and the action of zinc is closely linked to functions and pathological processes [9]. There is a large number of evidence on zinc-containing glutamatergic neurons that sequester zinc in the presynaptic vesicles and release it in a calcium- and impulse-dependent manner [7, 10]. Zinc concentration in the presynaptic vesicles is the highest in the giant boutons of hippocampal mossy fibers and is estimated to be approximately 300 μM there [11]. All giant boutons of mossy fibers contain zinc in the presynaptic vesicles, while approximately 45% of Schaffer collateral boutons is zinc-positive [12]. Vesicular zinc may serve as an endogenous neuromodulator of several important receptors including the α-amino-3-hydroxy-5-methyl-4-isoxazolepropionic acid (AMPA)/kainate receptor, N-methyl-D-aspartate (NMDA) and γ-amino butyric acid (GABA) receptors [13-15]. However, the extracellular concentration of zinc reached after the release is a matter of debate. Estimates after tetanic stimulation range between 10 and 100 μM [16, 17], even up to 300 μM under extreme conditions [18]. Excess of extracellular zinc become neurotoxic via the translocation of zinc to postsynaptic neurons [19-23].

This paper summarizes zinc action via crosstalk between zinc and calcium in both functional and pathological aspects and also enhanced glutamate excitotoxicity in zinc deficiency.

2. Zinc Transport into the Brain

Zinc-binding affinity for ligands in the plasma is important to understand the mechanism of zinc transport into the brain through the brain barrier system (Figure 1) [4].Plasma zinc (approximately 15 μM) is partitioned between high molecular–weight and low molecular–weight fractions. The former is a protein–bound form (98 %) and the latter is a low molecular–weight ligand–bound form (1–2 %) and free zinc (Zn^{2+}), which is estimated to be as low as 10^{-9}–10^{-10} M.

The largest component of exchangeable zinc in the plasma is albumin. A brain autoradiogram with ^{65}Zn in the Nagase analbuminemic rat, which has a genetic mutation affecting albumin mRNA processing and lacks plasma albumin, demonstrated that ^{65}Zn distribution in the brain is similar to that in normal rats and that albumin is not essential for zinc transport into the brain [24].

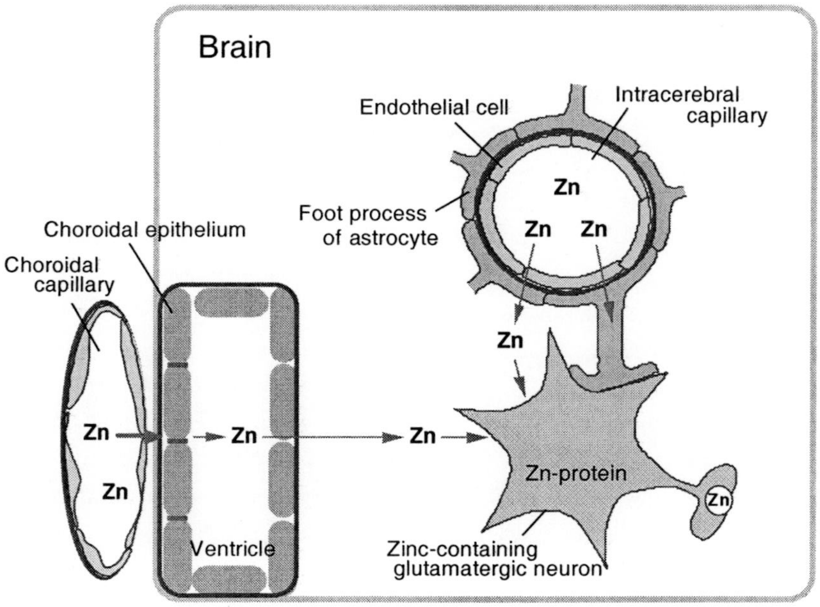

Figure 1. Zinc transport into the brain via the brain barrier system. Zinc bound to albumin and amino acids, *i.e.*, histidine and cysteine, serves as the exchangeable zinc pool in the plasma. Zinc is transported into the brain through the blood-brain and the blood-CSF barriers.

However, plasma albumin appears to participate in zinc transport as a large pool of exchangeable zinc in normal animals. Zinc is also known to bind to other plasma proteins such as transferrin and α_2–macroglobulin. Although zinc firmly binds to α_2-macroglobulin, its functional significance is unknown.

The next largest component of exchangeable zinc in the plasma is amino acids, *i.e.*, histidine and cysteine. Aiken *et al.* reported that the ratio of ^{65}Zn concentration in the brain, as well as in other tissues, to plasma ^{65}Zn concentration is enhanced by L-histidine infusion [25]. Brain distribution of ^{65}Zn-His is consistent with the data of an L-histidine infusion experiment [26]. It is possible that L-histidine is involved in zinc transport into the brain parenchyma through the brain barrier system [27]. A rat brain peptide/histidine transporter (PHT1) has been cloned [28]. PHT1 mRNA is intensely expressed in the choroid plexus. However, it is unknown whether histidine-bound forms actually pass across the plasma membranes of the choroidal epithelial cells (and brain capillary endothelial cells). On the other hand, DMT1, a divalent metal transporter, is expressed in brain capillary endothelial cells and choroidal epithelial cells [29]. Alternatively, it is possible that histidine serves to transfer zinc to DMT1. Other zinc transporters, *e.g.*, hZIP (ZRT1, IRT1-like protein), are also candidates for zinc transport across the brain barrier system [30].

Zinc is transported into the brain across the blood-CSF barrier, in addition to the blood-brain barrier, the main supply path of zinc (Figure 1); the choroid plexus might participate in slow supply of zinc to the brain. ^{65}Zn was highly distributed in the choroid plexus of mice and rats 1 h after intravenous injection of $^{65}ZnCl_2$ and then distributed in the brain parenchyma with a decrease in choroidal ^{65}Zn [8]. The maximum concentrations of ^{65}Zn in the rat brain were observed 6–10 days after the injection [31]. In the brain, ^{65}Zn was concentrated in the hippocampus, especially the hippocampal CA3 and dentate gyrus, and also in the amygdala, especially the amygdalopiriform transition and the amygdalo-hippocampal transition areas.

Zinc concentration in the CSF is approximately 0.15 μM. To study zinc uptake by the brain parenchyma cells *via* the CSF, $^{65}ZnCl_2$ and ^{65}Zn-His were injected intracerebroventricularly in rats [32, 33]. ^{65}Zn was distributed in the brain parenchyma, *i.e.*, the hippocampus and hypothalamus, after intracerebroventricular injection of $^{65}ZnCl_2$, while the radioactivity from ^{65}Zn-His was distributed extensively in the brain compared with that from $^{65}ZnCl_2$. PHT1 mRNA is widely expressed in the brain [28]. Especially intense hybridization signals are found in the hippocampus, cerebellum, and pontine nucleus. It is possible that PHT1 is involved in zinc uptake in neurons and glial cells. On the other hand, the finding that histidine decreases ^{65}Zn uptake in the synaptosomal preparation suggests that histidine does not participate in a carrier-mediated uptake in neuron terminals [34]. Therefore, the mechanism of zinc uptake by neurons and glial cells remains to be clarified. On the other hand, calcium channels are involved in zinc uptake in excitation (depolarization) of zinc-containing glutamatergic neurons as described later (Figure 6) [35,36]; zinc influx via calcium channels is linked to not only the modulation of zinc-containing glutamatergic neuron activity but also neurodegeneration.

Zinc turnover in the brain is very slow in contrast to zinc dynamics in the synaptic activity. The half-time for elimination of ^{65}Zn from the rat brain is in the range of 16–43 days; the longest is observed in the amygdala, followed by those in the piriform cortex and perirhinal cortex [31].

3. Differential Action of Extracellular Zinc in Extracellular Neurotransmitter Concentrations

Zinc taken up by neurons is transported anterogradely and retrogradely *via* the axonal transport system [37,38]. In zinc-containing glutamatergic neurons, the zinc may be transported into synaptic vesicles (Figure 1). Zinc sequestered in the synaptic vesicles is released with glutamate into the synaptic cleft [7]. However, the role of vesicular zinc is poorly understood. To examine the action of extracellular zinc in extracellular glutamate concentration, *in vivo* microdialysis experiments were carried out in the rat hippocampus. In the hippocampal CA3 region innervated by mossy fibers of dentate granule cells, glutamate concentration in the extracellular fluid was decreased by perfusion with 10–300 μM $ZnCl_2$, while increased by perfusion with 1 mM CaEDTA, a membrane-impermeable zinc chelator, suggesting that endogeneous zinc released into the synaptic cleft attenuates glutamate release in the CA3 region [39]. On the other hand, GABA concentration in the extracellular fluid was increased by perfusion with zinc, while decreased by perfusion with CaEDTA. Extracellular zinc may differentially act on glutamatergic and GABAergic neurotransmitter systems.

The hippocampal CA1 is connected with CA3 pyramidal cells *via* Schaffer collaterals (Figure 3). Extracellular glutamate concentration was also decreased by perfusion with zinc in the CA1 region. Thus, zinc action in the postsynaptic response (glutamate release) was examined by using the CA1–entorhinal connection (Figure 2) [40]. Perfusion of the CA1 with 1 μM tetrodotoxin, a sodium channel blocker, significantly decreased extracellular glutamate concentration in the entorhinal cortex, while perfusion of the CA1 with 50 μM glutamate significantly increased it, indicating that the response of postsynaptic CA1 pyramidal cells can be monitored by the level of glutamate released from the neuron terminals in the entorhinal cortex. When the CA1 region was perfused with 50 μM $ZnCl_2$, extracellular glutamate concentration was decreased not only in the CA1 but also in the entorhinal cortex (Figure 2a). Furthermore, the increase in extracellular glutamate concentration in the entorhinal cortex during perfusion of the CA1 with 50 μM glutamate was completely suppressed by addition of 50 μM $ZnCl_2$ to the CA1 (Figure 2b). Zinc could potentiate AMPA/kainate receptors, a subtype of glutamate receptors, on CA1 pyramidal cells [41]. However, in addition to the attenuation of glutamate release from presynaptic neurons, the inhibition of NMDA receptors on the CA1 pyramidal cells and the enhancement of GABA release from interneurons, which are both mediated by zinc, appear to attenuate the above potentiation, resulting in the attenuation of glutamate release from the postsynaptic neuron terminals.

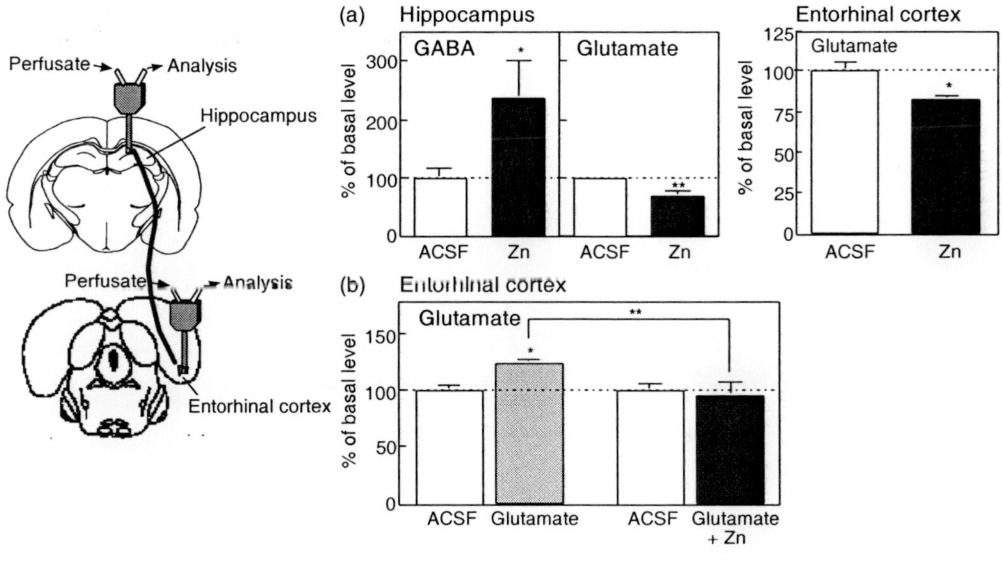

Figure 2. Zinc-dependent response of hippocampal CA1-entorhinal cortex connection. Position of microdialysis probes in the hippocampal CA1 and the entorhinal cortex is shown in the brain maps. a, When the hippocampal CA1 was perfused with ACSF for 60 min and then with 50 μM $ZnCl_2$ in ACSF for 30 min, the entorhinal cortex was perfused with ACSF alone. Each bar and line represent the mean ± s.e.m. The asterisks indicate significant differences (*, $p<0.05$; **, $p<0.01$) from the basal level (ACSF). b, When the hippocampal CA1 was perfused with ACSF for 60 min and then with 50 μM glutamate in ACSF or 50 μM glutamate + 50 μM $ZnCl_2$ in ACSF for 30 min, the entorhinal cortex was perfused with ACSF alone. Each bar and line represents the mean ± s.e.m. The asterisks indicate significant differences (*, $p<0.05$; **, $p<0.01$) from ACSF or glutamate group.

4. Zinc Dynamics along with Neuronal Activity

In the brain extracellular fluid, zinc seems to consist of two forms mainly. One is the form bound to low molecular weight ligands such as amino acids and another is free zinc (Zn^{2+}) [4]. Free zinc concentration in the extracellular fluid is estimated to be approximately 20 nM [42]. ZnAF-2, a membrane-impermeable zinc indicator, has a low K_d value of 2.7 nM for zinc and its fluorescence is minimally changed in the presence of calcium, magnesium, cadmium, nickel or other heavy metals [43,44]. ZnAF-2 also has no apparent toxicity to living cells. To observe zinc dynamics during excitation of mossy fiber synapses, intracellular zinc in hippocampal slices was stained with membrane-permeable ZnAF-2 DA, diacetylated form of ZnAF-2. ZnAF-2 DA taken up by cells is hydrolyzed to ZnAF-2, which cannot permeate the cell membrane (Figure 3). Intracellular ZnAF-2 signal was highly detected in the dentate hilus and stratum lucidum, in which mossy fibers exist.

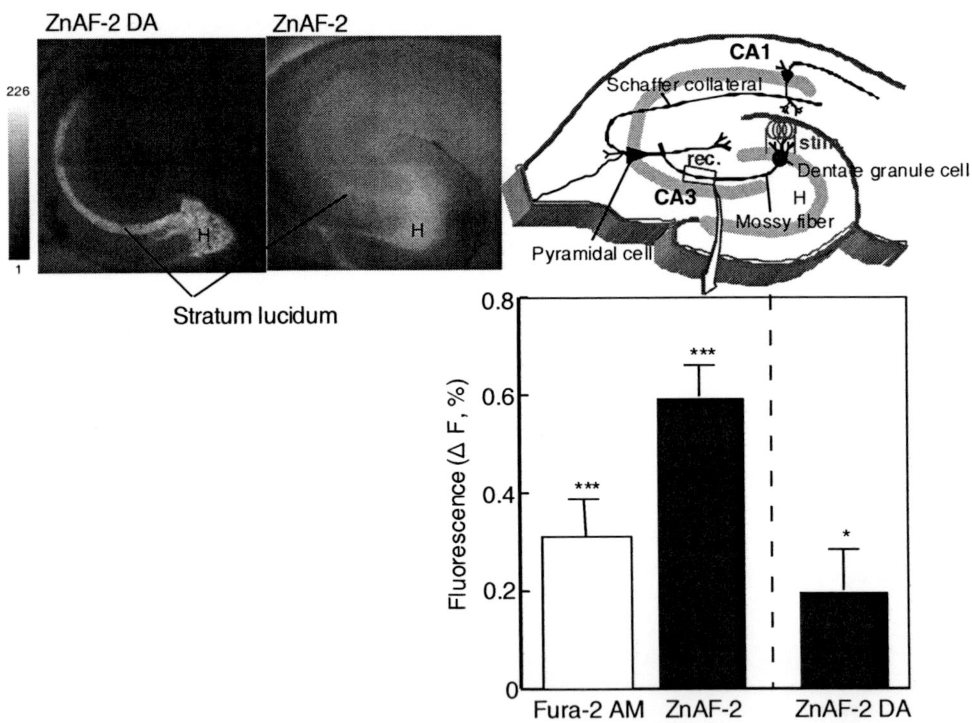

Figure 3. Zinc dynamics in intracellular and extracellular compartments in the hippocampus. ZnAF-2 DA and ZnAF-2 was applied to rat hippocampal slices to image intracellular and extracellular zinc, respectively. The fluorescent signal was imaged with an Argus-50/CA. H, hilus. In another experiment, hippocampal slices were double-stained with Fura-2 AM and ZnAF-2. Hippocampal slices were also stained with ZnAF-2 DA. Tetanic stimuli at 100 Hz for 1 s were delivered to the dentate granule cell layer and the fluorescent signals were monitored with a confocal laser-scanning microscopic system LSM 510 at the rate of 1 Hz through a 10 × objective. Region of interest was set in the stratum lucidum. The data represent the increment (%) of fluorescent signal during the stimulation (1 s) to the basal fluorescent signal just before the stimulation (the mean of 3 s). Each bar and line represents the mean ± S.E.M. *, $p<0.05$, **, $p<0.001$, vs. the basal level before the stimulation.

On the other hand, when ZnAF-2 was added to hippocampal slices, extracellular ZnAF-2 signal was immediately detected in the dentate hilus and stratum lucidum. The fluorescent signal was also detected around their regions.

When tetanic stimuli at 100 Hz for 1 s were delivered to the dentate granule cell layer, both extracellular and intracellular ZnAF-2 signals were increased with intracellular calcium (Fura-2) signal in the stratum lucidum during the stimulation (Figure 3) [45]. It is possible that the increase in intracellular ZnAF-2 signal is due to the binding of cytosolic ZnAF-2 to zinc retaken up after the release and that the increment in cytosolic signal is more than the decrement in vesicular signal. This possibility is supported by the data that the increase in intracellular ZnAF-2 signal in mossy fiber terminals during the tetanic stimulation was significantly decreased in the presence of 1 mM CaEDTA (Figure 4) [45]. Zinc released from mossy fiber terminals during tetanic stimulation may be immediately retaken up by the mossy fibers, probably via calcium channels such as VDCC, and also taken up into postsynaptic CA3 neurons.

5. Zinc Action via Crosstalk between Zinc and Calcium in Functional Aspect

Neural circuits of the zinc-containing glutamatergic neurons are considered to be associated with the episodic memory function and are important for behavior, emotional expression and cognitive-mnemonic operations [4]. Lu et al. [46] demonstrated that endogenous zinc is required for the induction of long-term potentiation (LTP) in hippocampal mossy fiber synapses. Li et al. [47] demonstrated that the induction of LTP in hippocampal mossy fiber synapses requires translocation of synaptically released zinc. On the other hand, the impairment of spatial learning, memory or sensorimotor function was not observed in zinc transporter-3-null mice, which lack the histochemically reactive zinc in synaptic vesicles [48,49]. There is also some evidence that zinc has no role in the CA3 mossy fiber LTP [16,50]. Thus, clarifying the physiological significance of vesicular zinc in synaptic plasticity is important to understand the hippocampal function.

The presynaptic action of zinc released from mossy fibers during tetanic stimulation was examined by using rat hippocampal slices (Figure 4). In mossy fiber terminals preferentially double-stained with zinc and calcium indicators, the increase in calcium orange signal during delivery of tetanic stimuli (100 Hz, 1 s) to the dentate granule cell layer is enhanced in the presence of 1 mM CaEDTA and attenuated in the presence of 100 μM zinc [45]. It is likely that zinc released from mossy fiber terminals suppresses the increase in calcium signal (Ca^{2+}) in the presynaptic terminals induced by tetanic stimulation, followed by inhibitory modulation of the presynaptic activity (Figure 5). Presynaptic calcium influx through voltage-dependent calcium cannel (VDCC) triggers vesicular exocytosis. FM4-64, a fluorescent indicator of synaptic vesicle recycling, is taken up into presynaptic vesicles in an activity-dependent manner. Subsequent rounds of exocytosis arising from depolarization lead to the release of the dye from the presynaptic terminals (destaining) [51, 52].

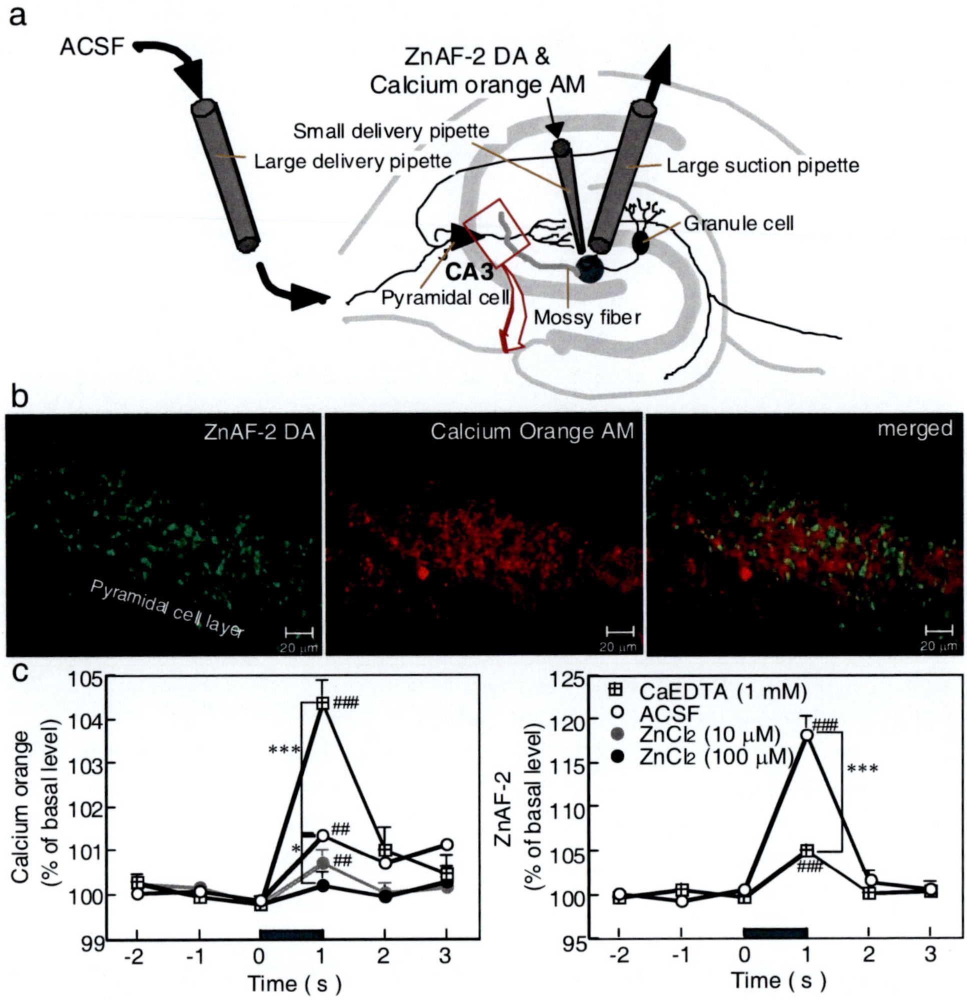

Figure 4. Inhibition of calcium signal by zinc in mossy fiber boutons during tetanic stimulation. a, Schematic illustration of the procedure. The stratum lucidum was regionally perfused with ZnAF-2 DA and calcium orange AM for mossy fiber terminal labeling. The red square was observed as shown in b. b, Giant boutons of mossy fibers were double-labeled with calcium orange and ZnAF-2 as shown in a merged image. c, Tetanic stimuli at 100 Hz for 1 s were delivered to the dentate granule cell layer in ACSF or reagents in ACSF. The fluorescent signals were monitored with a confocal laser-scanning microscopic system LSM 510 at the rate of 1 Hz through a 20 × objective. Region of interest (around 5 μm in diameter) was set in giant boutons of mossy fibers double-labeled with calcium orange and ZnAF-2. The data represent the ratio (%) of each fluorescent signal (1 s) to the basal fluorescent signal (the mean of 3 s) just before the stimulation. Each point and line represents the mean ± S.E.M. The shaded bar indicates the period of the stimulation. *, p<0.05, **, p<0.001, vs. ACSF group; **, p<0.01, ***, p< 0.001, vs. the basal level (the mean of 3 points just before the stimulation).

When tetanic stimuli at 10 Hz for 180 s, which induce mossy fiber LTP, are delivered to the dentate granule cell layer, the decrease in FM4-64 signal is enhanced by addition of 1 mM CaEDTA and suppressed by addition of 100 μM zinc [45]. Zinc released from mossy fiber terminals during tetanic stimulation may suppress vesicular exocytosis, via inhibitory modulation of presynaptic calcium mobilization (Figure 5).

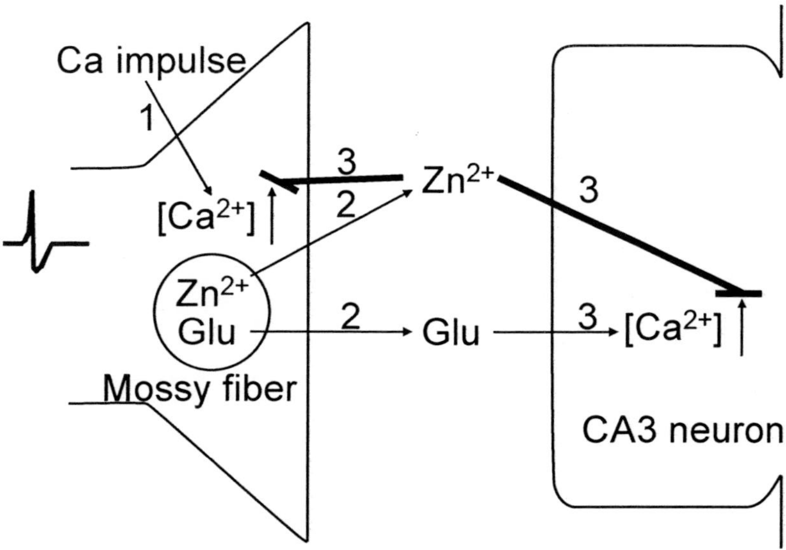

Figure 5. Negative zinc action against calcium mobilization in mossy fiber excitation. When the action potential is delivered to mossy fibers, calcium influx occurs via voltage-dependent calcium channel (VDCC) and Ca^{2+} concentration increases in the terminals (1), followed by exocytosis (2). Zinc released from mossy fiber boutons may negatively modulate the presynaptic activity via suppression of the increase in Ca^{2+} concentration (3). The negative modulation of the presynaptic activity by zinc leads to suppression of postsynaptic calcium mobilization. Zinc also negatively modulates the increase in Ca^{2+} concentration in postsynaptic CA3 neurons (3).

The hippocampal mossy fiber LTP is expressed by presynaptic mechanisms leading to persistent enhancement of neurotransmitter release. The induction of mossy fiber LTP is critically dependent on the increase in presynaptic calcium induced by stimulation of depolarization [53-55], which activates the calcium-calmodulin-sensitive adenyl cyclase I [56]. Mossy fiber zinc might be involved in the presynaptic mechanism leading to the LTP and be a negative neuromodulator against the presynaptic mechanism. It is possible that the release from mossy fibers leads to cross talk between extracellular free zinc and calcium via calcium channels such as VDCC, which is inhibited by zinc [57]. In an experiment using synaptosomal fraction of rat hippocampal CA3, on the other hand, zinc inhibits glutamate release via activation of presynaptic ATP-dependent potassium (K_{ATP}) channel [58]. Low micromolar concentrations of zinc inhibit cAMP signaling via inhibition of adenyl cyclase in N18TG2 neuroblastoma cells [59]. The IC_{50} of zinc against recombinant adenyl cyclase I is 1.4 μM. It is also possible that zinc taken up into the mossy fibers suppresses the cAMP-PKA signaling pathway via inhibition of adenyl cyclase I in mossy fiber boutons.

To examine zinc uptake by CA3 pyramidal cells and its significance, zinc uptake was checked in the CA3 pyramidal cell layer stained with ZnAF-2DA after tetanic stimulation to the dentate granule cell layer. The increase in intracellular ZnAF-2 signal induced by tetanic stimulation was completely inhibited in the presence of CNQX, an AMPA/kainate receptor antagonist [35]. These results suggest that zinc released from mossy fibers is taken up into CA3 pyramidal cells via activation of AMPA/kainate receptors. Activation of AMPA/kainate receptors provides the membrane depolarization and activates NMDA receptors and VDCC. Therefore, it is likely that calcium channels such as NMDA receptors and VDCC, in addition to calcium-permeable AMPA/kainate (GluR2-lacking) receptors, are involved in zinc uptake into CA3 pyramidal cells after mossy fiber excitation (Figure 6) [22].

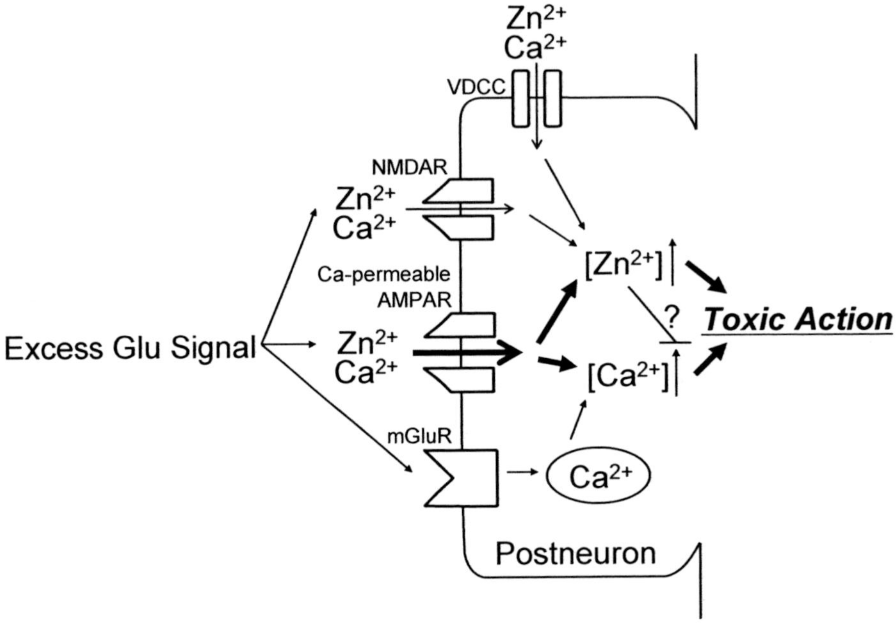

Figure 6. Zinc action in excessive excitation of mossy fibers. Excess of extracellular glutamate triggers excessive influx of zinc and calcium into postsynaptic CA3 neurons via ionotropic glutamate receptors and VDCC, followed by neurodegeneration. Calcium-permeable AMPA/kainite receptors play an important role in neurodegenerative processes. Extracellular zinc suppresses calcium mobilization into postsynaptic CA3 neurons via blockade of NMDA receptors and VDCC. The increase in intracellular zinc may suppress calcium release from internal stores, although it is harmful to postsynaptic CA3 neurons.

Mossy fiber activity evokes intracellular calcium mobilization via group I metabotropic glutamate receptors in CA3 pyramidal cells, in addition to calcium influx via NMDA receptors and VDCC, which is inhibited by zinc (Figure 6) [60]. Group I metabotropic glutamate receptors are coupled to phospholipase C and their activation evokes calcium release from internal stores via IP_3 [61,62]; *t*ADA stimulates phosphoinositide hydrolysis and calcium release from internal stores [60,63]. It is possible that the increase in free zinc concentration in CA3 pyramidal cells by mossy fiber activity influence calcium mobilization via group I metabotropic glutamate receptors. Group I metabotropic glutamate receptor activation is associated with memory formation. *t*ADA facilitates maintenance of LTP in the dentate gyrus in vivo and memory formation in the spatial alteration task [64,65]. On the other hand, group I metabotropic glutamate receptor activation is also associated with neurotoxic effects of glutamate in post-ischemic neuronal damage and epileptiform activity [61,62]. Because group I metabotropic glutamate receptor activation is involved in both functional and toxic aspects, it seems to be important to understand the effect of free zinc levels on intracellular calcium mobilization. Cytosolic concentration of basal free zinc is estimated to be subnanomlar [66]. The action of free zinc levels in CA3 pyramidal cells after group I metabotropic glutamate receptor activation was examined by regional delivery of *t*ADA to the stratum lucidum after blockade of AMPA/kainate receptor-mediated calcium and zinc influx [35]. Intracellular calcium orange signal in the CA3 pyramidal cell layer was increased by *t*ADA, suggesting that calcium release from internal stores is induced by group I metabotropic glutamate receptor activation in CA3 pyramidal cells (Figure 6). Furthermore, pyrithione was used to manipulate intracellular free zinc concentration [35]. The perfusion with 50 μM pyrithione decreased basal ZnAF-2 signal in the CA3 pyramidal cell layer without affecting basal calcium orange signal. The increase in calcium orange signal in the CA3 pyramidal cell layer by *t*ADA was enhanced in the presence of 50 μM pyrithione. These results suggest that calcium mobilization via group I metabotropic glutamate receptor pathway is facilitated by the decrease in free zinc concentration in CA3 pyramidal cells. The perfusion with 50 μM pyrithione and 100 μM zinc increased basal ZnAF-2 signal in the CA3 pyramidal cell layer and deceased basal calcium orange signal, suggesting that basal free calcium concentration is decreased by the increase in free zinc concentration in CA3 pyramidal cells. The increase in calcium orange signal by *t*ADA was blocked in the presence of 50 μM pyrithione and 100 μM zinc, suggesting that calcium mobilization via group I metabotropic glutamate receptor pathway is blocked by the increase in free zinc concentration in CA3 pyramidal cells (Figure 5 and 6). Therefore, zinc influx via AMPA/kainate receptor activation may negatively modulate calcium mobilization via group I metabotropic glutamate receptor pathway in CA3 pyramidal cells after mossy fiber excitation. It is possible that the negative modulation by zinc is linked to not only synaptic plasticity such as LTP and long-term depression (LTD) but also a neuroprotective function [61]; activation of group I metabotropic glutamate receptors amplifies NMDA-mediated neuronal toxicity and blockade of them is neuroprotective against NMDA-mediated damage. However, excessive zinc influx via AMPA/kainate receptor activation may be neurotoxic in postsynaptic neurons [19-23], even if the zinc negatively modulates intracellular calcium mobilization (Figure 5 and 6).

6. Zinc Action via Crosstalk between Zinc and Calcium in the Pathological Aspect

In global ischemia, CA1 pyramidal neurons degenerate, whereas CA3 pyramidal neurons remain intact [67-68]. Excess of extracellular glutamate is known as excitotoxicity. It is possible that mossy fiber zinc serves protectively against glutamate excitotoxicity elicited by excessive excitation of dentate granule cells, although excess of extracellular zinc is also harmful because of the translocation to the postneuron as described above. The response of hippocampal mossy fiber zinc to excessive glutamate release was examined by regional delivery of 1 mM glutamate to the stratum lucidum [69]. Extracelluar zinc signal was markedly increased in the stratum lucidum and intracellular calcium signal was persistently increased in the CA3 pyramidal cell layer. These results suggest that the excessive delivery of glutamate leads to the release of zinc and glutamate from mossy fibers and excites mossy fiber synapses. Presynaptic kainate receptors is involved in glutamate release [70]. Excessive delivery of glutamate activates presynaptic kainate receptors, followed by the release of zinc and glutamate from mossy fibers. Excessive delivery of glutamate may also facilitate glutamate release from astrocyres via activation of AMPA/kainate receptors [71].

The persistent increase in calcium signal in the CA3 pyramidal cell layer during stimulation with glutamate was significantly attenuated in the presence of 100 μM zinc, while significantly enhanced in the presence of 1 mM CaEDTA [69]. These results suggest that zinc released from mossy fibers attenuates the increase in intracellular calcium signal in mossy fiber synapses and postsynaptic CA3 neurons after excessive inputs to dentate granular cells (Figure 5). The blockade of calcium influx via NMDA receptors and VDCC by zinc may be involved in attenuating the increase in intracellular calcium. Furthermore, zinc may inhibit intracellular calcium mobilization via activation of group I metabotropic glutamate receptors as described above (Figure 6). An enhanced GABA release from interneurons by zinc might be also involved in attenuating the increase in calcium signal in mossy fibers and postsynaptic CA3 neurons [39]. Extracelullar and intracellular action of zinc seems to be linked to negative modulation of the increase in intracellular calcium in mossy fiber synapses (Figure 5). Therefore, excessive zinc release from mossy fibers is involved in both functional and toxic aspects.

The role of zinc in excitation of Schaffer collateral-CA1 pyramidal cell synapses was examined by using hippocampal slices stained with ZnAF-2 and ZnAF-2DA [36]. Extracellular and intracellular ZnAF-2 signals were increased in the stratum radiatum of the CA1, in which Schaffer collateral synapses exist, during tetanic stimulation (200 Hz, 1 s). Both the increases were completely blocked in the presence of 1 mM CaEDAT, suggesting that 1 mM CaEDTA is effective to chelate zinc released from Schaffer collaterals. The increase in intracellular fura-2 signal in the stratum radiatum of the CA1 during delivery of tetanic stimuli at 100 Hz for 1 s to the Schaffer collaterals is significantly enhanced in the presence of CaEDTA [72], suggesting that zinc released from Schaffer collaterals negatively modulates presynaptic and/or postsynaptic calcium signaling in the CA1. The presynaptic action of Schaffer collateral zinc was examined by using FM4-64 [36]. The decrease in FM4-64 signal during tetanic stimulation (10 Hz, 180 s) was enhanced in the stratum radiatum of the CA1 (Schaffer collaterals) in the presence of 1 mM CaEDTA, while suppressed in the presence of 5 μM $ZnCl_2$, suggesting that zinc released from Schaffer collaterals suppresses presynaptic activity during tetanic stimulation. When Schaffer collateral synapses were regionally stimulated with 1 mM glutamate, calcium orange signal was increased in the CA1 pyramidal cell layer. This increase was enhanced in the presence of 1 mM CaEDTA and attenuated in the presence of 10 μM $ZnCl_2$. These results suggest that zinc attenuates excitation of Schaffer collateral synapses elicited with glutamate via suppression of presynaptic activity and inhibition of postsynaptic NMDA receptors. Schaffer collateral zinc may also negatively modulate intracellular calcium signal in excessive excitation of Schaffer collateral synapses.

7. Zinc and Epilepsy

Zinc homeostasis in the brain is closely associated with epileptic seizures [9]; zinc concentration in the brain is decreased by epileptic seizures [73]. Kainate is an agonist of glutamate receptor subtypes and kainate-challenged mice and rats are experimental models of human temporal lobe epilepsy. They have been used to understand brain zinc movement in epileptic seizures. Zinc concentration in the hippocampus is significantly decreased in kainate-challenged mice [73]. A selective loss of Timm's stain is observed in the hippocampal mossy fibers after electrical stimulation of the perforant path, which evokes hippocampal granule spikes and epileptiform discharges [74,75]. Extracellular concentrations of zinc and glutamate are significantly increased in the hippocampus of young rats challenged with kainate [76]. The attenuation of Timm's stain is due to excessive excitation of zinc-containing glutamatergic neurons, followed by the translocation of zinc to postsynaptic neurons and neurodegeneration. Neuronal loss is observed in the CA1, CA2 and CA3 pyramidal cell layers after challenge with kainite [77] and seems to be linked to the loss of zinc from the hippocampus [73].

8. Zinc and Novelty Stress

Acute stress enhances hypothalamo-pituitary-adrenocortical (HPA) axis activity. Painful physical stress such as electro footshocks, restraint, forced swimming and tail pinch enhances the HPA axis activity, while a novel environment that an animal has not previously experienced can also influence the HPA axis activity [78,79]. An animal suddenly placed in a novel environment for a specified period shows a complex behavior for exploratory activity [80,81]. This placement is a novelty stress and is regarded as painless psychological stress. Exploratory activity involves several processes, including arousal, attention and locomotor activity and is associated with anxiety, fear and stress [82]. An extensive neuronal activity takes place in the hippocampus during exploratory behavior [83]. The hippocampal function is linked with the HPA axis activity and is involved in stress response [84].

To analyze the response of extracellular zinc in the hippocampus against novelty stress, rats were placed for 50 min in a novel environment once a day for 8 days [85]. Extracellular glutamate in the hippocampus was increased during exploratory behavior on day 1, whereas extracellular zinc was decreased. The same phenomenon was observed during exploratory behavior on day 2 and extracellular zinc returned to the basal level during exploratory behavior on day 8. To examine the significance of the decrease in extracellular zinc in exploratory activity, exploratory behavior was observed under perfusion with 1 mM CaEDTA. Locomotor activity in the novel environment was decreased by perfusion with CaEDTA. The decrease in extracellular zinc and the increase in extracellular glutamate in exploratory period were abolished by perfusion with CaEDTA. These results suggest that zinc uptake by hippocampal cells is linked with exploratory activity and required for the activation of glutamatergic neurotransmitter system. The zinc uptake may be involved in the response in painless psychological stress or the cognitive processes.

9. Enhanced Glutamate Excitotoxicity in Zinc Deficiency

Approximately 50% of the world population does not get adequate zinc [86]. Dietary zinc deficiency not only retards the growth of humans and animals, but also affects brain maturation [87,88]. Inadequate dietary zinc causes changes in behavior such as reduced activity and responsiveness. Periods of rapid growth such as pregnancy and infancy are particularly susceptible to dietary zinc deficiency.

Dietary zinc deficiency causes a decrease in the plasma zinc concentration [89], probably in the exchangeable zinc pool, followed by a decrease in the zinc concentration in the liver and femur. On the other hand, several researchers failed to find any change in brain zinc concentration in dietary zinc deficiency [9, 87]. Although zinc concentration in the hippocampus was not decreased after 4-week zinc deprivation, hippocampal extracellular zinc and histochemically reactive zinc detected by Timm's staining were decreased in young rats and mice fed a zinc-deficient diet for 4 weeks [76, 90]. Furthermore, zinc concentration in the hippocampus was significantly decreased after 12-week zinc deprivation [91]. The hippocampus is responsive to dietary zinc deficiency and vesicular zinc seems to be more responsive to dietary zinc deficiency than zinc metalloproteins.

Alteration of zinc homeostasis in the brain influences the etiology and manifestation of epileptic seizures [92-95]. Susceptibility to kainate-induced seizures was enhanced in mice and rats fed a zinc-deficient diet for 4 weeks [76,96]. Enhanced release of glutamate associated with a decrease in GABA concentration in the hippocampus is a possible mechanism for the enhanced seizure susceptibility in zinc deficiency. Neuronal loss and TUNEL-positive cells are more observed in the CA1, CA2 and CA3 pyramidal cell layers of zinc-deficient group than those of the control group after challenge with kainite [77]; kainate-induced excitotoxcity is enhanced by zinc deficiency (Figure 7). Glutamate excitotoxicity is a final common pathway for numerous pathological processes such as Alzheimer's disease and amyotrophic lateral sclerosis, in addition to stroke/ischemia, temporal lobe epilepsy [97,98]. It is likely that pathological processes associated with glutamate excitotoxicity are aggravated by zinc deficiency. Therefore, adequate zinc supply to the brain is important for prevention of neurological diseases.

10. Activation of HPA Axis in Early Zinc Deficiency

Dietary zinc deficiency causes anorexia, followed by weight loss and growth retardation [99-101]. It is possible that the stress of severe food restriction leads to the increase in serum corticosterone concentration via activation of the HPA axis; serum corticosterone concentration is significantly increased in young rats after 2-week zinc deprivation (Figure 7) [102]. The activation of HPA axis occurs prior to the decrease in extracellular zinc and histochemically reactive zinc in the hippocampus in zinc deficiency. Behavioral abnormality in the open-field test and the increase in anxiety-like behavior are observed in young rats after 2-week zinc deprivation, suggesting that the increase in serum corticosterone concentration is associated with behavioral abnormality in zinc deficiency.

It is possible that HPA axis activity in zinc deficiency alter hippocampal function. The increase in serum corticosterone concentration increases corticosterone concentration in the brain extracellular fluid [103], followed by the increase in cytosolic free calcium concentration in neurons [84] (Figure 7).

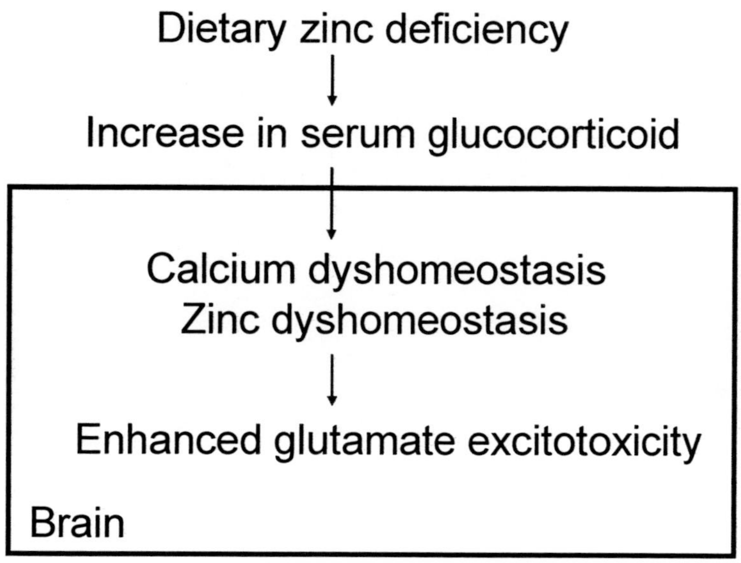

Figure 7. Enhanced glutamate excitotoxicity in the hippocampus in zinc deficiency. Zinc deficiency activates the HPA axis, followed by the increase in serum corticosterone, and increases the basal levels of intracellular Ca^{2+} in the hippocampus prior to the decrease in extracellular zinc. Intracellular calcium dyshomeostasis, in addition to zinc dyshomeostasis, may be linked to the enhanced glutamate excitotoxicity in the hippocampus in zinc deficiency.

In hippocampal slices from rats fed the zinc-deficient diet for 2 weeks, fura-2 signal was more than in the control hippocampal slices [102]. These results suggest that basal Ca^{2+} concentration in hippocampal cells is increased in association with the increase in serum corticosterone concentration. Glucocorticoids increase cytosolic Ca^{2+} concentration in cultured hippocampal neurons [104,105] and potentiate extracellular glutamate accumulation in the hippocampus [106,107]. Therefore, zinc-deficient mice and rats is a potential model to study pathological processes of psychological diseases.

References

[1] Vallee BL, Falchuk KH: Zinc and gene expression. *Philos. Trans R. Soc. Lond B Biol. Sci.* 294, 185-197, 1981.

[2] Coleman JE: Zinc proteins: enzymes, storage proteins, transcription factors, and replication proteins. *Annu. Rev. Biochem.* 61, 897-946, 1992.

[3] Vallee BL, Falchuk KH: The biological basis of zinc physiology. *Physiol. Rev.* 73, 79-118, 1993.

[4] Takeda A: Movement of zinc and its functional significance in the brain. Brain Res. *Rev.* 34, 137-148, 2000.

[5] Takeda A: Essenttial trace metals and brain function. *Yakugaki Zasshi* 124, 577-585, 2004.

[6] Haug FMS: Heavy metals in the brain. A light microscope study of the rat with Timms' sulphide silver method. Methodological considerations and cytological and regional staining patterns. *Adv. Anat. Embryol. Cell Biol.* 47, 1-71, 1973.

[7] Frederickson CJ: Neurobiology of zinc and zinc-containing neurons. *Int. Rev. Neurobiol.* 31, 145-238,1989.

[8] Takeda A, Akiyama T, Sawashita J, Okada S: Brain uptake of trace metals, zinc and manganese, in rats. *Brain Res.* 640, 341-344, 1994.

[9] Takeda A: Zinc homeostasis and functions of zinc in the brain. *BioMetals* 14: 343-352, 2001.

[10] Frederickson C.J, Koh JY, Bush AI: The neurobiology of zinc in health and disease. *Nat. Rev. Neurosci.* 6, 449-462, 2005.

[11] Frederickson CJ, Klitenick MA, Manton WI, Kirkpatrick JB: Cytoarchitectonic distribution of zinc in the hippocampus of man and rat. *Brain Res.* 273, 335-339, 1983.

[12] Sindreu CB, Varoqui H, Erickson JD, Perez-Clausell J: Boutons containing vesicular zinc define a subpopulation of synapses with low AMPAR content in rat hippocampus. *Cerebral Cortex* 13, 823-829, 2003.

[13] Harrison NL, Gibbons SJ: Zn^{2+}: an endogenous modulator of ligend- and voltage-gated ion channels. *Neuropharmacolgy* 33, 935-952, 1994.

[14] Smart TG, Xie X, Krishek BJ: Modulation of inhibitory and excitatory amino acid receptor ion channels by zinc. *Prog. Neurobiol.* 42, 393-441, 1994.

[15] Huang EP: Metal ions and synaptic transmission: think zinc. *Proc. Natl. Acad. Sci.* 94, 13386-13387, 1997.

[16] Vogt K, Mellor J, Tong G, Nicoll R: The actions of synaptically released zinc at hippocampal mossy fiber synapses. *Neuron* 26, 197-196, 2000.

[17] Li Y, Hough CJ, Suh SW, Sarvey JM, Frederickson CJ: Rapid translocation of Zn(2+) from presynaptic terminals into postsynaptic hippocampal neurons after physiological stimulation. *J. Neurophysiol.* 86, 2597-2604, 2001.

[18] Assaf SY, Chung S-H: Release of endogeneous Zn^{2+} from brain tissue during activity. *Nature* 308, 734-735, 1984.

[19] Koh JY, Suh SW, Gwag BJ, He YY, Hsu CY, Choi DW: The role of zinc in selective neuronal death after transient global cerebral ischemia. *Science* 272, 1013-1016, 1996.

[20] Choi DW, Koh JY: Zinc and brain injury. *Annu. Rev. Neurosci.* 21, 347-375, 1998.

[21] Lee JM, Zipfel G.J, Choi DW: The changing landscape of ischaemic brain injury mechanisms. *Nature* 399, A7-A14, 1999.

[22] Weiss JH, Sensi SL, Koh JY: Zn(2+): a novel ionic mediator of neural injury in brain disease. *Trends Pharmacol. Sci.* 21, 395-401, 2000.

[23] Suh SW, Garnier P, Aoyama K, Chen Y, Swanson RA: Zinc release contributes to hypoglycemia-induced neuronal death. *Neurobiol. Dis.* 16, 538-545, 2004.

[24] Takeda A, Kawai M, Okada S: Zinc distribution in the brain of Nagase analbuminemic rat and enlargement of the ventricular system. *Brain Res.* 769, 193-195, 1997.

[25] Aiken SP, Horn NR, Saunders NR: Effect of histidine on tissue zinc distribution in rats. *BioMetals* 5, 235-243, 1992.

[26] Takeda A, Suzuki M, Okada S, Oku, N: Influence of histidine on zinc transport into rat brain. *J. Health Sci.* 46, 209-213, 2000.

[27] Takeda A, Suzuki M, Oku N: Possible involvement of plasma histidine in differential brain permeability to zinc and cadmium. *BioMetals* 15, 371-375, 2002.

[28] Yamashita T, Shimada S, Guo W, Sato K, Kohmura E, Hayakawa T, Takagi T, Tohyama M: Cloning and functional expression of a brain peptide/histidene transporter. *J. Biol. Chem.* 272, 10205-10211, 1997.

[29] Gunshin H, Mackenzie B, Berger UV, Gunshin Y, Romero MF, Boron WF, Nussberger S, Gollan JL, Hediger MA: Cloning and characterization of a mammalian protein-coupled metal-ion transporter. *Nature*, 388, 482-488, 1997.

[30] Gaither LA, Eide DJ: Functional expression of the human hZIP2 zinc transporter. *J. Biol. Chem.* 275, 5560-5564, 2000.

[31] Takeda A, Sawashita J, Okada S: Biological half-lives of zinc and manganese in rat brain. *Brain Res.* 695, 53-58, 1995.

[32] Takeda A, Sawashita J, Okada S: Localization of rat brain of the trace metals, zinc and manganese, after intracerebroventricular injection. *Brain Res.* 658, 252-254, 1994.

[33] Takeda A, Suzuki M, Okada S, Oku N: ^{65}Zn localization in rat brain after intracerebroventricular injection of ^{65}Zn-histidine. *Brain Res.* 863, 241-244, 2000.

[34] Wensink J, Molenaar AJ, Woroniecka UD, Van den Hamer CJA: Zinc uptake into synaptosomes. *J. Neurochem.* 50, 782-789, 1988.

[35] Takeda A, Fuke S, Minami A, Oku N: Role of zinc influx via AMPA/kainate receptor activation in metabotropic glutamate receptor-mediated calcium release. *J. Neurosci. Res.* 85, 1310-1317, 2007.

[36] Takeda A, Fuke S, Tsutsumi W, Oku N: Negative modulation of presynaptic activity by zinc released from Schaffer collaterals. *J. Neurosci. Res.*, in press.

[37] Takeda A, Ohnuma M, Sawashita J, Okada S: Zinc transport in the rat olfactory system. *Neurosci. Lett.* 225, 69-71, 1997.

[38] Takeda A, Kodama Y, Ohhuma M, Okada S: Zinc transport from the striatum and substantia nigra. *Brain Res. Bull.* 47, 103-106, 1998.

[39] Takeda A, Minami A, Seki Y, Oku N: Differential Effects of Zinc on Glutamatergic and GABAergic Neurotransmitter Systems in the Hippocampus. *J. Neurosci. Res.* 75, 225-229, 2004.

[40] Takeda A, Minami A, Seki Y, Oku N: Inhibitory function of zinc against excitation of hippocampal glutamatergic neurons. *Epilepsy Res.* 57, 169-174, 2003.

[41] Rassendren FA, Lory P, Pin JP, Nargeot J: Zinc has opposite effects on NMDA and non-NMDA receptors expressed in Xenopus Oocytes. *Neuron* 4, 733-740, 1990.

[42] Frederickson CJ, Giblin LJ, Krezel A, McAdoo DJ, Muelle RN, Zeng Y, Balaji RV, Masalha R, Thompson RB, Fierke CA, Sarvey JM, Valdenebro M, Prough DS, Zornow MH: Concentrations of extracellular free zinc (pZn)e in the central nervous system

during simple anesthetization, ischemia and reperfusion. *Exp. Neurol.* 198, 285-293, 2006.

[43] Hirano T, Kikuchi K, Urano Y, Higuchi T, Nagano T: Highly zinc-selective fluorescenct sensor molecules suitable for biological applications. *J. Am. Chem. Soc.* 122, 12399-12400, 2000.

[44] Hirano T, Kikuchi K, Urano Y, Nagano T: Improvement and biological applications of fluorescent probes for zinc, ZnAFs. *J. Am. Chem. Soc.* 124, 6555-6562, 2002.

[45] Minami A, Sakurada N, Fuke S, Kikuchi K, Nagano T, Oku N, Takeda A: Inhibition of presynaptic activity by zinc released from mossy fiber terminals during tetanic stimulation. *J. Neurosci. Res.* 83, 167-176, 2006.

[46] Lu YM, Taverna FA, Tu R, Ackerley CA, Wang YT, Roder J: Endogenous Zn(2+) is required for the induction of long-term potentiation at rat hippocampal mossy fiber-CA3 synapses. *Synapse* 38, 187-197, 2000.

[47] Li Y, Hough CJ, Frederickson CJ, Sarvey JM: Induction of mossy fiber→CA3 long-term potentiation reguires translocation of synaptically released Zn2+. *J. Neurosci.* 21, 8015-8025, 2001.

[48] Cole TB, Wenzel HJ, Kafer KE, Schwartzkroin PA, Palmiter RD: Elimination of zinc from synaptic vesicles in the intact mouse brain by disruption of the ZnT-3 gene. *Proc. Natl. Acad. Sci. USA* 96, 1716-1721, 1999.

[49] Cole TB, Robbins CA, Wenzel HJ, Schwartzkroin PA, Palmiter RD: Seizures and neuronal damage in mice lacking vesicular zinc. *Epilepsy Res.* 39, 153-169, 2000.

[50] Xie X, Smart TG: Modulation of long-term potentiation in rat hippocampal pyramidal neurons by zinc. *Pflugers Arch.* 427, 481-486, 1994.

[51] Klingauf J, Kavalali ET, Tsien RW: Kinetics and regulation of fast endocytosis at hippocampal synapses. *Nature* 394, 581-585, 1998.

[52] Zakharenko SS, Zablow L, Siegelbaum SA: Visualization of changes in presynaptic function during long-term synaptic plasticity. *Nat. Neurosci.* 4, 711-717, 2001.

[53] Castillo PE, Weisskopf MG, Nicoll RA: The role of Ca^{2+} channels in hippocampal mossy fiber synaptic transmission and long-term potentiation. *Neuron* 12, 261-269, 1994.

[54] Nicoll RA, Malenka RC: Contrasting properties of two forms of long-term potentiation in the hippocampus. *Nature* 377, 115-118, 1995.

[55] Breustedt J, Vogt KE, Miller RJ, Nicoll RA, Schmitz D: Alpha1E-containing Ca^{2+} channels are involved in synaptic plasticity. *Proc. Natl. Acad. Sci. USA* 100, 12450-12455, 2003.

[56] Wang H, Storm DR: Calmodulin-regulated adenylyl cyclases: cross-talk and plasticity in the central nervous system. *Mol. Pharmacol.* 63, 463-468, 2003.

[57] Takeda A: Inhibitory modulatuon of glutamatergic neuron activity by zinc in the hippocampus. *Biomed. Res. Trace Elements* 17, 399-405, 2006.

[58] Bancila V, Nikonenko I, Dunant Y, Bloc A: Zinc inhibits glutamate release via activation of pre-synaptic K_{ATP} channels and reduces ischaemic damage in rat hippocampus. *J. Neurochem.* 90, 1243-1250, 2004.

[59] Klein C, Sunahara RK, Hudson TY, Heyduk T, Howlett AC: Zinc inhibition of camp signaling. *J. Biol. Chem.* 277, 11859-11865, 2002.

[60] Kapur A, Yeckel MF, Johnston D: L-type calcium channels are required for one form of hippocampal mossy fiber LTP. *J. Neurophysiol.* 79, 2181-2190, 1998.

[61] Bordi F, Ugolini A: Group I metabotropic glutamate receptors: implications for brain diseases. *Prog. Neurobiol.* 59, 55-79, 1999.

[62] Nicoletti F, Bruno V, Catania MV, Battaglia G, Copani A, Barbagallo G, Cena V, Sanchez-Prieto J, Spano PF, Pizzi M. Group-I metabotropic glutamate receptors: hypotheses to explain their dual role in neurotoxicity and neuroprotection. *Neuropharmacol.* 38, 1477-1484, 1999.

[63] Favaron M, Manev RM, Candeo P, Arban R, Gabellini N, Kozikowski AP, Manev H: Trans-azetidine-2,4-dicarboxylic acid activates neuronal metabotropic receptors. *NeuroReport* 4, 967-970, 1993.

[64] Riedel G, Manahan-Vaughan D, Kozikowski AP, Reymann KG: Metabotropic glutamate receptor agonist trans-azetidine-2,4-dicarboxylic acid facilitates maintenance of LTP in the dentate gyrus in vivo. *Neuropharmacol* 34, 1107-1109, 1995.

[65] Riedel G, Wetzel W, Reymann KG: Metabotropic glutamate receptors in spatial and nonspatial learning in rats studied by means of agonist and antagonist application. *Learn Mem.* 2, 243-265, 1995.

[66] Sensi SL, Canzoniero LMT, Yu SP, Ying HS, Koh JY, Kerchner GA, Choi DW: Measurement of intracellular free zinc in living cortical neurons: routes of entry. *J. Neurosci.* 15, 9554-9564, 1997.

[67] Schmidt-Kastner, R, Freund TF: Selective vulnerability of the hippocampus in brain ischemia. *Neuroscience* 40, 599-636, 1991.

[68] Liu S, Lau L, Wei J, Zhu D, Zou S, Sun H, Fu Y, Liu F, Lu Y: Expression of Ca2+-permeable AMPA receptor channels primes cell death in transient forebrain ischemia. *Neuron* 43, 43-55, 2004.

[69] Takeda A, Minami A, Sakurada N, Nakajima S, Oku N: Response of hippocampal mossy fiber zinc to excessive glutamate release. *Neurochem. Int.* 50, 322-327, 2007.

[70] Rodriguez-Moreno A, Sihra TS: Presynaptic kainate receptor facilitation of glutamate release involves protein kinase A in the rat hippocampus. *J. Physiol.* 557, 733-745, 2004.

[71] Levi G, Patrizio M: Astrocytes heterogeneity: endogenous amino acid levels and release evoked by non-N-methyl-D-aspartate receptor agonists and by potassium-induced swelling in type-1 and type-2 astrocytes. *J. Neurochem.* 58, 1943-1952, 1992.

[72] Takeda A, Nakajima S, Fuke S, Sakurada N, Minami A, Oku N: Zinc release from Schaffer collaterals and its significance. *Brain Res. Bull.* 68, 442-447, 2006.

[73] Takeda A, Hirate M, Tamano H, Oku N: Zinc movement in the brain under kainate-induced seizures. *Epilepsy Res.* 54, 123-129, 2003.

[74] Sloviter R: A selective loss of hippocampal mossy fiber Timm stain accompanies granule cell seizure activity induced by perforant path stimulation. *Brain Res.* 330, 150-153, 1985.

[75] Frederickson CJ, Hernandez MD, Goik SA, Morton JD, McGinty JF: Loss of zinc staining from hippocampal mossy fibers during kainic acid induced seizures: a histofluorescence study. *Brain Res.* 446, 383-386, 1988..

[76] Takeda A, Hirate M, Tamano H, Nishibaba D, Oku N: Susceptibility to kainate-induced seizures under dietary zinc deficiency. *J. Neurochem.* 85, 1575-1580, 2003.

[77] Takeda A, Tamano H, Nagayoshi A, Yamada K, Oku N: Increase in hippocampal cell death after treatment with kainate in zinc deficiency. *Neurochem. Int.* 47, 539-544, 2005.

[78] Biondi M, Picardi A: Psychological stress and neuroendocrine function in humans: the last two decades of research. *Psychother Psychosom.* 68, 114-150, 1999.

[79] Negrao AB, Deuster PA, Gold PW, Singh A, Chrousos GP: Individual reactivity and physiology of the stress response. *Biomed. Pharmacother.* 54, 122-128, 2000.

[80] Hennessy MB, Levine S: Sensitive pituitary-adrenal responsiveness to varying intensities of psychological stimulation. *Physiol. Behav.* 21, 295-297, 1978.

[81] Armario A, Lopez-Calderon A, Jolin T, Castellanos JM: Sensitivity of anterior pituitary hormones to graded levels of psychological stress. *Life Sci.* 39, 471-475, 1986.

[82] Dai H, Krost M, Carey RJ: A new methodological approach to the study of habituation: the use of positive and negative behavioral indices of habituation. *J. Neurosci. Methods* 62, 169-174, 1995.

[83] Whishaw IQ, Vanderwolf CH: Hippocampal EEG and behavior: changes in amplitude and frequency of RSA (theta rhythm) associated with spontaneous and learned movement patternsin rats and cats. *Behav. Biol.* 8, 461-484, 1973.

[84] Lee AL, Ogle WO, Sapolsky RM: Stress and depression: possible links to neuron death in the hippocampus. *Bipolar Disorders* 4, 117-128, 2002.

[85] Takeda A, Sakurada N, Kanno S, Minami A, Oku N: Response of extracelluar zinc in the ventral hippocampus against novelty stress. *J. Neurochem.* 99, 670-676, 2006.

[86] Brown KH, Wuehler SE, Peerson JM: The importance of zinc in human nutrition and estimation of the global prevalence of zinc deficiency. *Food Nutri. Bull.* 22, 113-125, 2001.

[87] Golub MS, Keen CL, Gershwin ME, Hendrickx AG: Developmental zinc deficiency and behavior. *J. Nutr.* 125, 2263-2271, 1995.

[88] Sandstead HH, Frederickson C.J, Penland JG: History of zinc as related to brain function. *J. Nutr.* 130, 496S-502S, 2000.

[89] Prohaska JR, Luecke RW, Jasinski R: Effect of zinc deficiency from day 18 of gestation and/or during lactation on the development of some rat brain enzymes. *J. Nutr.* 104, 1525-1531, 1974.

[90] Takeda A, Hirate M, Tamano H, Oku N: Release of glutamate and GABA in the hippocampus under zinc deficiency. *J. Neurosci. Res.* 72, 537-542, 2003.

[91] Takeda A, Minami A, Takefuta S, Tochigi M, Oku N. Zinc homeostasis in the brain of adult rats fed zinc-deficient diet. *J. Neurosci. Res.* 63, 447-452, 2001.

[92] Sterman MB, Shouse MN, Fairchild MD: Zinc and seizure mechanisms. In Morley, J.E., Sterman M.B., and Walsh J.H., (Eds.), *Nutritional modulation of neural function*, Academic Press, San Diego, pp.307-319, 1988.

[93] Takeda A, Hanajima T, Ijiro H, Ishige A, Iizuka S, Okada S, Oku N: Release of zinc from the brain of El (epilepsy) mice during seizure induction. *Brain Res.* 828, 174-178, 1999.

[94] Hirate M, Takeda A, Tamano H, Enomoto S, Oku N: Distribution of trace elements in the brain of EL (epilepsy) mice. *Epilepsy Res.* 51, 109-116, 2002.

[95] Takeda A, Hirate M, Oku N: Elimination of zinc-65 from the brain under kainate-induced seizures. *BioMetals* 17, 141-144, 2004.

[96] Takeda A, Tamano H, Oku N: Involvement of Unusual Glutamate Release in kainate-induced seizures in zinc-deficient adult rats. *Epilepsy Res.* 66, 137-143, 2005.

[97] Lipton SA, Rosenberg PA: Excitatory amino acids as a final common pathway for neurologic disorders. *N. Engl. J. Med.* 330, 613-622, 1994.

[98] Obrenovitch TP, Urenjak J: Altered glutamatergic transmission in neurological disorders: from high extracellular glutamate to excessive synaptic efficacy. *Prog. Neurobiol.* 51, 39-87, 1997.

[99] Reeves PG, Frissell SG, O'Dell BL: Response of serum corticosterone to ACTH and stress in the zinc-deficient rat. *Proc. Soc. Exp. Biol. Med.* 156, 500-504, 1997.

[100] Gaetke LM, Frederick RC, Oz HS, McClain CJ: Decreased food intake rather than zinc deficiency is associated with changes in plasma leptin, metabolic rate, and activity levels in zinc deficient rats. *J. Nutr. Biochem.* 13, 237-244, 2002.

[101] Chu Y, Mouat MF, Harris RBS, Coffield JA, Grider A: Water maze performance and changes in serum corticosterone levels in zinc-deprived and pair-fed rats. *Physiol. Behav.*, 78, 569-578, 2003.

[102] Takeda A, Tamano H, Kan F, Itoh H, Oku N: Anxiety-like Behavior of Young Rats after 2-Week Zinc Deprivation. *Behav. Brain Res.* 177, 1-6, 2007.

[103] Cook CJ: Glucocorticoid feedback increases the sensitivity of the limbic system to stress. *Physiol. Behav.* 75, 455-464, 2002.

[104] Elliott EM, Sapolsky RM. Corticosterone enhances kainic acid-induced calcium elevation in cultured hippocampal neurons. *J. Neurochem.* 59, 1033-1040, 1992.

[105] Elliott EM, Sapolsky RM. Corticosterone impairs hippocampal neuronal calcium regulation – possible mediating mechanisms. *Brain Res.* 602: 84-90, 1993.

[106] Stein-Behrens BA, Elliott EM, Miller CA, Schilling JW, Newcombe R, Sapolsky RM. Glucocorticoids exacerbate kainic acid-induced extracellular accumulation of excitatory amino acids in the rat hippocampus. *J. Neurochem.* 58, 1730-1735, 1992.

[107] Stein-Behrens BA, Lin WJ, Sapolsky RM. Physiological elevation of glucocorticoids potentiates glutamate accumulation in the hippocampus. *J. Neurochem.* 63, 596-602, 1994.

In: Neural Pathways Research
Editor: F. L. Pichler, pp. 95-113
ISBN: 978-1-60456-214-9

Chapter IV

Approaches to the Visualization of Neural Pathways in Chick Imprinting

***Fumihiko Maekawa*[1,2,*,#], *Katsushige Sato*[3] and *Hiroko Ohki-Hamazaki*[1#]**

[1]Laboratory of Molecular Neuroscience,
School of Biomedical Science and Medical Research Institute,
Tokyo Medical and Dental University, Bunkyo-ku,
Tokyo 113-8519, Japan
[2]Department of Physiology, Division of Integrative Physiology,
Jichi Medical University, Shimotsuke, Tochigi 329-0498, Japan
[3]Department of Physiology and Cell Biology,
Tokyo Medical and Dental University Graduate School and
Faculty of Medicine, Bunkyo-ku, Tokyo 113-8519, Japan

Abstract

Among precocious Aves, day-old chicks show a preference for the moving object to which they were first exposed. This behavior, called "imprinting," is known as a form of learning. Visual and acoustic clues are naturally used by chicks as guides to their mothers. This imprinting phenomenon can be reproduced in the laboratory using such clues as an imprinting stimulus. We established a system of inducing visual imprinting behavior in chicks (*Gallus gallus domesticus*) using moving 2-dimensional images on a thin-film transistor liquid crystal display. Since this system enables us to control the components of the image presented to the chicks, we are now able to discriminate those

[*]Address correspondence to: F. Maekawa (fmaekawa@nifty.com) and H. Ohki-Hamazaki (hamazaki.aud@mri.tmd.ac.jp) Laboratory of Molecular Neuroscience, School of Biomedical Science and Medical Research Institute, Tokyo Medical and Dental University, 1-5-45 Yushima, Bunkyo-ku, Tokyo 113-8519, Japan. TEL:+81-3-5803-4693 or 5844 ; FAX:+81-3-5803-4693.
[#] These authors equally contributed to this work.

components to which the chicks direct their attention from among all the components of the stimulus, such as shapes, colors, textures and directions of movement.

By applying this system, we examined the neural pathways related to color recognition and memory. We adopted two methods of visualizing these pathways: the optical imaging of the intrinsic signal, which is suitable for mapping gross changes in activity on the surface of the brain, and c-FOS immunohistochemistry, which is suitable not only for examining the deep structure of the brain but for determining what types of neurons respond to specific stimuli within the layers and nuclei of the brain. Using the first approach, we found that the visual wulst, one of the higher visual centers in Aves, is related to both color recognition and the acquisition of memory. Using the second approach, we found that the cholecystokinin neurons in the visual wulst and adjacent areas are indispensable for the acquisition of memory.

In this review, we describe the technical approaches that led us to these findings. Imprinting in chicks might be a good model of infant and child learning in humans, because infants and chicks share a specific time-window called the "critical period," during which the infant is sensitive to certain experiences. Further study using our techniques might enable a more detailed characterization of this "critical period."

1. Imprinting as a Rapid and Highly-Reproducible Tool for Investigating the Infant and Child Learning Process

Most newborn mammals and birds grow up under the protected circumstances created by their parents and/or conspecifics. In this period, newborns discriminate their parents and conspecifics from predators with the aid of clues such as shape, color, odor and voice, and memorize and imitate the behaviors of their parents and conspecifics. This "instinct" function inherent in newborns gives them 1) increased chances of survival as a result of protection from conspecifics, and 2) a social model after maturation [1]. The process through which newborns discriminate their conspecifics from other entities begins in the uterus in the case of mammals. The human fetus can discriminate the mother's voice from other voices, indicating that the fetus can memorize voices within the intra-uterine environment [2, 3]. After birth, the human newborn can memorize the mother's face in a short period of time [4]. This is not a phenomenon specific to humans; it is reported that newborn lambs can also distinguish their mothers by means of visual and auditory clues [5].

Among precocious Aves, newly hatched chicks can recognize their mother's shape and voice. There are thought to be two underlying brain mechanisms that make this recognition possible: 1) predisposition and 2) imprinting [6]. Predisposition is a mechanism of inborn perceptual preference for conspecifics. Newly hatched chicks have a natural preference for conspecifics, a preference that does not derive from any experience. One excellent example of predisposition in chicks is described in a previous report of Long *et al.* [7]. They have found that the transplantation of midbrain tissue from quail to chicken increases the preference for the maternal voice produced by the adult quail female. This indicates that the neural circuit which makes chicks prefer the voice of their own species is localized in the midbrain in chicks. It has been reported that chicks have an innate visual preference for the maternal shape as well. On the other hand, newly hatched chicks can learn the mother's characteristics by means of visual and auditory clues [6]. Lorenz was the first to observe that the chicks of precocious birds can chase the object to which they are first exposed after hatching [8], and he named this phenomenon "imprinting." The period during which the memory of the chicks is sensitive to the object is called the "critical period" [9]. The critical period in ground-nesting birds extends from hatching to a few days thereafter, although that in tree-nesting birds generally continues for about 2 weeks [10]. Other than this filial imprinting, the birds keep the characteristics of their conspecifics in their memories and use this knowledge to choose a partner for copulation after sexual maturation. This is called "sexual" imprinting [11]. While the interaction between predisposition and imprinting is still under discussion, both play important roles in inducing optimal social interaction among conspecifics.

To study the imprinting process of precocious birds might serve to better understand the neural circuit mechanism which enables human infants to learn, because precocious birds and humans share some common features. Both are endowed with higher brain function accompanied by superior visual and auditory resolutions. In addition, in both cases, the critical period limits the duration of learning. To elucidate the brain function underlying this behavior, imprinting has been explored in the laboratory using real objects as surrogate mothers instead of hens. We conceived that simplified computer-generated images on a thin-film transistor liquid crystal display (TFT-LCD) can also be used as stimuli for imprinting, and established a system of inducing visual imprinting behavior in chicks. By applying this technique, we can easily control the factors of the image exhibited to chicks and determine precisely which features chicks utilize as cues for imprinting. In addition, this system enables us to simplify the presentation of the stimulus when we record either the electrical or the optical brain signal. In this review, we focus on introducing our method and our new findings regarding the neural pathways of imprinting which we have obtained by applying this method.

2. Use of Symbolic Images as Stimuli of Imprinting: Can Chicks Recognize Simplified 2-Dimensional Images as Imprintable Stimuli?

There are many reports on systems which enable the examination of the imprinting behavior of domestic chickens in the laboratory [6, 12, 13, 14]. In most cases, newly hatched chicks are exposed to visual and/or auditory stimuli, as an initial input, over a certain period (the so-called "training phase"). To assess whether chicks really memorize the object/acoustic stimuli or not, the chicks are sequentially exposed, after a certain period of time, to either the objects/acoustic stimuli which they have already experienced or objects/acoustic stimuli which they have never encountered. If the memory was formed at the first exposure, the chicks chase the familiar object/acoustic stimulus and do not chase or avoid the object/acoustic stimulus which they encounter for the first time ("evaluation phase"). Among the apparatuses which have been used to evaluate imprinting, a circular pool with a ball rolling along the wall, a running wheel and a rectangular box allowing the chicks to see rotating objects at one or both ends have been used [6, 12, 15, 16, 17]. Stuffed animals, balls, cylinders and boxes with texture and/or color are used as artificial stimuli for imprinting, and chicks can memorize all such objects.

Meanwhile, it is known that images on CRT display or LCD can serve as stimuli of learning in the laboratory [18]. Among Aves, pigeons can discriminate complex images such as paintings on CRT display [19]. Ikebuchi *et al.* have also reported that male songbirds can emit songs as a response to the image of a female on the display [20]. This indicates that birds can recognize images that are projected on the screen and treat them as real counterparts. It has been reported that imprinting was achieved in newly hatched chicks using biological motion projected on a CRT screen [21, 22]. All these studies using birds suggest the possibility that images projected on a screen can be utilized as stimuli for imprinting. Therefore, as a first step, we planned on capturing on videotape a moving conventional object which would normally be used for imprinting studies and show the video to chicks.

To compare the effects of real objects with those of video stimuli, we started by evaluating the imprinting process using real objects. For imprinting training, chicks at 12–24 h after hatching (defined as PD0) were placed on a running wheel with a see-through wire net as the floor. The chicks could see a moving object (either a red cylinder or a blue box with a black stripe) on the turntable during training, according to the schedule shown in Figure 1A.

On the day after training (PD1), we examined the chick's preference for objects. The representative results for individual chicks are shown in Figure 1B. In this figure, the Y-axis indicates the difference between revolution toward (shown as a scale above 0) and away from (shown as a scale below 0) the object over a period of 1 min. We changed the stimuli every 5 min and presented them sequentially. When training was performed using the red cylinder, the chick chased the red cylinder and avoided the blue box with a black stripe (Figure 1B-a). On the other hand, when training was performed using the blue box with a black stripe, the chick chased the blue box with a black stripe and avoided the red cylinder (Figure 1B-b). If we did not show any object to the chick during training, the chick did not demonstrate any preference (Figure 1B-c). These findings indicate that the red cylinder and the blue box with a black stripe were appropriate stimuli for imprinting. We also examined the elapsed time until the chick could chase the object memorized at PD0. The chick trained to prefer the red cylinder at PD0 was able to chase it at PD1 and PD2 but not after PD3 (Figure 2). Although the reasons why the chicks did not chase the object after PD3 are obscure, we speculate the following: 1) there may be a predisposition which overcomes the imprinting effect through group-breeding, 2) there may be a brain mechanism of forgetting, or 3) a reward which might reinforce the association in the wild may be absent under laboratory conditions. Next, we used a movie showing a moving red cylinder as a stimulus of imprinting (Figure 3). The training achieved by this movie could not elicit a specific preference at the evaluation phase (Figure 3A). However, interestingly, when these chicks were exposed to the real objects, 3 of 4 chicks exhibited a preference for the blue box with a black stripe (Figure 3B). We do not have a clear explanation why the chicks mixed up the red cylinder on the LCD and the real blue box with a black stripe, but there are several possible causes for this misidentification: the chicks' inability to estimate the disparity between the object shown on the screen and the real object, the difference in color balance between the LCD and the real object, and the difference between the backlight emission on the LCD and the environmental illumination on the real object. This was an unsuccessful attempt, but we became convinced that we could use

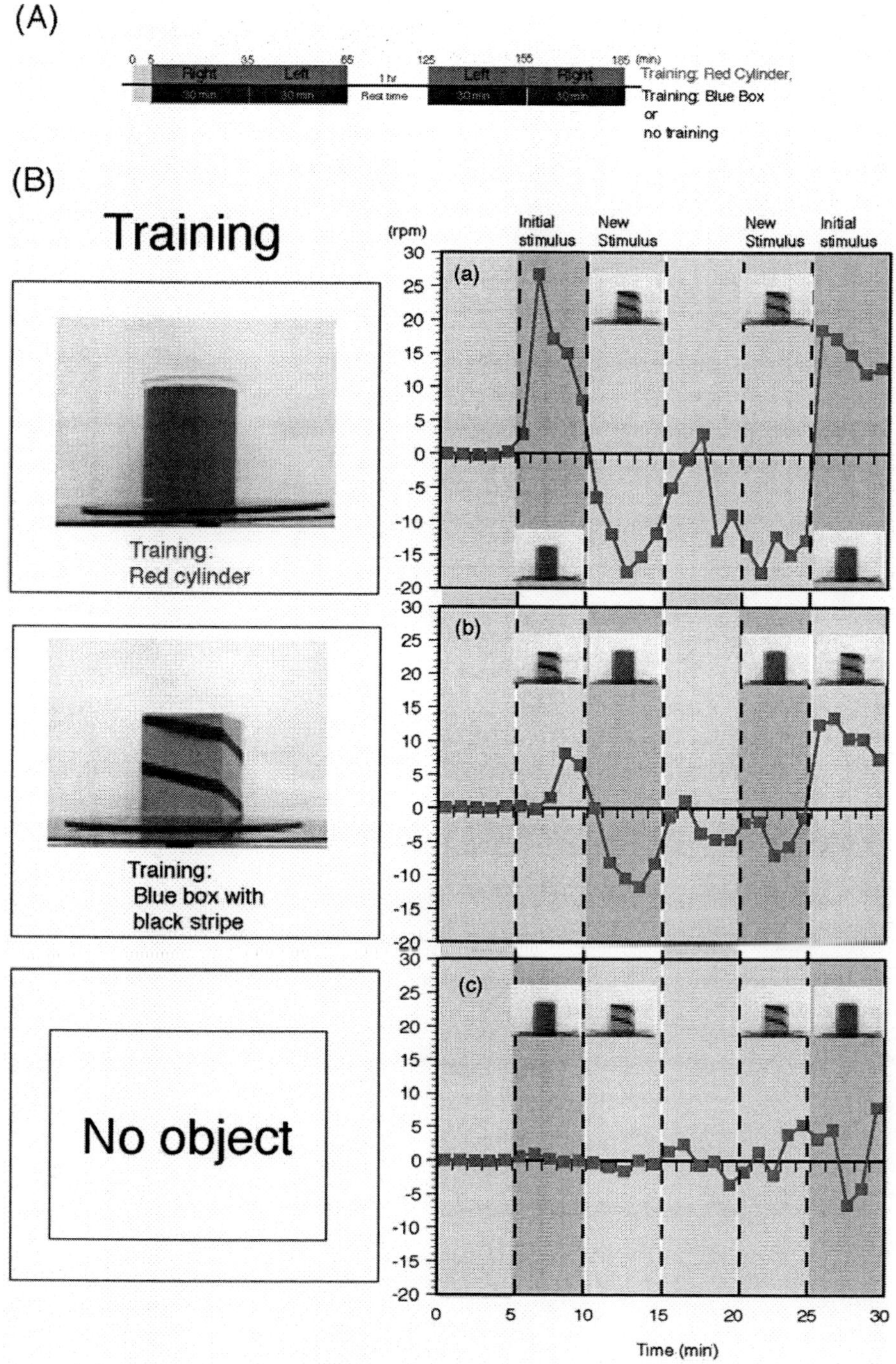

Figure 1. Imprinting using real objects. (A) Schedule of imprinting training; (B) results of the evaluation. (a) Chicks trained with a red cylinder chased the red cylinder at the evaluation phase and avoided the blue box with a black stripe. (b) Chicks trained with a blue box with a black stripe chased the blue box at the evaluation phase and avoided the red box. (c) The chicks which received no training did not show any preference.

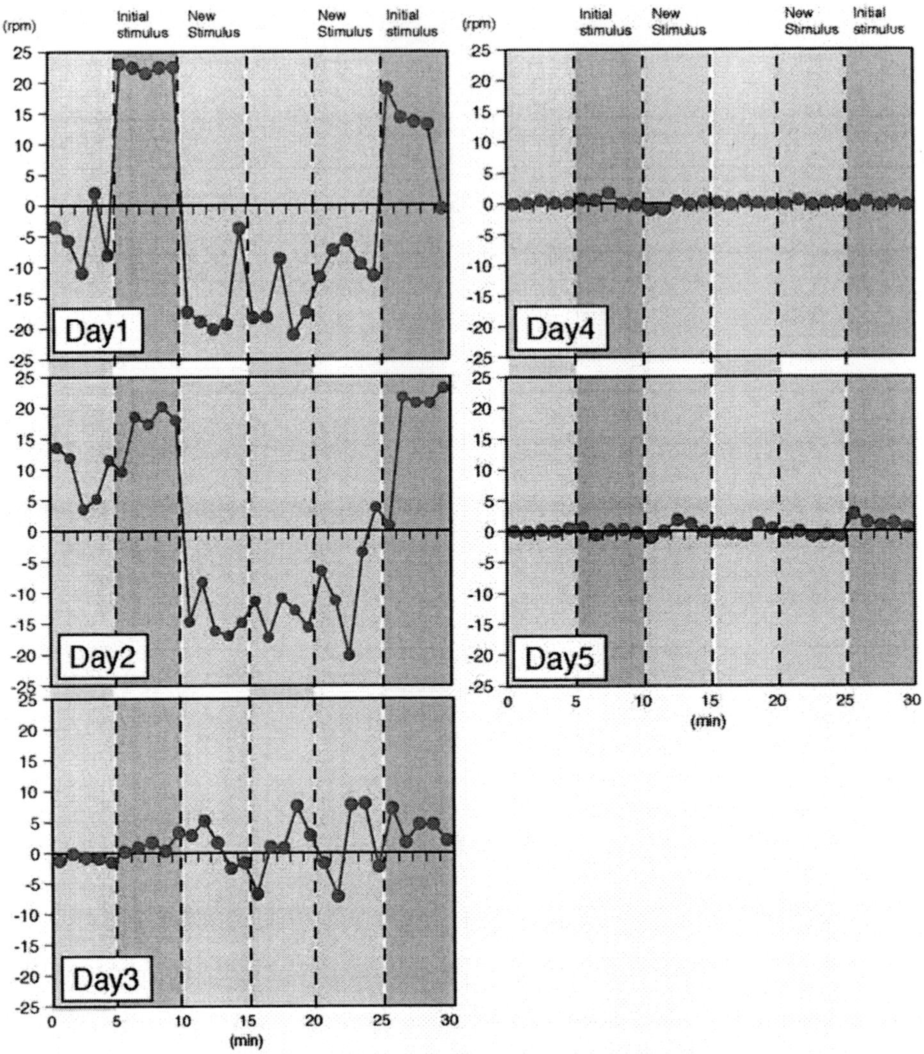

Figure 2. Elapsed time until the chick could chase the memorized object. Chicks trained with a red cylinder chased the red cylinder until PD2 but not after PD3.

video images displayed on an LCD as a stimulus for imprinting, because the chick could memorize the stimuli on the LCD, even if they mistook the target.

We changed our strategy following this attempt. We decided to use symbols generated by computer software, because it is complicated for us to modulate the image parameters of captured images, but not those of computer-generated symbols. We generated the images of a circle and a square, either red (RGB: 255, 0, 0) or blue (RGB: 0, 0, 255), that moved horizontally across the screen, using either MovieWorks ver. 4.7 (Interactive Solutions, Inc., Pleasanton, CA) or visual stimulus generator software (VSG, Cambridge Research Systems, Ltd., Kent, UK) (Figure 4, details in [23]). When the chicks were trained with the red circle on the LCD at PD0, they chased the red circle and square but avoided the blue circle and square (representative example in Figure 4A). Similarly, when the chicks were trained with the blue square on the LCD, they chased the blue square and circle but avoided the red symbols (Figure 4B).

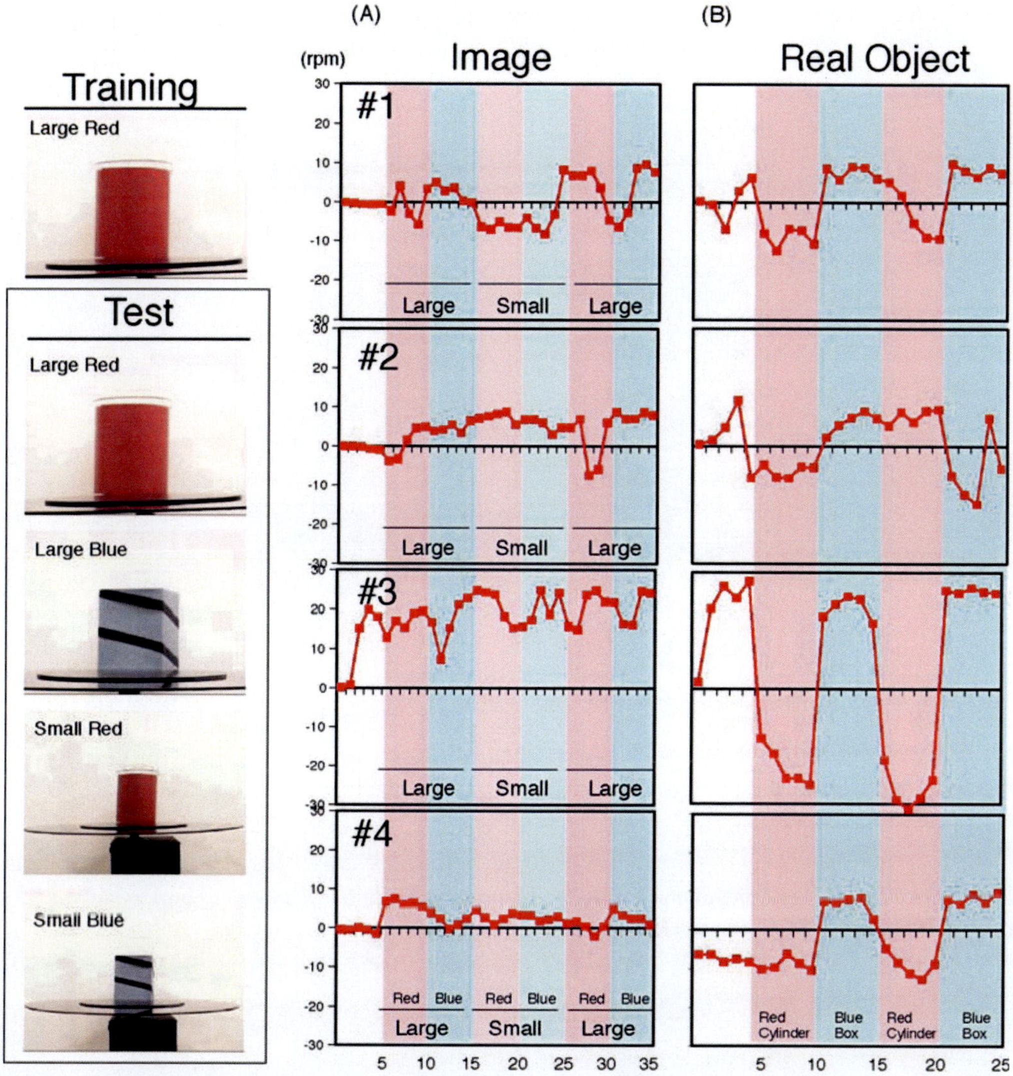

Figure 3. Unsuccessful imprinting attempt using the projection of real objects on an LCD. (A) The training using this movie could not elicit any specific preference at the evaluation phase in any chick. (B) When these chicks were exposed to the real objects, 3 of 4 chicks showed a preference for the blue box with a black stripe, indicating that the chicks mixed up the red cylinder on the LCD and the real blue box with a black stripe.

These facts indicate that the chick utilizes color rather than shape as a cue for recognition. We briefly examined the effect of vertical movement of the images (Figure 5). The vertically moving blue square could elicit imprinting (Figure 5A). In addition, chicks trained with a horizontally moving red circle could chase a vertically moving circle as well as a horizontally moving one (Figure 5B), indicating that the direction of movement is not an important factor once the chicks memorize the imprinted image. From these data, we concluded that color is the most important factor of the chicks' image recognition in our system. We are now interested in elucidating the necessary conditions which an image must fulfill in order for it to be imprinted in the chick's memory. Therefore, we are now searching for a moving image which cannot be imprinted in the chick's memory. This might help us deepen our understanding of imprinting and recognition in chicks.

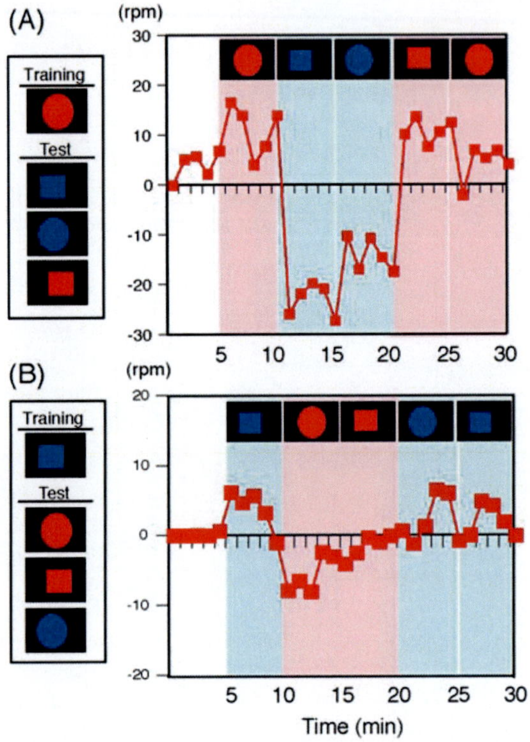

Figure 4. Successful imprinting attempt using computer-generated images moving horizontally. (A) Chicks trained with a red circle chased the red circle and box but not the blue box and circle at the evaluation phase. (B) Chicks trained with a blue box chased the blue box and circle at the evaluation phase.

3. *In Vivo* Optical Imaging of the Visual System in the Avian Brain

To circumvent the limitations of electrophysiology, which is only suitable for analyzing restricted areas, *in vivo* intrinsic optical recording of the brain has been developed as a technique of detecting response changes in wider areas at a time [24]. The optical signal is thought to emerge through the following steps: 1) increase of oxygen consumption in a local area of the brain, 2) transition of hemoglobin from oxy- to deoxy-form, and 3) changes in light scattering (decrease of light reflectance and increase of absorption in hemoglobin) [25]. Optical recording has been applied to the monitoring of visual processing areas in the brain, such as the ocular dominance column [26, 27], the orientation preference map [28] and the color preference map [29] in the visual cortex.

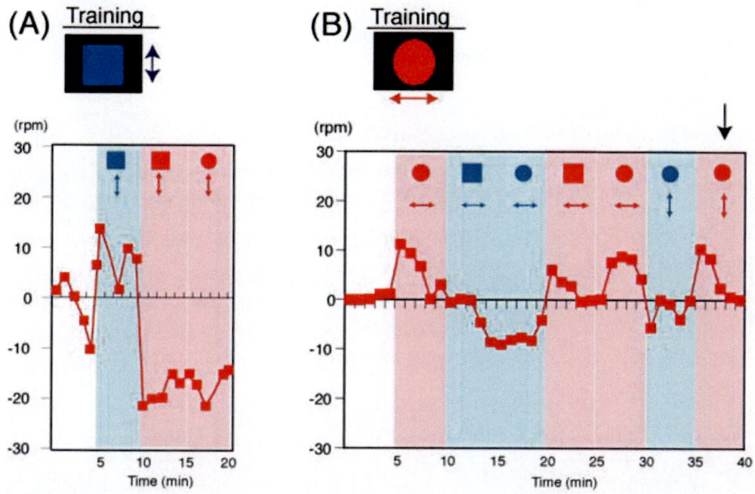

Figure 5. Vertical movement of images during imprinting. (A) Chicks trained with a blue box moving vertically chased the blue box, but not the red box or circle moving vertically. (B) Chicks trained with a red circle moving horizontally chased the red circle moving vertically (black arrow).

Birds have two parallel visual pathways. In the tectofugal pathway, visual information from the retina is transmitted to the following regions in the order listed: the optic tectum, the nucleus rotundus of the dorsal thalamus and the entopallium of the telencephalon [30]. In the thalamofugal pathway, visual information from the retina terminates in the visual wulst of the telencephalon via the nucleus dorsolateralis anterior thalami [30]. The two pathways are thought to play different roles in visual processing, but their detailed functions have not been fully elucidated yet. The entopallium is located within the deep pallium, whereas the visual wulst lies just beneath the surface of the brain. Anatomical and electrophysiological studies have demonstrated that both receive visual inputs from the retina, even in newly hatched chicks [31-33]. Our lesion study suggests that the visual wulst is related to the acquisition of memory in visual imprinting behavior [23]. In a study using barn owl, an orientation preference map comparable to the V2 of the cat and monkey was found in the wulst, indicating that the visual wulst is a structure homologous to the extrastriate visual area [34]. So, we decided to investigate the optical signal in the visual wulst in chicks under either electrical stimulation of the optical papilla or the presentation of computer-generated images on a LCD.

Schematic representations of the settings for electrical stimulation of the optical papilla and the computer-generated images on the LCD are shown in Figures 6A and 6B, respectively. For the optical recording, the chick was fixed in the stereotaxic device under urethane anesthesia. The body, except the head, was wrapped in cotton and placed on a heating pad. The body temperature was kept at 35 °C using a temperature controller (FST, Foster City, CA). The skull and dura mater at the surface of the telencephalon were removed. The surface of the visual wulst was maintained in the horizontal position by adjusting the angle of the head. A plastic chamber was fixed onto the skull with dental cement (Quick

Resin, Shofu, Kyoto, Japan) and filled with silicone oil (KE106, Shin-Etsu Chemical Co., Ltd., Tokyo, Japan). The dorsal surface of the visual wulst was illuminated by two adjustable light guides attached to 100-W tungsten-halogen lamps driven by a DC power supply (model 66184, Oriel Co., Darmstadt, Germany). The filter used for visualizing the surface of the visual wulst and its vascular pattern had a band pass of 540±30 nm, and the filters used for intrinsic imaging had a band pass of 630±30 nm (Asahi Spectra Co., Tokyo, Japan). Intrinsic imaging was performed using the differential video acquisition system Imager 2001 (Optical Imaging, Inc., Germantown, NY), via a charge-coupled device camera (CCD-502 4N, Bischke, Ltd., Neumunster, Germany) fitted to the tandem lens of the microscope. The data were collected with the data acquisition software VDAQ (Optical Imaging) and analyzed using WinMix software (Optical Imaging). The details of the data acquisition and analysis were presented in the previous report [23].

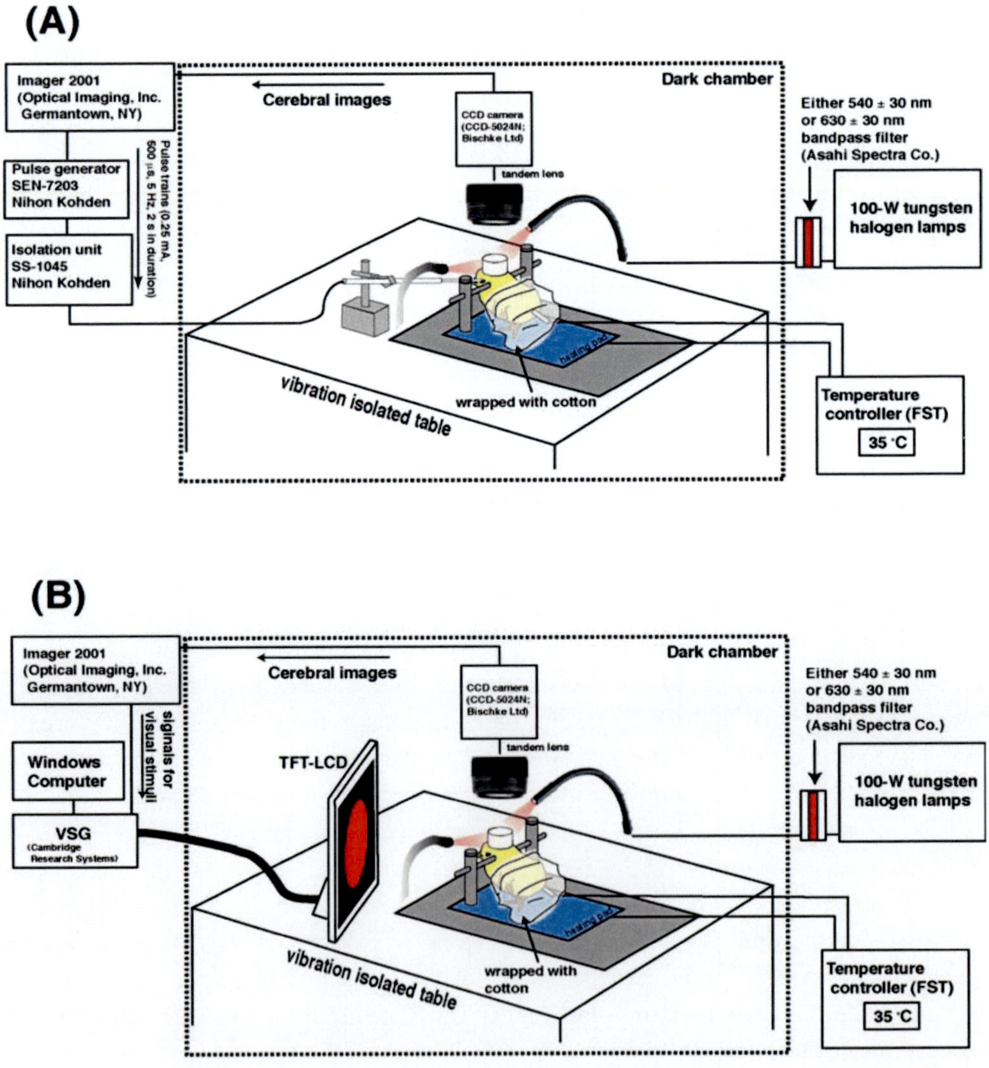

Figure 6. Settings of *in vivo* optical imaging for (A) electrical stimulation of the papilla and (B) visual stimulation via an LCD.

For electrical stimulation of the optic nerve, the left eyeball was removed and the optic papilla was exposed. Bipolar tungsten electrodes (TN201; Unique Medical Co., Ltd., Tokyo, Japan) were attached to the left optic papilla. Pulse trains (0.25 mA, 500 μs, 5 Hz, 2 s in duration) were delivered using a pulse generator (SEN-7203; Nihon Kohden Co., Tokyo, Japan) through an isolation unit (SS-1045; Nihon Kohden). The settings of visual stimulation by LCD were basically the same as those of electrical stimulation, except for the use of an LCD (Figure 6B). A 15-inch TFT-LCD (EIZO FlexScan L367; Nanao, Ishikawa, Japan) was placed 20 cm from the left eye of the chick. The program, written in Stimulus Description Language (Cambridge Research Systems), was sent to the visual stimulus generator (VSG, Cambridge Research Systems) in advance, and the VSG could work concurrently with Imager 2001.

When the left optical papilla was excited by electrical stimulation, the optical signal appeared in the right visual wulst (Figure 7). The response reached a peak at 2-3 sec after the start of stimulation and continued for at least 5 sec.

The localization of the response within the wulst obtained by the intrinsic optical recording was identical to that obtained by extrinsic signals using voltage-sensitive dye [23]. We also found that the intrinsic signal was glutamatergic receptor-dependent, because the application of both NMDA and non-NMDA glutamate receptor antagonists diminished the intrinsic signal [23]. All these results prove that the intrinsic signal in the visual wulst reflects actual neural activity and is not an artifact.

Next, we examined the response of the visual wulst to colored moving images on an LCD using the settings shown in Figure 6B. Interestingly, there were differences in the areas which responded to red-, green- and blue-colored squares, although the regions partially overlapped each other [23].

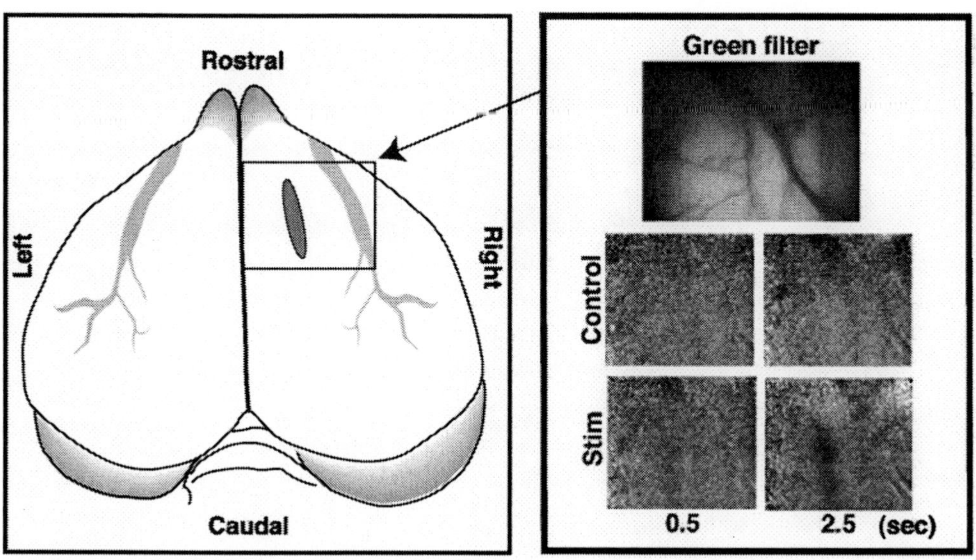

Figure 7. Typical optical signal obtained in the visual wulst after electrical stimulation of the optical papilla. Left: black frame showing the area chosen for optical imaging. Upper right: vascular image of the visual wulst. Lower right: optical signal appearing 2.5 sec after electrical stimulation of the optical papilla.

The area which responded to the blue square was in the rostral part of the wulst, whereas that which responded to the red square was in the caudal part. The area which responded to the green square was between the areas which responded to the blue and red squares. These results indicate that a functional map corresponding to color stimuli exists in the visual wulst. Following the finding of the color-oriented map, we examined the possibility that imprinting training affects the functional map. In the chicks trained with a blue square at PD0, the area which responded to the blue square shifted to the rostral part of the visual wulst. Similarly, in the chick trained with the red square at PD0, the area which responded to the red square also shifted to the rostral part [23]. These results suggest that a plastic change in the visual wulst occurs during the experience of imprinting training.

In mammals, retinal ganglion cells that receive input from cones send information to the V1 region of the visual cortex via the color-selective parvoceullar layer of the lateral geniculate nucleus [35]. In V1, the cells responding to various colors are thought to form a scattered blob pattern, as by detected cytochrome oxidase staining [36]. Hubel *et al.* have also described that the color cells were clustered in thin stripes in the V2 region [37]. Recently, Xiao *et al.* have found a color-oriented map in V2 using *in vivo* optical recording [29]. In addition, V4 has color-selective neurons and plays an important role in color information processing [38]. In contrast to the great number of reports on mammals, color information processing in the avian telencephalon has not been examined yet. As in our findings of a color-orient map and plastic map change in the chick visual wulst, further improvement of the presentation of computer-generated images combined with *in vivo* optical recording might open the way to the elucidation of the avian neural pathway in the telencephalon related to visual processing.

4. Identification of Neurons Related to Visual Imprinting Using FOS Immunohistochemistry

To investigate the activity of the deep structure of the brain, techniques such as 2-deoxy-d-glucose uptake and electrophysiology have been employed as classical methods [39]. In the mid-1980's, neuroscience researchers started to use the expressions of immediate early genes such as c-FOS, c-JUN and ZENK as indices of neural firing [40, 41]. Owing to its high resolution, this technique enables us not only to readily examine which sites are activated after stimulation, but also to determine what types of neurons become active, by means of immunohistochemistry or a combination of immunohistochemistry and *in situ* hybridization. c-FOS immunohistochemistry has been applied in the study of avian visual learning [42, 43]. By applying this method, McCabe *et al.* have found a good correlation between the degree of learning and the increase of c-FOS-immunopositive cells in the intermediate medial mesopallium (IMM), which was formerly known as the intermediate medial hyperstriatum ventrale [42]. The importance of the IMM has been established by a study in which lesions in the IMM were found to disturb the formation of preference during imprinting [44]. It has been reported that c-FOS-expressing cells in this region contain a gamma protein kinase C isoenzyme and GABA to some extent [45, 46]. However, the wulst and IMM also contain a variety of neurons other than GABA neurons, and their roles in relation to imprinting still remain unclear. Neuropeptides and their receptors are known to modulate synaptic plasticity in mammals. Although there are some reports showing that endogenous opioids affect imprinting behavior [47], little has been found regarding the interaction with other neuropeptides. In accordance with the evidence obtained by *in vivo* optical imaging, we examined the correlation between the expression of neuropeptide-immunoreactive cells and c-FOS expression in some nuclei of the dorsal pallium containing the hyperpallium intercalatum (HI), mesopallium dorsale (MD) and mesopallium ventrale (MV) 1 hour after imprinting training. The HI constitutes one of the layers of the visual wulst, and the wulst subdivisions and the MV include the IMM. Among the bombesin, cholecystokinin (CCK), choline acetyltransferase, enkephaline, neuropeptide Y, somatostatin and substance P-like immunoreactivities, the distribution of CCK-like immunoreactive cells was strongly correlated with that of c-FOS immunoreactivity. Therefore, we performed double staining with c-FOS immunohistochemistry and CCK *in situ* hybridization in order to confirm that the CCK neurons are really activated by imprinting training. We summarized the protocol of this method applied to chick brain sections in Table 1.

One hour after imprinting training, both c-FOS-positive and CCK-mRNA-positive cells in these three areas were increased, and the percentage of double-positive cells in the chicks which received training was 3 times higher than that in chicks without training [48]. In addition, semi-quantitative PCR analysis revealed that CCK mRNA was significantly increased in the trained chicks. The CCK neurons were possibly glutamatergic but not GABAergic in these areas, because the CCK neurons were positive for VGLUT2, which is a molecular marker of glutamatergic neurons, and negative for GAD65 and parvalbumin, which are markers of GABAergic neurons. Furthermore, we found that CCK receptor inhibitor disturbed the formation of memory. All these findings suggest that glutamatergic CCK neurons play an important role in visual imprinting.

As we mentioned in the section regarding *in vivo* optical recording, the interruption of glutamatergic input by the receptor antagonists diminished the optical signal in the visual wulst. NMDA receptor antagonist treatment has been reported to inhibit the acquisition of memory during imprinting [49]. In addition, chicks that are imprinted release more glutamate in the IMM than those that are not imprinted [17]. Although we cannot exclude the possibility that neural projection from the lateral geniculate nucleus to the telencephalon plays the main role in the increase of glutamate in this region, the increase of glutamate might also be at least partly accounted for by the activation of glutamatergic CCK neurons within the visual wulst and IMM. Taking this possibility into consideration, it is reasonable to speculate that CCK might cooperate with glutamate to form memory during the critical period of imprinting.

Table 1. Protocol for simultaneous detection of mRNA and c-FOS protein in the chick brain

1. Imprinting training: expose for 30 min to the red circle presented on the TFT-LCD.

2. Perfusion of Zamboni's fixative (4% paraformaldehyde and 15% saturated picric acid in 0.05 M phosphate buffer (PB))

3. post-fixation for O/N in the Zamboni's fixative .

4. Cryoprotect with 30% sucrose in 0.05 M PB

5. Cut at 30-50 μm thinkness by cryostat

6. *in situ* hybridization
   a) wash with 0.1 M PBS
   b) proteinase K treatment (1 μg/ml) in Tris-EDTA buffer (pH8.0) at 37 °C for 10 min
   c) acetylation with 0.25 % acetic anhydride in 0.1 M Triethanolamine for 10 min
   d) digoxigenin(DIG)-labeled cRNA probes generated using Roche DIG RNA labeling kit at 60 °C for 14-16 hrs in a solution that contained 50% formamide, 2% blocking reagent (Roch Applied Science), 5X sodium chloride/sodium citrate buffer (SSC), 0.1% N-lauroylsarcosine, 0.1% sodium lauryl sulfate.
   e) wash twice in 50% formamide plus 2X SSC for 15 min at 50 °C
   f) RNase treatment (20 μg/ml) in Tris-EDTA buffer (pH8.0) at 37 °C for 30 min
   g) wash 2X SSC and 0.2 X SSC (each for 30 min) at 50 °C
   h) wash by DIG-1 buffer (0.1 M Tris-HCl (pH7.5) , 0.15 M NaCl and 0.1% Tween 20)
   i) blocking of antibody by DIG1-buffer containing 5% BSA and 5% normal goat serum for 1 h at 25 °C
   j) alkaline phosphatase-conjugated antibody against DIG (1:2,000, Roche Applied Science) in blocking buffer O/N at 4 °C
   k) wash three times with DIG-1 buffer for 15 min
   l) rinse by DIG-3 buffer (0.1 M Tris-HCl, pH9.5, 0.1 M NaCl, 50 mM MgCl2)
   m) Incubation in nitro blue tetrazolium chloride plus 5-bromo-4-chloro-3-indoly phosphate (1:50 in DIG-3 buffer Roche) for 2-10 hrs at 25 °C.
   n) wash in 0.1M Tris-HCl (pH8.0)

7. c-FOS Immunohistochemistry
   a) incubate with 0.6 % H2O2 in 0.05 M PBS for 30 min
   b) Blocking by 5% normal horse serum in 0.05 M PBS for 30 min
   c) anti-c-FOS antibody treatment (1: 10 000, Santa Cruz Biotechnology) in blocking buffer O/N at 25 °C
   d) wash 3 times by 0.05 M PBS
   e) EnVision+/Rabbit-horseradish peroxidase (DAKO) for 1 h
   f) wash 3 times by 0.05 M PBS
   g) diamino benzidine liquid system (DAKO)
   h) wash 3 times by 0.05 M PBS

8. Mounting in aquerous media that contained 14% polyvinyl alcohol and 30% glycerol in PBS

9. Microscopic observation (blue: mRNA , red: c-FOS)

Concluding Remarks

We established an imprinting system using an LCD screen. Since chicks use color as a cue for preferential imprinting, we decided to examine the relationship between color vision and imprinting behavior. The preferential usage of color vision by chicks in our system might be a natural consequence. The chicken retina is cone-dominant, and cones outnumber rods 6 to 1 in the central retina and 3 to 1 in the peripheral retina in the chicken [50]. In addition, the chicken retina has 4 different types of cones [51, 52]. On the contrary, the visual resolution of objects which are far away is not good in the chicken due to its natural shortsightedness [53]. These natural characteristics might lead chicks to rely on color vision more than on other features of vision. Therefore, chicks are thought to be a good model for investigating how color vision is consolidated; popular laboratory animals such as the mouse and the rat are not suitable for investigating color vision, because they have only two types of cones in the retina, and they are red-green color-blind [54, 55]. However, there have not been many reports on the relationship between color vision and chick behavior [51, 56, 57]. We hope our system will contribute to the further research of color vision in Aves.

As to the mechanism of color recognition, many issues remain to be examined. Firstly, to clarify the role of the wulst, a more detailed map should be made. In a reported primate experiment, areas functionally corresponding to at least 10 different colors have been found in V2 [29]. In Aves, the wulst is homologous to the neocortex in terms of the expression of homeobox genes and the projection from the thalamus [30, 58]. However, we do not have much functional evidence that the visual wulst has a function similar to that of the visual cortex. Since the chicken has ultraviolet vision in addition to the trichromatic color vision of primates and humans [52], we do not know how ultraviolet color is coded in the chick brain. To assess a functional color map containing ultraviolet vision might contribute not only to the understanding of the connection between vision and memory in Aves, but it may also be a first step toward a better understanding of the evolution of color vision in general. Another matter which should be clarified in the future is the relationship between the visual wulst and the entopallium in chicks at the imprinting stage. The entopallium is the center of another higher visual system called the "tectofugal pathway." In our previous study, the entopallium was also found to play an important role in the retention of memory during chick imprinting [23], in contrast to the visual wulst, which is only related to the acquisition but not to the retention of memory. This means that different visual processing steps are applied independently in retention and acquisition. On the other hand, it has been reported that there is no direct anatomical connection between the wulst and the entopallium. Similarly, we do not assume that the CCK neurons in the visual wulst and IMM influence the activity of the entopallium directly, because CCK receptor is absent in the entopallium [48]. However, an indirect connection via other pallial structures is supposed [59]. Knowing the precise connection between these two higher centers might help solve the problem of how memory is formed based on image recognition.

Concerning technical aspects for further study, it would be desirable to establish a system of optical imaging under free-moving conditions. Such a system would enable us to design a longitudinal study of chick development. However, it is difficult to keep the chamber glued on top of the skull for a long period, because the chick skull is too soft. Other than this problem, there are some limitations of *in vivo* intrinsic optical imaging. For example, *in vivo* optical recording cannot be used to probe the deep structure of the brain. As far as chicks are concerned, there is the additional problem that the chick wulst is not a flat structure with simple layers like the mammalian visual cortex. This might cause disturbances in the acquisition of higher-resolution images, even of the brain surface. Further technical advances are required to improve the signal intensity and resolution.

References

[1] Goth, A and Hauber, M.E. (2004). Ecological approaches to species recognition in birds through studies of model and non-model species. *Ann. Zool. Fennici*, 41, 823-842.

[2] Eimas, P.D., Siqueland, E.R., Jusczyk, P. and Vigorito, J. (1971) Speech perception in infants. *Science*, 171, 303–306.

[3] Kuhl, P.K., Williams, K.A., Lacerda, F., Stevens, K.N. and Lindblom, B. (1992) Linguistic experience alters phonetic perception in infants by 6 months of age. *Science*, 255, 606–608.

[4] Bushnell, I.W.R. (2001) Mother's face recognition in newborn infants: learning and memory. *Inf. Child Dev.* 10, 67-74.

[5] Nowak, R., Keller, M., Val-Laillet D., Levy, F. (2007) Perinatal visceral events and brain mechanisms involved in the development of mother-young bonding in sheep. *Horm. Behav.* 52, 92-98.

[6] Bolhuis, J.J., (1999) Early learning and the development of filial preferences in the chick. *Behav. Brain Res.*, 98, 245-52.

[7] Long, K.D., Kennedy, G., Balaban, E., (2001) Transferring an inborn auditory perceptual predisposition with interspecies brain transplants. *PNAS* 98,5862-7.

[8] Lorenz, K. (1937) The companion in the bird's world. *Auk* 54, 245-273.

[9] Hensch, T.K. (2004) Critical period regulation. *Ann. Rev. Neurosci.*, 27, 549-579.

[10] Jones, M.P. (2001) Behavioral aspects of captive birds of prey. Vet. *Clin. North. Am Exotic Anim. Pract.*, 4, 613-632.

[11] Mills, A.D., Crawford, L.L., Domjan, M., Faure, J.M. (1997) The behavior of the japanese or domestic quail Coturnix japonica. *Neurosci. Biobehav. Rev.*, 21, 261-281.

[12] Horn, G. (2004) Pathways of the past: the imprint of memory. *Nat. Rev. Neurosci.*, 5, 108-120.

[13] Matsushima, T., Izawa, E., Aoki, N., Yanagihara, S. (2003) The mind through chick eyes: memory, cognition and anticipation. *Zoolog Sci.*, 20, 395-408.

[14] Braun, K., Bock, J., Metzger, M., Jiang, S., Schnabel, R. (1998) The dorsocaudal neostriatum of the domestic chick: a structure serving higher associative functions. *Behav. Brain Res.,* 98, 211-218.

[15] Hess EH (1959) Two conditions limiting critical age for imprinting. *J. Comp. Physiol.* Psychol., 52, 515-518.

[16] Vallortigara, G., Regolin, L., Pagni, P. (1999) Detour behaviour, imprinting and visual lateralization in the domestic chick. *Brain Res. Cogn. Brain Res.*, 7, 307-320.

[17] Tsukada, Y., Kanamatsu, T., Takahara, H. (1999) Neurotransmitter release from the medial hyperstriatum ventrale of the chick forebrain accompanying filial imprinting behavior, measured by *in vivo* microdialysis. *Neurochem. Res.*, 24, 315-320.

[18] Miyashita, Y. (1988) Neuronal correlate of visual associative long-term memory in the primate temporal cortex. *Nature*, 335, 817-820.

[19] Watanabe, S., Sakamoto, J., Wakita, M. (1995) Pigeons' discrimination of paintings by Monet and Picasso. *J. Exp. Anal. Behav.*, 63, 165-174.

[20] Ikebuchi, M., Okanoya, K. (1999) Male zebra finches and bengalese finches emit directed songs to the video images of conspecific females projected onto a TFT display. *Zool. Sci.* 16:63-70.

[21] Yamaguchi, M.K., Fujita, K. (1999) Perception of biological motion by newly hatched chicks and quail. *Perception 28 ECVP* Abstract Supplement.

[22] Regolin, L., Tommasi, L., Vallortigara, G. (2000) Visual perception of biological motion in newly hatched chicks as revealed by an imprinting procedure. *Anim. Cogn.*, 3, 53-60.

[23] Maekawa, F., Komine, O., Sato, K., Kanamatsu, T., Uchimura, M., Tanaka, K., Ohki-Hamazaki, H. (2006) Imprinting modulates processing of visual information in the visual wulst of chicks. *BMC Neuroscience* 7:75.

[24] Bonhoeffer, T., Grinvald, A. (1996). Optical Imaging based on intrinsic signals: the methodology. In *Brain mapping; the methods*. A.W. Toga and J.C. Mazziotta, Eds. Academic Press, 55-97.

[25] Malonek, D., Grinvald, A. (1996) Interactions between electrical activity and cortical microcirculation revealed by imaging spectroscopy: implications for functional brain mapping. *Science*, 272, 551-4.

[26] Ts'o, D.Y., Frostig, R.D., Lieke, E.E., Grinvald, A. (1990) Functional organization of primate visual cortex revealed by high resolution optical imaging. *Science*, 249, 417-420.

[27] Grinvald, A., Frostig, R.,D., Siegel, R.M., Bartfeld, E. (1991) High-resolution optical imaging of functional brain architecture in the awake monkey. *PNAS*, 88, 11559-63.

[28] Weliky, M., Bosking, W.H., Fitzpatrick, D. (1996) A systematic map of direction preference in primary visual cortex. *Nature,* 379, 725-8.

[29] Xiao, Y., Wang, Y. and Felleman, D.J. (2003) A spatially organized representation of colour in macaque cortical area V2. *Nature* 421, 535–539.

[30] Shimizu, T., Bowers, A.N. (1999) Visual circuits of the avian telencephalon: evolutionary implications. *Behav. Brain Res.*, 98, 183-191.

[31] Wilson, P. (1980). The organization of the visual hyperstriatum in the domestic chick. I. Topology and topography of the visual projection. *Brain Res.* 188, 319-32.

[32] Wilson, P. (1980) The organization of the visual hyperstriatum in the domestic chick. II. Receptive field properties of single units. *Brain Res.* 188, 333-45.

[33] Koshiba, M., Yohda, M., Nakamura, S. (2005) Topological relation of chick thalamofugal visual projections with hyper pallium revealed by three color tracers. *Neurosci. Res.*, 52, 235-42.

[34] Liu, G.B., Pettigrew, J.D.(2003) Orientation mosaic in barn owl's visual Wulst revealed by optical imaging: comparison with cat and monkey striate and extra-striate areas. *Brain Res.*, 961, 153-8.

[35] Conway, B.R. (2003) Colour vision: a clue to hue in V2. *Curr. Biol.*, 13, R308-10.

[36] Livingstone, M.S. and Hubel, D.H. (1984). Anatomy and physiology of a color system in the primate visual cortex. *J. Neurosci.* 4, 309–356.

[37] Hubel, D.H. and Livingstone, M.S. (1987). Segregation of form, color and stereopsis in primate area 18. *J. Neurosci.* 7, 3378–3415.

[38] Bartels, A., Zeki, S. (2000) The architecture of the colour centre in the human visual brain: new results and a review. *Eur. J. Neurosci*., 12, 172-93.

[39] Kohsaka, S.I., Takamatsu, K., Aoki, E., Tsukada, Y. (1979) Metabolic mapping of chick brain after imprinting using [14C]2-deoxyglucose technique. *Brain Res*. 172, 539-44.

[40] Herrera, D.G., Robertson, H.A. (1996) Activation of c-fos in the brain. *Prog. Neurobiol.* 50, 83-107.

[41] Knapska, E., Kaczmarek, L. (2004) A gene for neuronal plasticity in the mammalian brain: Zif268/Egr-1/NGFI-A/Krox-24/TIS8/ZENK? *Prog Neurobiol*., 74, 183-211.

[42] McCabe, B.J., Horn, G. (1994) Learning-related changes in Fos-like immunoreactivity in the chick forebrain after imprinting. *PNAS*, 91, 11417-21.

[43] Anokhin, K.V., Mileusnic,R., Shamakina, I.Y., Rose, S.P.R. (1991) Effects of early experience on c-fos gene expression in the chick forebrain, *Brain Res*. 544 ,101-107.

[44] McCabe, B.J., Horn, G., Bateson, P.P.G. (1981) Effects of restricted lesions of the chick forebrain on the acquisition of filial preferences during imprinting. *Brain Res*., 205, 29-37.

[45] Ambalavanar, R., Van der Zee, E.A., Bolhuis, J.J., McCabe, B.J., Horn, G. (1993) Co-expression of Fos immunoreactivity in protein kinase (PKC gamma)-positive neurones: quantitative analysis of a brain region involved in learning. *Brain Res*. 606:315-8.

[46] Ambalavanar, R., McCabe, B.J., Potter, K.N., Horn,G. (1999)Learning-related fos-like immunoreactivity in the chick brain: time-course and co-localization with GABA and parvalbumin. *Neuroscience*. 93, 1515-24.

[47] Baldauf, K., Braun, K., Gruss, M. (2005) Opiate modulation of monoamines in the chick forebrain: possible role in emotional regulation? *J. Neurobiol*., 62, 149-63.

[48] Maekawa, F., Nakamori, T., Uchimura, M., Fujiwara, K., Yada, T., Tsukahara, S., Kanamatsu, T., Tanaka, K., Ohki-Hamazaki, H. (2007) Activation of cholecystokinin neurons in the dorsal pallium of the telencephalon is indispensable for the acquisition of chick imprinting behavior. *J. Neurochem*. 102, 1645-1657.

[49] McCabe, B.J., Davey, J.E., Horn, G. (1992) Impairment of learning by localized injection of an N-methyl-D-aspartate receptor antagonist into the hyperstriatum ventrale of the domestic chick. *Behav. Neurosci*., 106, 947-53.

[50] Morris, V. B., and Shorey, C. D. (1967) An electron microscope study of types of receptor in the chick retina. *J. Comp. Neurol*., 129, 313-340.

[51] Osorio, D., Vorobyev, M., Jones, C.D. (1999) Colour vision of domestic chicks. *J. Exp. Biol*. 202, 2951-2959.

[52] Prescott, N.B., Wathes, C.M. (1999) Spectral sensitivity of the domestic fowl (Gallus g. domesticus). *Br. Poult. Sci*. 40:332-9.

[53] Fernandez-Juricic, E., Erinchsen, J.T., Kacelnik, A., (2004) Visual perception and social foraging in birds. *Trends Ecol. Evol*. 19:25-31.

[54] Conway, B.R. (2007) Color vision: mice see hue too. *Curr. Biol*., 17, R457-60.

[55] Onishi, A., Hasegawa,J., Imai , H., Chisaka, O. , Ueda, Y., Honda, Y. , Tachibana, M., Shichida, Y. (2005) Generation of knock-in mice carrying third cones with spectral sensitivity different from S and L cones, *Zoolog. Sci*. 22, 1145–1156.

[56] Baddeley, R.J., Osorio, D., Jones, C.D. (2007) Generalization of Color by Chickens: Experimental Observations and a Bayesian Model. *Am. Nat.* 169, S27-41.

[57] Izawa, E., Zachar, G., Aoki, N., Koga, K., Matsushima, T. (2002) Lesions of the ventro-medial basal ganglia impair the reinforcement but not the recall of memorized color discrimination in domestic chicks. *Bahav. Brain Res.* 136, 405-14.

[58] Medina, L., Reiner, A. (2000) Do birds possess homologues of mammalian primary visual, somatosensory and motor cortices? *Trends. Neurosci.*, 23, 1-12.

[59] Alpar, A., Tombol, T. (1998) Telencephalic connections of the visual system of the chicken: tracing the interrelation of the efferents of the visual Wulst and the hyperstriatum ventrale. *Ann. Anat.* 180, 529-36.

In: Neural Pathways Research
Editor: F. L. Pichler, pp. 115-128
ISBN: 978-1-60456-214-9

Chapter V

Modulation of Neuronal Signal by Cross-Talk within Sensory Ganglia: Contribution to Pathological Pain

Mamoru Takeda* and Shigeji Matsumoto
Department of Physiology, School of Life Dentistry at Tokyo,
Nippon Dental University, 1-9-20., Fujimi-cho,
Chiyoda-ku, 102-8159, Tokyo, Japan

Abstract

The transmission of sensory information from the peripheral to the central nervous system is modified at each relay station (e.g. spinal cord, thalamus and sensory cortex) by a local circuit and descending and ascending pathways. Peripheral tissue injury/inflammation can also alter the properties of somatic sensory pathways, causing behavioral hypersensitivity and resulting in increased responses to pain caused by noxious stimulation (hyperalgesia) and normally innocuous stimulation (allodynia). Although no synaptic transmission has been found in the primary sensory ganglia, it has been discovered that the activity of neighboring neurons elicits a functional cross-excitation in the somata of affected sensory neurons under normal conditions, indicating that non-synaptically released diffusible chemical messengers modify the neuronal excitability of the sensory ganglia. Non-synaptically released chemical mediators (substance P, ATP and cytokine) were derived from both the neurons and the satellite glia (e.g. neuron-neuron and neuron-glia interactions). More recent studies have suggested that modulation of neuronal excitability within sensory ganglia may trigger chronic pain (hyperalgesia/allodynia) via the autocrine/paracrine mechanism, and this augmented excitability of the primary afferent neurons may also cause the development of changes in the central pain-signaling neurons (central sensitization). Therefore, the present review focuses on the modulation of the neuronal signal by cross-talk in the sensory ganglia, particularly with regard to its contribution to inflammatory and

* Corresponding Author : Mamoru Takeda, Ph.D., Department of Physiology, School of Life Dentistry at Tokyo, Nippon Dental University, 1-9-20, Fujimi-cho, Chiyoda-ku, Tokyo, 102-8159, Japan. Tel/Fax: -81-3-3261-8740; E-mail:m-takeda@tokyo.ndu.ac.jp

neuropathic pain, and discusses the potential therapeutic target for the prevention of hyperalgesia/allodynia

Keywords: Sensory ganglia; satellite glia paracrine; allodynia; hyperalgesia.

1. Introduction

Sensory signals, including nociception from peripheral tissue to the central nervous system, are modulated at each relay station (at the level of the spinal cord, thalamus and cortex). It is well known that nociceptive transmission in the spinal and medullary dorsal horns is modulated by the descending inhibitory/facillitatory control system [1,2], and also that local inhibitory interneurons, such as GABA/glycinergic inhibitory interneurons, are controlled by the supraspinal activities of periaqueductal gray matter (PAG) [3.4], nucleus raphe magnus (NRM) and locus cerulus (LC) [5,6] (Figure 1). These observations suggest that within the ascending sensory pathway, the site of synaptic transmission at each relay station works to amplify and attenuate the sensory signal.

Generally, peripheral tissue injury and inflammation can alter the properties of somatic sensory pathways, causing behavioral hypersensitivity and resulting in increased responses to pain caused by both noxious stimulation (hyperalgesia) and normally innocuous stimulation (allodynia) [7, 8]. Changes in the excitability of primary afferents (peripheral sensitization) and/or altered information processing in the spinal cords/higher center contribute to the enhancement of pain sensations [9].

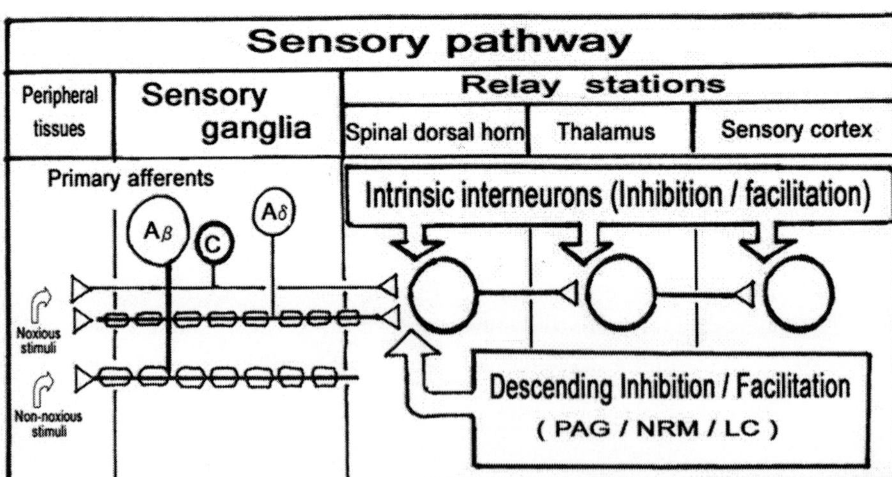

Figure 1. A schematic drawing of the modulation sites of sensory signals from the peripheral to the central nervous system. Both noxious and innoxious information is conveyed independently through primary afferent fibers to the secondary neurons at the spinal level. At each relay station, the sensory signal is attenuated or augmented by intrinsic interneurons and/or descending inhibition/facilitation (PAG, NRM and LC). PAG: periaqueductal gray matter; LC: locus coeruleus; NRM: nucleus raphe magnus.

Recent electrophysiological studies indicate that the activity of neighboring neurons causes functional cross-excitation of the somata of affected sensory neurons under normal conditions [18,19], suggesting that a non-synaptically released diffusible chemical messenger may modify the somatic excitability of neurons in the sensory ganglia. In fact, a combination of microdialysis and radioimmunoassay techniques provided evidence that ATP and substance P (SP) are released concurrently from trigeminal ganglion (TRG) neurons within the trigeminal ganglia (TRGs) under *in-vivo* conditions [20]. The release of SP in the TRG neuron increases predominantly after peripheral inflammation, indicating that the local paracrine mechanism in the sensory ganglia contributes to the development of inflammation-induced sensory abnormalities. In addition, there are an increasing number of reports indicating that spinal dorsal horn glial cells (astroglia and microglia) play an important role in the maintenance of pathological pain [21-22]. For example, behavioral hyperalgesia by peripheral inflammation is reportedly attenuated by the administration of either glial metabolic inhibitors or cytokine inhibitors [23], suggesting that the spinal glia cells may contribute to the development of hyperalgesia. Sensory ganglia containing satellite glia cells (SGC), as well as the astroglia in the the central nervous system (CNS), play an important role in pathological pain. Recent studies have demonstrated the importance of satellite glial cells in modulating the function of sensory transmission, including nociception [24-27].

The present review focuses on a possible site for the modulation of neuronal signals by cross-talk within sensory ganglia (e.g. neuron-neuron, neuron-glia interactions) through the ascending sensory pathway. In particular, we address the question of whether intraganglionic sensory modulation contributes to pathological pain, such as hyperalgesia and allodynia, and we also discuss a potential therapeutic target for chronic pain prevention.

2. Overview of Sensory Ganglia

Sensory ganglia contain the cell bodies of primary afferent neurons, which transmit sensory information from the periphery into the CNS [28, 29]. Most sensory signals from the cell body are conveyed to the CNS by DRG, which are neurons consisting of a single axon with a T-shaped bifurcation. The long branch of this axon extends into the periphery and forms sensory endings in the skin, muscles, viscera/joints and other organs, and the short branch enters the spinal cord. The second major type of sensory ganglia are the trigeminal ganglia (TRGs), which provide sensory innervation to most of the head, including tooth pulp and the orofacial organ. The third major type of sensory ganglia are the nodose ganglia (NGs), which receive sensory inputs via the vagus nerves from internal organs such as the heart, respiratory tract and stomach.

Additionally, sensory ganglia contain satellite glial cells (SGCs) that outnumber and completely envelope the sensory neurons [30, 31]. In contrast to the large amount of information that has been published on glial cells in the CNS, little is known about SGCs in the sensory ganglia. Morphologically, SGCs are laminar and have no process, unlike CNS astroglia [26]. In general, each sensory neuron has its own SGCs, and thus the neurons and their surrounding SGCs form a distinct morphological functional unit [26]. Like other glial cels, such as schwan cells and astrocytes, SGCs are characterized as having numerous receptors for neurotransmitters and other bioactive molecules [26,32]. It can be argued that SGCs are analogus to astrocytes because both cell types are closely related to neuronal somata, carry transporters for neurotransmitters, and regulate ionic concentration in extracellular space [26]. It is thus possible that the cross-talk within sensory ganglia (neuron-neurons, neurons to glia) serves as a neuronal signal modification site in the sensory pathway.

3. Cross-Communication within Sensory Ganglia

3.1. Cross-Communication: Neurons-Neurons

It has been found that autonomic ganglia, such as sympathetic and parasympathetic ganglion neurons, have dendrites and chemical synapses, while both absent in sensory ganglia. Although primary afferent neurons are generally thought to constitute independent sensory communication channels, Devor and Wall [33] reported cross-excitation with the dorsal root ganglia (DRGs) in intact animals. As shown in Figure 2*A*, they found that most DRG neurons are transiently depolarized when the axons of neighboring neurons of the same ganglions are stimulated repeatedly (*"cross-excitation"*) [18,34], and suggested that activity-dependent changes in the extracellular concentration of K^+ and the non-synaptic release of transmitters might mediate mutual depolarization among DRG neurons [33]. Matsuka et al. [26] reported evidence that a TRG neurons release ATP and substance P (SP) concurrently under *in-vivo* conditions. Oh and Weinreich [35] observed that the cross-excitation of the NG neurons (vagal primary afferent neurons) is qualitatively similar to that of the DRG neurons. Thus, we can conclude that cross-talk is a general physiological property of manmmalian primary sensory neurons, including both sensory and visceral sensory afferents (Figure 2*A*).

CNS astrocytes and retinal Muller cells regulate extracellular K^+ concentration; they are able to take up this ion and transport it away from the site of its accumulation [36]. This process, called "spatial buffering," is considered a major function of glial cells. In fact, it has also been found that the mean volume of nerve cell bodies corresponding to an SGC is lower for small neurons than for large ones [26]. Although there is evidence that some chemical signals, such as substance P (SP), affect the cultured function of SGC under normal conditions, little is known about the chemical messengers derived from sensory neurons that affect the excitability of SGC cells. More recently, Zhang et al. [36] reported that electrical stimulation of DRG neurons elicits robust vesticular ATP release from their somata, and the rate of release events increases with the frequency of nerve stimulation; external Ca^{2+} entry is required for release.

Additionally, the released ATP activates P2X7 receptors in the SGCs that envelope each DRG neuron and triggers communication between the neuronal soma and glial cells. The researchers concluded that the somata of DRG neurons release transmitters actively and play a crucial role in the bidirectional communication between neurons and their surrounding SGCs [37] (Figure 2B). Many previous studies also shown that glial cells in most regions of the CNS are mutually coupled by gap junctions, which allow the passage of ions and small molecules [38,39]. The degree of dye-coupling depends upon changes in extracellular pH.

Since relatively few DRG neurons and SGC cells exhibit dye-coupling (5%)[25], the physiological importance of dye-coupling between DRG neurons and SGC cells under normal conditions is currently unclear.

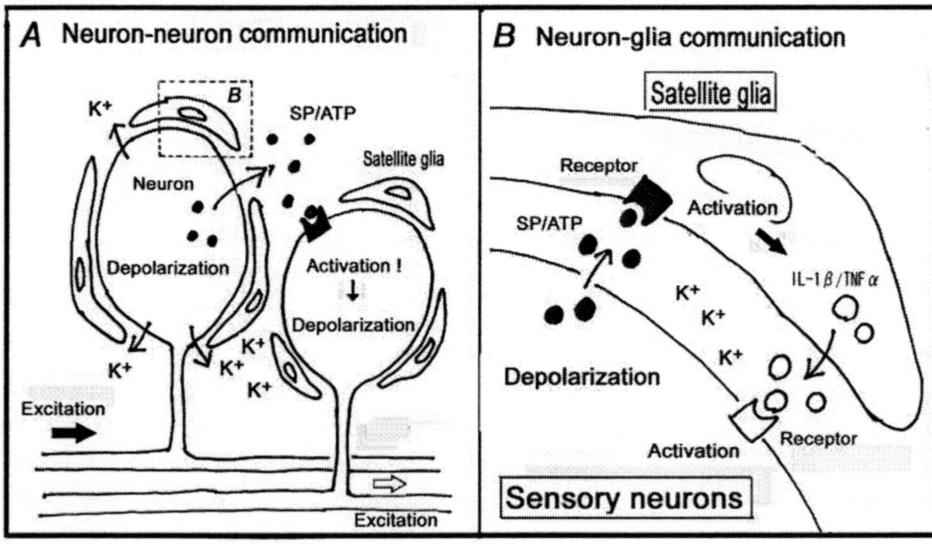

Figure 2. A possible mechanism for the intraganglionic modulation of sensory signals. *A*: Neuron-neuron communication. Activated neurons release K^+ ions and neurotransmitters into the extracellular space within the sensory ganglia. The elevated potassium concentration depolarizes the neighbor neurons and/or the neurotransmitter activates neighbor neurons via their receptors. *B*: Neuron-glial communication. Activated neurons release neurotransmitters into the extracellular space within the sensory ganglia. The neurotransmitters released also activate satellite glia via their receptors. Activated satellite glial cells release cytokins (IL-1β, TNFα), which may further potentiate neuronal excitability. IL-1β: interleukin 1β; TNFα: tumor necrosis factor α.

4. A Possible Mechanism for the Sensory Ganglionic Communication of Pathological Pain

4.1. The Contribution of Neuron-Neuron Communication to the Development of Allodynia

Changes in the excitability of primary afferent neurons (peripheral sensitization) and alteration of the information processing of the spinal cord or higher centers contribute to intensifying the pain sensation [8, 9]. Peripheral inflammation induces the production of many inflammatory mediators, such as prostaglandin E_2 (PGE_2), bradykinin (BK) and 5-hydroxytriptamine (5HT), and neurotrophic factors increase the excitability of the peripheral terminals of small-diameter DRG neurons [40]. Neurotrophic factors, including nerve growth factor (NGF), are up-regulated in the inflamed tissues and transported back to the DRG, and this increase in NGF in the peripheral nerves generates inflammatory pain [41]. In actuality, the production of NGF in the inflamed tissues leads to an alternation between up-regulation of preprotachykinin-A mRNA [42] and increase in SP content in the DRG [43].

Recent electrophysiological studies have demonstrated that a non-synaptically released diffusible chemical messenger may modify the somatic excitability of neurons within the sensory ganglia [18,19,33,34]. In fact, there is evidence that SP is released from the somata of DRG neurons *in vitro* [44], suggesting that neuronal cell cross-communication within the sensory ganglia contributes to the development of mechanical allodynia at a site far from the inflammation site. A model for temporomandibular joint (TMJ) inflammatory mechanical allodynia in the rat trigeminal system has recently been developed, as has a complete Freund's adjuvant (CFA) model for inflammation that produces persistent behavioral hyperalgesia and allodynia in the orofacial region of rats [45,46]. TMJ inflammation is associated with the spread of pain and hyperalgesia/allodynia [47], and patients suffering from TMJ disorders complain about pain due to innoxious vibrotactile stimulation [48].

We have recently reported that under *in-vivo* and *in-vitro* conditions, TMJ inflammation modulates the excitability of the Aβ-TRG neurons, which innervate the facial cutaneous area, via a paracrine mechanism when substance P (SP) is released from small-diameter (Aδ-/ C-) TRG cell bodies [49]. Such a release may be an important factor in understanding the trigeminal inflammation that induces mechanical allodynia [49-51]. As shown in Figure 3, we hypothesized that the possible mechanism underlying mechanical allodynia was an increase in the excitability of the Aβ-TRG neurons that innervate the facial skin after TMJ inflammation. Under normal conditions, when innoxious mechanical stimulation is applied to the skin, the low-threshold mechanical receptor is activated and generates action potentials. The discharge of action potentials is conveyed through the Aβ-fibers to the central terminal (e.g. trigeminal spinal nucleus/spinal dorsal horn) in secondary neurons such as low threshold mechanoreceptive (LTM) neurons. The LTM neurons then transmit the non-nociceptive signal (tactile sensation) to the higher center (thalamus and cortex). Since we recently found that TMJ inflammation potentiated the number of SP positive neurons in inflamed rats [49], we can speculate that the production of NGF due to TMJ inflammation leads to increased SP content in the TRG neuronal soma. TMJ inflammation causes the release of SP, and this SP release activates up-regulated NK1 receptors in the TRG neurons that innervate the facial skin via a paracrine mechanism. Under these conditions, when innoxious mechanical stimulation was applied to the skin, strong membrane depolarization was evoked in the peripheral terminals of low-threshold mechanoreceptors. This depolarization caused an increase in action potential discharges and traveled through the axons of Aβ-fibers to central terminals in the secondary neurons (e.g. trigeminal spinal nucleus/spinal dorsal horn), such as LTM and/or wide dynamic range (WDR) neurons. It is known that the central terminal of Aβ-neurons projects onto the secondary neurons in the III-IV layer of spinal dorsal horns [41,52], and some of these secondary neurons also respond to noxious stimuli (Aδ-/C-fibers); these are WDR neurons.

The augmentation of central pain signaling neurons' responsiveness to inputs from low-threshold mechanoreceptors, often termed "central sensitization," has been previously documented [53]. Since we found that TMJ inflammation lowered the mechanical response threshold in this study, it can be speculated that "central sensitization" contributes to the generation of inflammation-induced mechanical allodynia. Although LTM neurons still transmit tactile sensations to the higher center, the increased excitability of WDR neurons may result in the additional transmission of noxious signals (pain) to the higher center.

Therefore, it is possible that the potentiation of the excitability of Aβ-neurons activates the nociceptive neurons may trigger allodynia as a result.

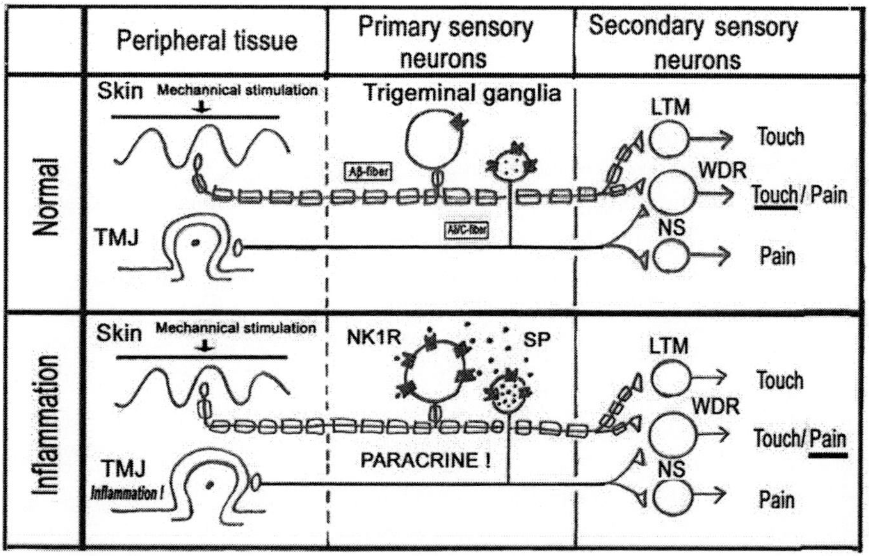

Figure 3. A possible mechanism for mechanical allodynia following inflammation and nerve injury. *A. normal*: Application of an innocuous mechanical stimulation to the skin activates the low-threshold mechanoreceptor and generates action potentials. The discharge of the action potential is conducted through the axons of the Aβ-fiber to the central terminal (e.g. trigeminal spinal nucleus) in the secondary neurons, such as the low-threshold mechanical (LTM) neuron. The LTM neurons further transmit the non-nociceptive signal (touch: tactile sensation) to a higher center (thalamus and cortex). *B. Inflamed:* After TMJ inflammation, the production of nerve growth factor (NGF) in the inflamed tissue leads to an increase in the SP content of the TRG neuronal soma. TMJ inflammation causes the release of SP, which activates up-regulated NK1 receptors in the TRG neurons that innervate the facial skin via a paracrine mechanism. Under this condition, when an innoxious mechanical stimulation was applied to the skin, strong membrane depolarization excitation was evoked in the peripheral terminals of the low-threshold mechanical receptors. This depolarization caused an increase in the action potential discharges, which were conducted through the axons of Aβ-fibers to the central terminal (e,g. trigeminal spinal nucleus) via the secondary neurons, such as LTM and/or WDR neurons. Although the LTM neurons still transmit tactile sensations to the higher center (thalamus and cortex), WDR neurons with increased excitability may transmit noxious signals (pain) to the higher center. Therefore, this phenomenon may be a component of the mechanism for the inflammatory allodynea that accompanies TMJ disorder. LTM: Low-threshold mechanoreceptive; WDR: wide dynamic range; NS: nociceptive specific.

4.2. The Contribution of Neuron-Satellite Glial Cell Communication to Hyperalgesia

It has recently been found that spinal cord microglia and astrocytes are responsible for the creation of an exaggerated pathological state of pain [21-22]. Recent studies have demonstrated that the importance of satellite glial cells lies in their ability to modulate the function of sensory transmission, including nociception [24-27]. Indeed, Stephenson and Byers [24] discovered an increased in the presence of glial fibrillary acidic protein (GFAP) in satellite glial cells of the TRGs following tooth pulp injury. GFAP is a marker of glial cells as well as of central nervous system astrocytes. Following inflammation, glial cells are activated and produce proinflammatory cytokines, such as interleukin-1β (IL-1β) [23,27,54]. The trigeminal neuropathic model was used to demonstrate that axotomy could effect changes not only in neurons but also in the satellite glial cells of the TRG [25, 55], implicating these cells in the development of neuropathic hyperalgesia (abnormally intense pain elicited by noxious stimuli).

Recently, using the patch-clamp technique combined with immunohistochemistry, we investigated the basis for both mechanical hyperalgesia and changes in the sensitivity of neurons following peripheral inflammation. We found that *(1)* significantly more TRG neurons were encircled by GFAP/IL-1β-immunoreactive cells in inflamed rats than in control rats; *(2)* inflammation upregulated interleukin type I receptor (IL-1RI) expression in the small-diameter TRG neurons innervating the facial skin; and (*3*) inflammation potentiated the excitability of acutely dissociated small-diameter (Aδ-/C-types) TRG neurons innervating the facial skin [27]. Taken together, these findings suggest that the activation of SGCs modulates the excitability of small diameter (nociceptive) TRG neurons via the IL-1β following inflammation, and that the up-regulation of IL1RI in the soma may be a part of the underlying mechanism for inflammatory hyperalgesia [27].

Alternatively, it has been reported that axotomy and inflammation induced profound changes not only in neurons but also in glial cells, causing greatly augmented coupling among SGCs [25,56]. Although the precise functional significance of this phenomenon is yet unknown, we can speculate that the increased excitability of TRG neurons may lead to an increase in extracellular K^+ concentration, and that the augmented coupling may enable the SGCs to redistribute K^+ ions and other harmful substances more effectively. We can assume that such reactions alter glia-neuron interactions and participate in the production of chronic pain.

5. The Functional Significance of Intraganglionic Cross-Talk to Pathological Pain

Mechanical allodynia and their related neuronal responses were attenuated by systemic and local pretreatment with a specific neurokinin 1 receptor blocker [55, 57-60]. Intrathecal administration of an NK1 receptor antagonist can attenuate mechanical allodynia in rats with neuropathic pain [59]. Moreover, using the rats with the TMJ inflammation model, we found that iontophoretic application of an NK1 receptor antagonist to the trigeminal sensory ganglia decreased the mechanical threshold in the inflamed rats and reversed it in the control rats [55]. It has previously been reported that after several weeks of the peripheral nerve injury procedure, tactile afferent Aβ-fibers sprout into substantia gelatinosa (lamina II in the spinal dorsal horn) [61]. Two days of peripheral inflammation results in the faclilitation of Aβ-fiber mediated synaptic inputs to substantia gelatinosa [62], but no Aβ-fiber sprouting is detected at two days following inflammation. Therefore, we can postulate that for a short period after inflammation, mechanical allodynia resulting from peripheral sensitization may trigger the "central sensitization" without sprouting of the central afferent. Therefore under chronic pain conditions, enhancement of the specific NK1 receptor expressed in the primary afferent fibers (including peripheral, soma and central terminals) plays a significant role in the induction of mechanical allodynia. Since stimulation of Aβ-fiber induced wind-up was inhibited by the superfusion of an NK1 receptor antagonist [63], this antagonist may serve as a useful therapeutic tool for the alleviation of chronic pain associated with mechanical allodynia. In this review, we indicated that neuronal cell cross-talk within the sensory ganglia contributes to central sensitization, and this mechanism may trigger the exaggeration of chronic mechanical allodynia at the level of the spinal dorsal horn (sensitization and disinhibition).

Since damage to the trigeminal nerve, represented by axotomy, nerve crush and tooth-injury, induces activation of satellite ganglionic cells [24,25,40], our recent study suggests that SGCs play an important role not only in trigeminal neuropathic pain, but also in trigeminal inflammatory pain. It has already been discovered that IL-1β also plays an important role in the generation of hyperalgesia [63-65]. Indeed, administration of the IL-1 receptor antagonist significantly reduces mechanical hyperalgesia produced by peripheral inflammation [66]. Our recent study has also discussed small-diameter nociceptive TRG neurons [27]. These observations suggest that inflammation-induced satellite glial activation may cause the development of trigeminal hyperalgesia, suggesting that IL-1 receptor blockers are a potential therapeutic agent for the prevention of trigeminal inflammatory hyperalgesia.

Finally, concerning the functional significance of transganglionic signal communication to pathological pain, we suggested the following two points. First, nociceptive information conveyed by small-diameter primary afferents terminates in the superficial layers of the spinal dorsal horn, and this signal is modified by the descending inhibitory system (neuronal networks that arise from the PAG, NRN and LC). In contrast to sensory transmission in the secondary neurons of the spinal dorsal horn, nociceptive and non-nociceptive signals are conveyed to the secondary neurons independently by small- and large-diameter fibers, respectively, at the level of primary afferents. Thus, we can speculate that transganglionic communication between sensory ganglia may allow the different types of sensory information transmission to interact, and this interaction may trigger the development of dysesthesia/pathological pain (e.g. allodynia). Second, it is generally believed that each relay station in the synaptic transmission of sensory signals from peripheral tissue, as well as from the CNS (ascending sensory pathway), works as a sensory signal amplification/attenuation site. It is possible that sensory signals are effectively amplified or attenuated within the downstream of the ascending sensory pathway and then sent to the higher center (CNS). Therefore, we can speculate that trans-ganglionic communication is one of the phenomena that can trigger the central sensitization that contributes to the development of pathological pain.

6. Conclusion and Perspective

In conclusion, several advances have been made toward understanding the peripheral and central mechanisms for pathological pain, such as hyperalgesia and allodynia. We recently provided evidence that a paracrine mechanism in the trigeminal ganglia may explain the development of acute inflammatory mechanical allodynia without the reorganization of synaptic connections (e.g. sprouting non-nociceptive fibers) in the spinal dorsal horn. The enhanced excitability of peripheral afferent discharges, including neuronal cross-communication with nociceptive and non-nociceptive neurons in the sensory ganglia, may trigger the central sensitization and disinhibition associated with mechanical allodynia. This senstization in the primary afferent neurons (nociceptive and non-nociceptive) causes both the facilitation of WDR-type neurons (e.g. the wind-up phenomenon) and the disinhibition of the GABA/glycinergic local circuit in the spinal cord. Since peripheral inflammation and nerve injury-induced up-regulation of NK 1 receptors in primary and secondary neurons may contribute to the peripheral and/or central sensitization of pain pathways, we can speculate that treatment with an NK1 receptor antagonist may effectively alleviate the mechanical allodynia associated with peripheral inflammation.

Recently, there has been increasing evidence that glia cells within the spinal dorsal horn contribute to the maintenance of a pathological state of pain, including hyperalgesia and allodynia [21, 22]. Thus, the glial-neuronal interaction has received much attention with regard to pain mechanisms. Our recent study suggests that the activation of satellite glial cells modulates the excitability of small-diameter TRG neurons via a cytokine paracrine mechanism following inflammation, and that the up-regulation of cytokine receptors in the soma may be a component of the mechanism for inflammatory hyperalgesia [27]. Therefore, further studies are needed to clarify this question, which we believe may contribute to the development of therapeutic agents for the prevention of trigeminal inflammatory/neuropathic pain (allodynia/hyperalgesia).

Acknowledgments

This study was supported by a Grant-in-Aid for Scientific Research (C) from the Japanese Society for the Promotion of Science (*No.15591980 and No17591953)* to M.T.

References

[1] Gebhart GF and Randich A, Brainstem modulation of nociception, In: *Brainstem Mechanisms of Behavior*. In Klemm WR, Vertes RP, Eds. New York, Wiley-Intersciece Publication, 1990.

[2] Millan MJ, The induction of pain: an integrative review, *Prog. Neurobiol.* 57:1-164,1999.

[3] Lin Q, Peng YB, Willis WD, Role of GABA receptor subtypes in inhibition of primate spinothlamic tract neurons: difference between spinal and periaqueductal gray inhibition. *J. Neurophysiol.* 75:109-123, 1996.

[4] Peng YB, Lin PB, Willis WD, Effects of GABA and glycine receptor antagonists on the activity and PAG-induced inhibition of rat dorsal horn neurons, *Brain Res.* 189-201, 1996.

[5] Field HL, Anderson SD, Evidence that raphe-spinal neurons mediate opiate and midbrain stimulation–produced analgaesia. *Pain* 5:333-349, 1978.

[6] Ren K, Randich A, Gebhart GF, vagal afferent modulation of a nociceptive reflex in rats: Involvement of spinal opioid and monoamine receptors. *Brain Res.* 446:285-294, 1989.

[7] Merskey H, Bogduk, N, :In IASP Task Force on Taxtomy, IASP press, Seatles, 1994.

[8] Sholtz J, Woolf CJ, Can we coquer pain? *Nat. Neurosci.* 5(Suppl.) 1062-1067, 2002.

[9] Millan MJ, The induction of pain: an integrative review, *Prog. Neurobiol.* 57:1-164,1999

[10] Bereiter DA, Benetti AP, Excitatory amino acid release within spinal trigeminal nucleus after mustard oil injection into the tremporomandibular joint region of the rat. *Pain* 67:451-459, 1996.

[11] Carleson J, Alsergren P, Appelgren A, Applegren B, Knopp, S, Theodorsson E, Lundeberg T, A model for experimental induction of acute temporomandibular joint inflammation in rats: effect of substance P (SP) on neuropeptides-like immunoreactivity, *Life Sci.* 59:11931201, 1996.

[12] Garry MG, Hagegeraves Enhanced release of immunoreactive CGRP and substance P from spinal dorsal horn slices occurs during carageenan inflammation. *Brain Res.* 582:139-142,1992

[13] Coderre TJ, Katz J, Vaccarino AL and Melzack R. Contribution of central neuroplasticity to pathological pain : review of clinical and experimental evidence. *Pain* 52: 259-285, 1993

[14] Dubner R and Ruda MA. Activity-dependent neuronal plasticity following tissue injury and inflammation. *Trends Neurosci.* 15: 96-103, 1992

[15] Ren K, Williams GM, Ruda MA and Dubner R. Inflammation and hyperalgesia in rats neonatally treated with capsaicin: effects on two classes of nociceptive neurons in the superficial dorsal horn. *Pain* 59: 287-300, 1994.

[16] Schible HG and Schmidt RF. Effects of an experimental arthritis on the sensory properties of fine articular afferent units. *J. Neurophysiol.* 54: 1109-1122, 1985.

[17] Xu GY and Zhao ZQ, Changes in excitability and phenotype of substance P and its receptor in cat Aβ sensory neurons following peripheral inflammation. *Brain Res.* 92: 112-119, 2001.

[18] Amir R and Devor M Chemically mediated cross-excitation in rat dorsal root ganglia. *J. Neurosci.* 16:4733-4741, 1996.

[19] Amir R and Devor M, Functional cross-excitation between afferent A- and C-neurons in dorsal root ganglia. *Neuroscience* 95:189-195, 2000.

[20] Matsuka Y, Neubert JK, Maidment NT and Spigelman I. Concurrent of ATP and substance P within guinea pig trigeminal in vivo. *Brain Res.* 915: 248-255, 2001.

[21] Watkins LR, Martin D, Ulrich P, Tracey KJ, Maier SF, Evidence for the involvement of spinal cord glia in subcutaneous formalin induced hyperalgesia in the rat. *Pain* 71:225-235,1997.

[22] Watkins LR, Milliga ED, Maie SF, Glial activation: a driving force for pathological pain. *Trends. Neurosci.* 4:450-455, 2001.

[23] Watkins LR, Maier, SF, Beyond neurons: evidence that immune and glial cells contribute to pathological pain states. *Physiol. Rev.* 82:981-1011, 2001.

[24] Stephenson JL, Byers, MR, GFAP immunoreactivity in trigeminal ganglion satellite cells after tooth injury in rats. *Exp. Neurol.* 131:11-22, 1995.

[25] Cherkas PS, Huang, T-Y, Pannicke, T, Tal, M, Reichenbach, A., Hanai, M, The effects of axotomy on neurons and satellite glial cells in mouse trigeminal ganglion. *Pain* 110:290-298, 2004.

[26] Hanani M, Satellite glial cells in sensory ganglia: from form to function. *Brain Res. Rev.* 48:457-476. 2005.

[27] Takeda M, Tanimoto T, Kadoi J, Nasu M, Takahashi M, Kitagawa J, Matsumoto S, Enhanced excitability of nociceptive trigeminal ganglion neurons by satellite glial cytokine following peripheral inflammation. *Pain* 29:155-166, 2007.

[28] Matthews MR, Cuello AC, Substance P–immunoreactive peripheral branches of sensory neurons innervate guinea pig sympathetic neurons. *Proc. Natl. Acad. Sci. USA.* 79:1668-1672,1982.

[29] Aldskogius H, Elfvin LG, Forsman CA, Primary sensory afferents in the inferior mesenteric ganglion and related nerves of guinea pig. An experimental study with anterogradely transported wheat germ agglutin-horseradish peroxidase conjugate. *J. Auton. Nerv. Syst.* 15:179-190, 1986.

[30] Pannese E,The satellite cells of sensory ganglia. *Adv. Anat. Embryol.* 65:1-111, 1981.

[31] Panesse E, Ledda M, Cherkas PS, Huang TY, Hanani M, Satellite cell reactions to axon injury of sensory ganglion neurons: increase in gap junctions and formation of bridges connecting previously separate perineuronal sheaths. *Anat. Embryol.* 206:337-347, 2003.

[32] Kettenmann H, Ransom BR, (Eds) *Neuroglia*, Oxford University, Press, New York, 1995, 1079, 1995.

[33] Devor M, Wall PD, Cross-excitation in dorsal root ganglia of nberve injured and intact rats. *J. Neurophysiol.* 64:17331746, 1990.

[34] Utzschneider D, Kocsis J and Devor M. Mutual excitation among dorsal root ganglion neurons in the rat. *Neurosci. Lett.* 146:53-56, 1992.

[35] Oh EJ and Weinreich D. Chemical communication between vagal afferent somata in nodose ganglia of the rat and guinea pig in vitro. *J. Neurophysiol.* 87: 2801-2807, 2001.

[36] Reichenbach A, Glial K^+ permeability and CNS K^+ clearance by diffusion and spatial buffering. *Ann. N.Y. Acad. Sci.* 633:272-286, 1991.

[37] Zhang X, Chen Y, Wang C, Huang L-YM, Neuronal somatic ATP release tribggers neuron-satellite glial cell communication in dorsal root ganglia. *Proc. Natl. Acad. Sci. USA*.104:9864-9869, 2007.

[38] Rouach N, Avignone F, Méme W, Koulakoff A, Venance L, Blomstrand F, Giame C, Gap junctions and connexin expression in the normal and pathological central nervous system. *Biol. Cell* 94: 457-475, 2002.

[39] Rozental R, Giaume C, Spray DC, Gap junctions in the nervous system, *Brain Res. Rev.* 32:11-15, 2002.

[40] Dray A. Imflamatory mediators of pain. *Br. J. Anaesth.* 75:125-131,1995.

[41] Woolf CJ, Doubell TP, The pathology of chronic pain –increased sensitivity to low threshold Aβ-fiber inputs. *Curr. Opin. Neurobiol.* 4: 525-534,1994.

[42] Leslie TA, Emson PC, Dowd PM and Woolf CJ. Nerve growth factor contributes to the up-regulation of growth-associated protein 43 and preprotachykinin A messenger RNAs in primary sensory neurons following peripheral inflammation. *Neuroscience* 67: 753-761, 1995.

[43] Donnerrer J, Shuligoi R, Stein C, Increased content and transport of substance P and carcitonin-gene related peptide in sensory nerves innervating inflamed tissue: evidence for regulatory function of nerve growth factor in vivo. *Neuroscience* 49:693-698, 1992.

[44] Huang LYM, Nehr E, Ca^{2+} dependent exocitosis in the somata of dorsal root ~~gn~~ganglion neuron. *Neuron* 17:135-145, 1996.

[45] Imbe H, Iwata K, Zhou QQ, Zou S and Dubner R. Orofacial deep and cutaneous tissue inflammation and trigeminal neuronal activation. *Cells Tissues Organs* 169: 238-247, 2001.

[46] Iwata K, Tashiro A, Tsuboi Y, Imai T, Sumino R, Morimoto T, Dubner R, and Ren K. Medullary dorsal horn neuronal activity in rats with persistant temporomandibular joint and perioral inflammation. *J. Neurophysiol.* 82: 1244-1253, 1999.

[47] Sessle BJ and Hu JW. Mechanisms of pain arising from articular tissues. Can J *Physiol. Pharmacol.* 69: 617-626, 1991.

[48] Filligim RB, Filligim LA, Hollins M, Sigurdsson A and Maixner W. Generalized vibrotactile allodynia in a patient with temporomandibular disorder. *Pain* 78: 75-78, 1998.

[49] Takeda M, Tanimoto T, Ikeda M, Nasu M, Kadoi J, Shima Y, Ohta H, Matsumoto S, Temporomandibular joint inflammation potentiates the excitabilities of trigeminal root ganglion neurons innervating the facial skin in rats. *J. Neurophysiol.* 93:2723-2738, 2005a.

[50] Takeda M, Tanimoto T, Nasu M, Ikeda M, Kadoi J, Matsumoto S, Activation of NK1 receptor of trigeminal root ganglion via substance P paracrine mechanism contributes to the mechanical allodynia in the temporomandibular joint inflammation in rats. *Pain* 116:375-385, 2005b.

[51] Takeda M, Masumoto S, Peripheral and central mechanisms of mechanicall allodynia: neurokinin receptor for therapeutic target. *Central Nervous System Agents in Medicinal Chemistry* 6:311-320, 2006.

[52] Light AR, Perl ER. Spinal termination of functionally identified primary afferent neurons with slowly conducting myelinated fibers. *J. Comp. Neurol* 186:133-150,1979.

[53] Raja SN, Meye RA, Ringkamp M, Chapbell JN, *Peripheral mechanisms of nociception.* In: Wall PD, Merzack R, editors. Text Book of Pain. 4th ed, London; Churchill Livingston; 1999.

[54] Li M, Jiang SHI, Tang J-R, Chen D, Ai B, Chen J, Wang L-N, Cao F-Y, Li L-I, Lin C-Y, Guan X-M. Effects of complete Freund's adjuvant on immunohistochemical distribution of IL-1β and IL-1RI in neurons and glia cells of dorsal root ganglion. *Acta Pharmacol Sin.* 26:192-198, 2005.

[55] Chudler EH, Anderson LC, Byers, MR, Trigeminal ganglion neuronal activity and glial fibrillary acidic protein immunoreactivity after inferior alveolar nerve crush in the adult rat. *Pain* 73:141-149, 1997.

[56] Dublin P, Hanani M, Satellite glial cells in sensory ganglia: Their possible contribution to inflammatory pain. *Brain Behav. Immunity* 21:592-598, 2007.

[57] Carlton SM, Zhou, S, Coggeshall R.E, Localization and activation of substance P receptors in unmyelinated axons of rat glabrous skin. *Brain Res.,* 734:103-108, 1996.

[58] Cahill CM, Coderre TJ, Attenuation of hyperalgesia in a rat model of neuropathic pain after intrathecal pre- or post-treatment with a neurokinin-1 antagonist. *Pain* 95:277-285, 2002.

[59] Meert TF, Vissers K, Greenen F, Kontinen VK, Functional role of exogenous administration of substance P in chronic constriction injury model of neuropathic pain in gerbils. *Pharmacol. Biochem. Behav.* 76:17-25, 2003.

[60] Pitcher GM, Henry JL, Nociceptive response to innocuous mechanical stimulation is mediated via myelinated afferents and NK-1 receptor activation in a rat model of neuropathic pain. *Exp. Neurol.*186:173-197, 2004.

[61] Woolf CJ, Shortland P, Coggeshall RE, Peripheral nerve injury triggers central sprouting of myelinated afferents. *Nature* 355:75-78, 1992.

[62] Baba H, Doubell TP, Woolf CJ, Peripheral inflammation facilitates Abeta fiber-mediated synaptic input to the substantia gelatinosa of the adult rat spinal cord. *J. Neurosci.* 15:859-867, 1999.

[63] Thompson SW, Dray A, Urban L, Injury-induced plasticity of spinal reflex activity: NK1 neurokinin receptor activation and enhanced A- and C-fiber mediated responses in the rat spinal cord in vitro. *J. Neurosci.* 14:3672-3678, 1994.

[64] Bianchi M, Dib B, Paneral AE. Interleukin 1 and nociception in the rat. *J. Neurosci. Res.* 53:645-650, 1998.

[65] Woolf G, Yirmiya R, Goshen I, Iverfeldt K, Holmlund L, Takeda K, Shavit Y. Impairment of interleukin-1 (IL-1β) signaling reduces basal pain sensitivity in mice: genetic, pharmacological and developmental aspects. *Pain* 104:471-480, 2003.

[66] Safieh-Garabedian B, Poole S, Allchorne A, Winter J, Woolf C, Contribution of interleukin-1β to the inflammation-induced increase in nerve growth factor and inflammatory hyperalgesia. *Br. J. Pharmacol.* 115:1265-127, 1995.

In: Neural Pathways Research
Editor: F. L. Pichler, pp. 129-155

ISBN: 978-1-60456-214-9

Chapter VI

White Matter Damage in the Developing and Adult Brain

C. Kaur**, *V. Sivakumar and E. A. Ling
Department of Anatomy, Yong Loo Lin School of Medicine,
Blk MD10, 4 Medical Drive, National University of Singapore,
Singapore 117597

Abstract

Brain white matter, especially in the periventricular region, is vulnerable to damage in hypoxic-ischemic and inflammatory conditions in the developing and adult brain. In the developing brain, damage to the axons and death of immature oligodendrocytes along with myelination disturbances has been widely reported. Release of excess glutamate and enhanced production of nitric oxide, free radicals and inflammatory cytokines have been implicated in this damage. Although microglial cells, the immune effector cells in the central nervous system, have been detected at sites of lesion in the periventricular white matter, their role in the development or amelioration of such lesions is not known. Another important factor which has remained elusive is the vascularity of this region. The blood vessels are thin-walled and are not surrounded completely by astrocyte end feet. In hypoxic-ischemic conditions, they dilate and their permeability is increased due to enhanced production of vascular endothelial growth factor (VEGF) which is known to cause fenestration of endothelial cells. Studies from our laboratory have shown increased production of VEGF and vacuolation of endothelial cells in the periventricular white matter in neonatal rats subjected to hypoxia.

The White matter damage in the adult brain has been reported in many pathological conditions such as multiple sclerosis, Alzheimer's disease, Parkinson's disease, experimental allergic encephalomyelitis, stroke, traumatic brain injury and due to aging. Factors similar to those observed in the developing brain have been implicated in this damage. The disruption of blood-brain barrier (BBB) may be an important factor underlying production of damage to the white matter. Although many strategies for

*Corresponding author: E-mail : antkaurc@nus.edu.sg; Fax: 65-67787643; Phone: 65-65163209

ameliorating the damage are being tested in animal models or in clinical trials, further research is needed to discover therapeutic agents to combat or prevent the damage. Methods of discovering successful drug delivery through the BBB may be timely, exciting and helpful in mitigating the damage.

Abbreviations

AD	Alzheimer's disease
AMPA	α-amino-3-hydroxy5-methyl-4-isoxazolepropionic acid
BBB	blood brain barrier
CNS	central nervous system
eNOS	endothelial nitric oxide
ICAM-1	intercellular adhesion molecule-1
IL-1β	interleukin-1 β
iNOS	inducible nitric oxide synthase
mGluRs	metabotropic glutamate receptors
MHC	major histocompatibility complex
NMDA	N-methyl-d-aspartate
NO	nitric oxide
NOS	nitric oxide synthase
nNOS	neuronal nitric oxide synthase
PWMD	periventricular white matter damage
TNF- α	tumor necrosis factor- α
VCAM-1	vascular cell adhesion molecule-1
VEGF	vascular endothelial growth factor
WMD	white matter damage

Introduction

White matter occupies a central position in the brain and comprises nearly half the brain volume (Zhang et al., 2000; Filley, 2005). It consists of axons and glial cells and plays a key role in development, aging, and many neurologic and psychiatric disorders (Filley, 2005). The axons are arranged in tracts, known as white matter fiber tracts, which connect widely separated cortical regions and are important for neural information processing (Dougherty et al., 2005). The disruption of these tracts results in disturbances of the function of neural networks leading to neurobehavioural syndromes (Filley, 2005), or neurological abnormalities (Mulhern et al., 2001; O'Sullivan et al., 2001a).

Damage to the white matter in the developing brain contributes significantly to neonatal mortality and long-term neurodevelopmental deficits such as cerebral palsy, impaired vision, hearing impairments and mental retardation (Volpe, 2003). In adults, white matter damage (WMD) is associated with neurological impairment, particularly cognitive function and gait disorders as described in many studies (Ylikoski et al., 1993; Pantoni and Garcia, 1996;

Baloh et al, 1995a, b). The white matter peripheral to the lateral ventricles, known as periventricular white matter, is vulnerable to damage in the developing (Volpe, 2001; Folkerth, 2006) and adult brain (van Swieten et al, 1991).

Periventricular white matter injury is the predominant form of brain injury during the perinatal period (Volpe, 2001, 2003). Multiple factors such as hypoxia-ischemia, cerebral hypoperfusion, a developmental lack of autoregulation of the cerebral blood flow, maternal intrauterine infection and prematurity are thought to be associated with the pathogenesis of periventricular white matter injury which is also known as periventricular leukomalacia. Several hypotheses, largely relating to hypoxia-ischemia, about the pathogenesis of white matter injury in the perinatal period have been proposed. Hypoxia can occur in association with many maternal causes such as diabetes, asthma, multiple gestation pregnancy, anemia, smoking, use of alcohol or drugs, decreased utero-placental blood flow, premature onset of labor or prolonged labor. Fetal hypoxia can also result as a consequence of in utero viral or bacterial infections (Debillon et al., 2000), endotoxins (Froen et al., 2002), general anesthesia, severe maternal or fetal heart disease (Takashima et al., 1978). Pulmonary and/or cardiac dysfunction (Bruder et al., 2004) or neonatal stroke (Jiang et al., 2003) also results in hypoxia in the neonates. The periventricular white matter is selectively vulnerable to a hypoxic-ischemic event prior to 32 weeks of gestation (Johnston, 1997).

In the adult, cerebral white matter lesions are associated with cardiovascular disease (Bots et al., 1993), aging and stroke (Breteler et al., 1994; Lindgren et al., 1994; Pantoni and Garcia, 1996; Ylikoski et al., 1995), ischemia (Pantoni et al., 1996), traumatic brain injury (Onaya, 2002; Hortobagyi et al., 2007), schizophrenia (Brambilla et al., 2005; Brambilla and Tansella , 2007; Whitford et al., 2007) multiple sclerosis (Fu et al., 1998; De Stefano et al., 1999; Suhy et al., 2000; Pleasure et al., 2006) and Parkinson disease (Piccini et al., 1995). Alzheimer's disease (AD) involves both gray and white matter pathology, the latter is reflected by atrophy of the corpus callosum (Gootjes et al., 2007). A notable decrease in the corpus callosum area has been reported in AD patients compared with age matched healthy subjects (Vermersch et al., 1996; Hampel et al., 1998; Bozzali et al., 2002). White matter lesions have been correlated with cognitive impairment in AD patients (Fazekas et al., 1987; Bondareff et al.., 1988; Kozachuk et al., 1990; Leys et al., 1990; Mirsen et al., 1991; Diaz et al., 1991; Harrell et al., 1991; Schmidt, 1992; Lopez et al., 1992). White matter damage (WMD) associated with AD has been detected by magnetic resonance imaging (Tsiskaridze et al., 1998) and verified neuropathologically (Englund, 1998). It is also associated with other conditions such as cerebral autosomal dominant arteriopathy with subcortical infarcts, leucoencephalopathy (Kalimo et al., 1999; Harris and Filley, 2001) and subcortical vascular dementia (Chui, 2001). Demyelination and damage of the cerebral white matter has also been reported in carbon monoxide poisoning (Grinker, 1925; Nuebuerger and Clarke, 1945; Gorman et al., 2001; Kim et al., 2003). Evidence of neurologic involvement in patients infected with human immunodeficiency virus was indicated by increased numbers of microglia, macrophages, astrocytes, and multinucleated giant cells in white matter and basal ganglia (Navia et al., 1986; Bell, 2004). White matter alterations are considered to contribute to cognitive decline in elderly patients (Skoog et al., 1996). A connection between white matter lesions and cognition has been demonstrated by many studies investigating effects of aging (O'Sullivan et al., 2001; Grade et al., 2000; Rose et al., 2000), depression (Steffens et al., 2002), multiple sclerosis (Edwards et al., 2001; Rovaris et al., 2002), Parkinson disease (Beyer et al., 2006) or other neurological abnormalities (Mulhern et al., 2001; O'Sullivan et al., 2001b).

WMD in the adult has been associated with hypoxia (Fernando et al., 2006). Increased expression of hypoxia inducible factor-1α and various factors upregulated by it such as vascular endothelial growth factor (VEGF) receptors have been reported (Fernando et al., 2006). Degenerative changes in small blood vessels resulting in cerebral hypoperfusion, altered vascular permeability and BBB dysfunction have been proposed as some of factors leading to the development of white matter lesions in the elderly (Pantoni and Garcia, 1997; Tomimoto et al., 1997).

Pathological Changes

Developing Brain

Axonal swelling, activated microglia, astrocytosis and oligodendroglial damage are hallmark features of the periventricular white matter damage (PWMD) (Takashima et al., 1995, Skoff et al., 2001; Ness et al., 2001; Folkerth, 2006). Swollen and degenerating axons (Kaur et al., 2006) along with cells undergoing necrosis or apoptosis (Kaur and You, 2000) have been demonstrated in the periventricular white matter (Figure 1).

The degenerating elements mentioned above were reported to be phagocytosed by the microglial cells. Besides apoptosis and necrosis, some cells which appeared to be glioblasts (Wu et al., 2006) considered as precursors of oligodendrocytes (Skoff and Knapp, 1991), showed ballooning of the nuclear envelope with splitting of the inner and outer nuclear membrane (Figure 2). Formation of cystic cavities accompanied by impaired or delayed myelination (Kurumatani et al., 1998; Jelinski et al., 1999; Cai et al., 2001) have been reported to occur in later stages. Reduced myelin basic protein after hypoxic injury suggested altered myelination process (Liu et al., 2002; Biran et al., 2006; Kaur et al., 2006). Several studies have shown the occurrence of oedema as a common feature in the periventricular white matter injury (Sridhar et al., 2001; Kadhim et al., 2003).

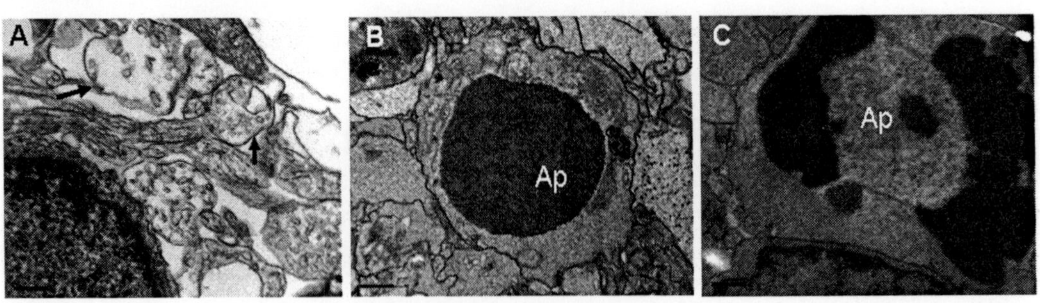

Figure 1. Swollen axons containing vesicular profiles (A, arrows) and apoptotic cells (Ap, B and C) are common in the periventricular white matter of a newborn rat after hypoxic injury. Scale bars, A-B = 0.5 μm, C = 1.0 μm.

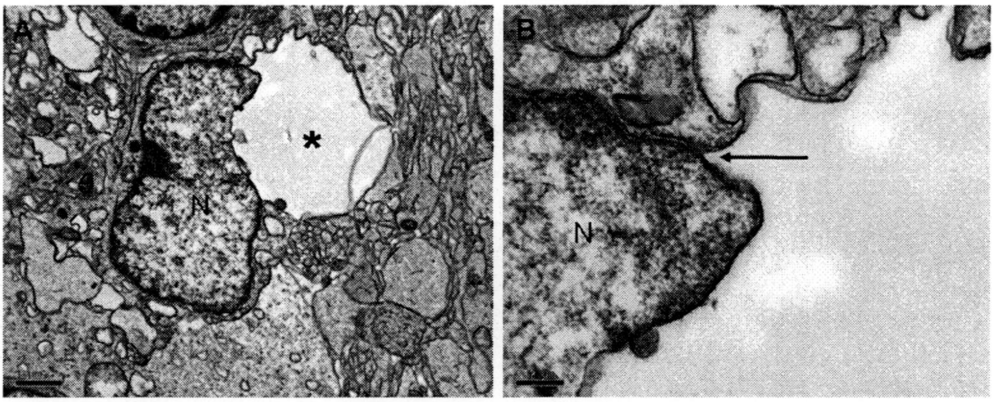

Figure 2. A. The nucleus (N) of a cell (glioblast) in the periventricular white matter of a newborn rat subjected to hypoxia showing ballooning (*) of the nuclear envelope. The arrow in B indicates the point of separation of the inner and outer layers of the nuclear envelope at high magnification. Scale bars, A = 2μm, B= 0. 2μm.

Adult Brain

Decreased myelin content in the periventricular white matter, gliosis, loss of ependyma, vacuolization of the white matter and enlarged perivascular spaces are some of the common pathological features of white matter lesions in adults (Sze et al., 1986; Leifer et al., 1990; Grafton et al., 1991; van Swieten et al., 1991; Chimowitz et al., 1992; Fazekas et al., 1993; Muñoz et al., 1993; Scarpelli et al., 1994; Moody et al., 1997). In addition to the above, thickening of the cerebral vascular walls has also been described (Révész et al., 1989 ; Leifer et al., 1990; van Swieten et al., 1991; Fazekas et al., 1993).

A significant decrease in number of oligodendrocytes is known to occur in white matter injury in AD (Sjöbeck and Englund, 2003). This is associated with an increase in astrocyte numbers, swelling of astrocytes and their end feet associated with the blood vessels. Activated microglial cells and clusters of macrophages have been observed in rarefied white matter (Akiguchi et al., 1997; Tomimoto et al., 2001). Arteriolar sclerosis has also been described as a feature of white matter damage besides myelin attenuation, axonal damage and oligodendrocyte loss (Brun and Englund, 1986)

Mediators of Periventricular White Matter Damage in the Developing Brain

As a significant association of PWMD with cardiopulmonary instability has been reported (Folkerth, 2006), the etiology of white matter injury is considered to be mainly hypoxic-ischemic. Infection and inflammation may also contribute to damage. Lack of cerebral blood flow autoregulation (Berger et al., 2002), glutamate toxicity (Volpe, 2001), impaired intracellular calcium turnover, release of inflammatory cytokines, iron accumulation (Adcock et al., 1996; Palmer et al., 1999) and overproduction of free-radicals (Amato and

Donati, 2000) have been considered to underlie pathogenesis of PWMD. Increased production of VEGF and nitric oxide (NO) also contribute the damage.

Vascular Factors

Immature Blood Vessels

It has been suggested that in view of the relatively low vascularity of white matter as compared to the gray matter, ischemia is likely to play a role in the pathogenesis of PWMD (Ballabh et al, 2004). The structural integrity of the cerebral blood vessels is believed to be maintained by astrocyte end-feet ensheathing them (El-Khoury et al., 2006). The blood vessels in the periventricular white matter of neonatal rats were reported to be thin walled and many of them were devoid of the astrocyte end-feet contacts. The vessels were extremely dilated in hypoxic rats (Figure 3). Increased expression of endothelial nitric oxide synthase (eNOS) in the blood vessels and VEGF in astrocytes associated with them was observed in hypoxic conditions. In addition, experimental and clinical data suggest that cerebrovascular autoregulation, the mechanism by which the cerebral blood flow is maintained and remains constant in many situations, is impaired in some neonates (Tsuji et al., 2000) and may be responsible for PWMD.

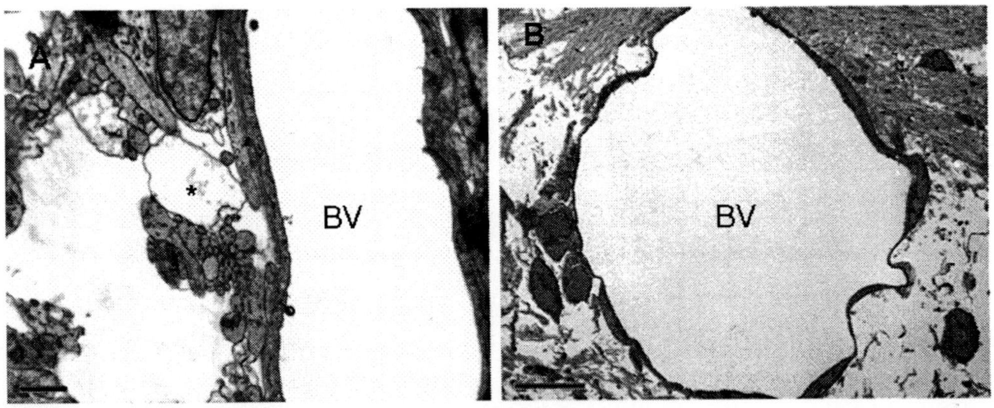

Figure 3. A and B. Extremely thin walled and dilated blood vessels (BV) occur in the periventricular white matter of a newborn rat subjected to hypoxia. The tissue around the blood vessels appears to be edematous. A swollen axon (*) is seen near the blood vessel in A. Scale bars, A = 0.2 μm and B = 1μm.

Vascular Endothelial Growth Factor

VEGF has been described as a vascular permeability enhancing factor (Croll et al., 2004) and has been shown to mediate increases in permeability and angiogenesis in response to chronic sublethal hypoxia in postnatal developing brain (Ment et al., 1997).

Our recent studies have shown increased VEGF concentration in the periventricular white matter and its specific localization in the astrocytes in this region following a hypoxic exposure (Kaur et al., 2006). VEGF has been reported to increase vascular permeability by altering the expression of zonula occludens-1 at the tight junctions of the endothelial cells (Fischer et al., 2002). It also induces fragmentation of the endothelium, fenestration of the endothelial cells and degenerative changes in endothelial basement membrane affecting the structural integrity of the microvessels and leading to extravasation of plasma proteins (Dobrogowska et al., 1998).

Nitic Oxide

NO is synthesized by the enzyme nitric oxide synthase (NOS) from L- arginine. Three isoforms of NOS are known to exist: neuronal (nNOS), eNOS and inducible (iNOS). The activities nNOS and eNOS are stimulated by increases in intracellular calcium whereas iNOS is calcium-independent. NO biosynthesis has been considered as a key factor in the pathophysiological response of the brain to hypoxia-ischemia.

Vasodilation occurring after hypoxic-ischemic episodes is mediated by eNOS (Bolanos and Almeida, 1999) leading to increased blood flow. Increased expression of eNOS occurs in the blood vessels in the developing periventricular white matter following a hypoxic insult (Kaur et al., 2006). Besides vasodilatation and increasing the blood flow, a predominant role for eNOS has been proposed in VEGF induced vascular hyperpermeability (Fukumura et al., 2001).

Expression of iNOS was induced in response to hypoxia in the microglial cells in the periventricular white matter. Excessive production of NO from iNOS is toxic to the immature oligodendrocytes resulting in delayed myelination in hypoxic injury (Baud et al., 2004; Kaur et al., 2006).

Excitotoxicity

Excitotoxicity due to excess release of glutamate has been implicated in the development of PWMD (Johnston 2001, 2005; Volpe, 2001, Tahraoui et al., 2001; Follet et al., 2004; Jensen et al., 2003). Glutamate is the predominant excitatory neurotransmitter in the central nervous system (CNS) and increased release of glutamate has been reported in hypoxic/ischemic conditions in the perinatal brain (Silverstein et al., 1991). Elevation of extracellular glutamate in white matter in hypoxia/ischemia has been described to be possibly due to release from damaged axons (Meng et al., 1997). Glutamate toxicity causes death of oligodendrocyte progenitors (Volpe, 2001) and has been described to occur by receptor and non-receptor mediated mechanisms (Inder and Volpe, 2000).

Receptor mediated glutamate toxicity occurs by activation of ionotropic [*N*-methyl-d-aspartate (NMDA), α-amino-3-hydroxy5-methyl-4-isoxazolepropionic acid (AMPA), kainate receptors] and metabotropic glutamate receptors (mGluRs). Oligodendrocytes are known to express ionotropic (Patneau et al., 1994; Pende et al., 1994; Jensen, 2005; Karadottir et al., 2005) as well as metabotropic glutamate receptors (Luyt et al., 2003; Deng et al., 2004). Non-receptor mediated glutamate toxicity occurs by glutathione depletion in oligodendrocytes leading to their death by free radical mediated mechanisms (Oka et al., 1993).

Activation of glutamate receptors is thought to initiate damage in the developing brain by a cascade of biochemical effects (Cai et al., 1995) such production of NO. High expression of AMPA receptors has been reported in immature oligodendrocytes in the cerebral white matter (Jensen, 2002). Enhanced activation of Ca^{2+} permeable AMPA/kainate receptors is thought to be a major mechanism producing hypoxic-ischemic injury to the immature oligodendrocytes by facilitating increased Ca^{2+} loading during excitotoxicity (Deng et al., 2003).

Inflammation

Overproduction of proinflammatory cytokines tumor-necrosis factor -α (TNF- α), interleukin-1β (IL-1β) and interleukin-6 (IL-6), interleukin-2 (IL-2) along with adhesion molecules such as intercellular adhesion molecule-1 (ICAM-1), vascular cell adhesion molecule-1 (VCAM-1) has been implicated in the pathogenesis of PWMD (Dammann and Leviton 1997; Yoon et al., 1997; Kadhim et al., 2001, 2002). The cytotoxic and inflammatory actions of TNF-α are mediated through two membrane receptors, the p55 and p75 TNF receptors (TNFRs), the expression of which has recently been reported on glial cells in PWMD (Kadhim et al., 2006).

Role of Microglial Cells in PWMD

Microglial cells, present in large numbers in the developing periventricular white matter, are resident immune cells in the nervous system (Ling and Wong, 1993). They are known to be active macrophages removing cellular debris during normal development as well as in pathological conditions. Although microglial cells are necessary for normal functions, their activation can cause bystander injury to other CNS cells. They play an important role in the development of an inflammatory response in the developing brain (Chew et al., 2006). Microglial cells have been detected in periventricular lesions in children with signs of periventricular leukomalacia (Deguchi et al., 1996; Kadhim et al., 2001) and have been shown to express both p55TNFαR1 and p75TNFαR2 whereas astrocytes and oligodendrocytes predominantly express p55TNFαR1 (Dopp et al., 1997). Aberrant TNF-α /p55TNFαR1 signaling in the CNS can have a potentially major role in pathologies in which oligodendrocyte death is a primary pathological feature (Akassoglou et al., 1998). Our recent studies have shown upregulation of TNF-α and IL-1β in the microglial cells in the periventricular white matter of hypoxic rats (Figures 4 and 5.).

Besides microglial cells, astrocytes in the periventricular white matter may also release cytokines (Sawada et al., 1992, 1995) such as TNF-α (Medana et al., 1997). It has been reported that microglia might contribute to the early phase of cytokine production and that astrocytes might contribute to it later (Sawada et al., 1995). TNF-α has toxic effects on developing oligodendrocytes (Merrill, 1991; Qi and Dal Canto, 1996; Ye and D'Ercole, 1999; Cammer, 2000) as it can decrease the number of oligodendrocyte progenitors (Cammer, 2000) by causing their apoptosis (Ladiwala et al., 1999). In addition to this, IL-1β and TNF- α may also be involved in transcriptional activation of iNOS gene (Lopez-Figueroa et al., 2000; Khadim et al., 2006) and NO generation (Betz et al., 1996; Bonmann et al., 1997). Microglial cells in the hypoxic periventricular white matter were observed to express iNOS (Kaur et al., 2006). Recent studies have suggested that peroxynitrite produced by iNOS in activated microglia may play an important role in the pathogenesis of white matter disorders (Li et al., 2005) as it is a highly reactive oxidant capable of inducing injury to a number of cell types.

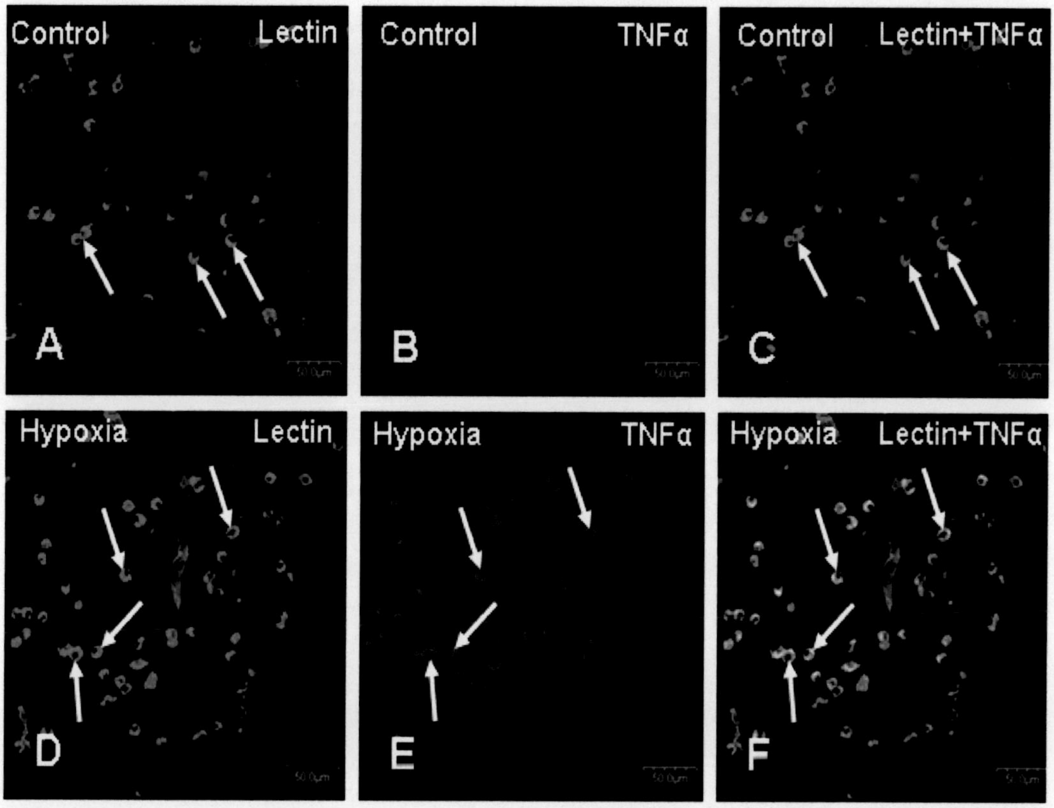

Figure 4. Amoeboid microglial cells (arrows) in the periventricular white matter labeled with microglial marker *Lycopersicon esculentum* lectin (green) in control (A) and a newborn hypoxic (D) rat, showing expression of TNF-α (B and E, red). C and F show co-localization of lectin and TNF-α (arrows). The expression of TNF-α is markedly increased in the amoeboid microglial cells in hypoxic rats. Scale bars, A-F = 50 μm.

In addition to iNOS and inflammatory cytokines, our studies have also shown the expression of NMDA receptors on the microglial cells (Kaur et al., 2006). Immunostimulated microglia are known to potentiate the NMDA receptor-mediated excitotoxicity through expression of iNOS (Kim and Ko, 1998). Microglia cells also express AMPA-KA receptors (Noda et al., 2000), the activation of which enhances the production of TNF- α which kills oligodendrocytes (Selmaj and Raine, 1988; Louis et al., 1993), resulting in destruction of myelin and the dysfunction of axons (Merrill and Benveniste, 1996).

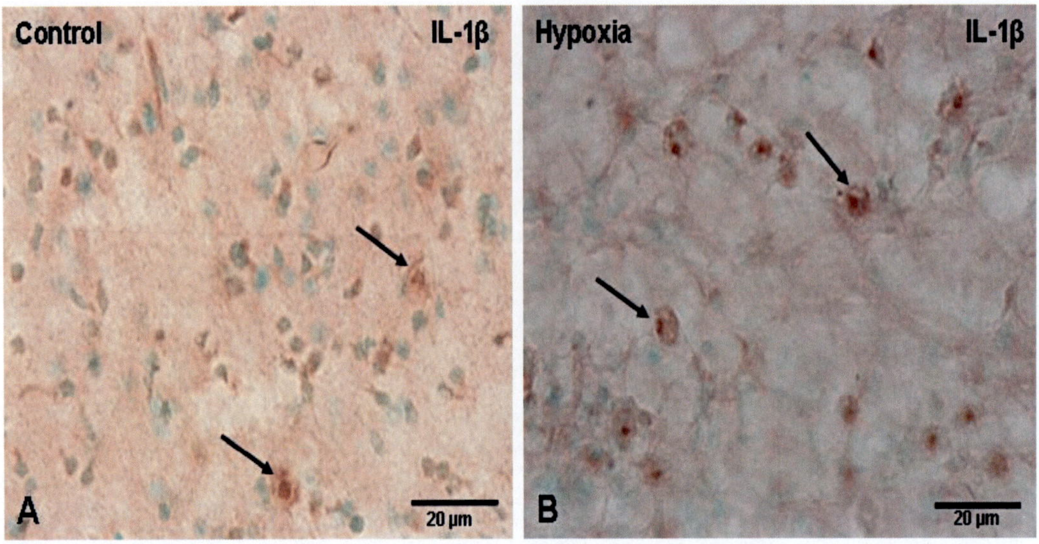

Figure 5. Periventricular white matter from a control (A) and a hypoxic (B) rat showing amoeboid microglial cells (arrows) expressing IL-1β immunoreactivity. The expression of IL-1β is enhanced in the cells from the hypoxic rat. Scale bars, A and B = 20 μm.

Mediators of White Matter Damage in Adult Brain

Hypoxia-ischemia has been put forward as a major factor resulting in white matter damage in the adults (Pantoni and Garcia, 1996; Goldberg and Ransom, 2003). Disruption of axonal ion homeostasis resulting in acute conduction failure and intra-axonal Ca^{2+} accumulation which may trigger subsequent cascades leading to the destruction of the axon structure have been demonstrated in hypoxic-ischemic or anoxic conditions (Stys, 1998; Tekkök and Goldberg, 2003). Studies from our laboratory have demonstrated disruption of myelin sheaths and axonal damage (Figures 6 and 7) in the periventricular white matter in hypoxic injury. In addition, swelling of astrocytes and their end feet in association with blood vessels was observed (Figure 8).

Excitotoxicity, as described above for the developing brain, plays a major role in the pathogenesis of white matter injury. Overactivation of AMPA/ kainate receptors damages the oligodendrocytes (Yoshioka et al., 1995; Matute et al., 1997; McDonald et al., 1998) and may be a common pathway for white matter injury in several conditions such as traumatic injury, (Wrathall et al., 1994), multiple sclerosis (Pitt et al., 2000) and experimental allergic encephalomyelitis (Smith et al., 2000; Matute et al., 2001). Expression of functional NMDA receptors in cells of the oligodendroglial lineage and myelin has been reported recently (Karadottir et al., 2005; Micu et al., 2006).

Activation of NMDA receptors mediates Ca^{2+} increase in myelin and causes loosening and vacuolation of the myelin sheath (Micu et al., 2006). These changes were attenuated by an NMDA-receptor antagonist (Micu et al., 2006).

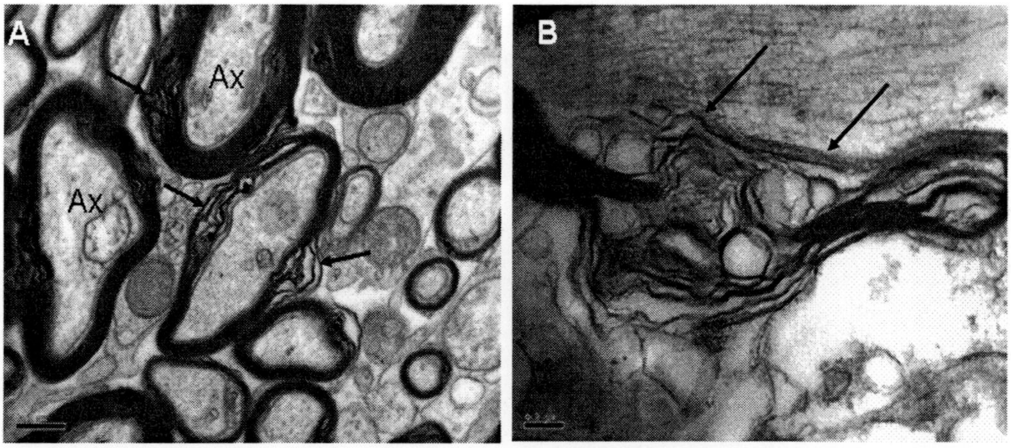

Figure 6. Axons (Ax) showing splitting of myelin sheaths (arrows) following a hypoxic injury in the periventricular white matter of an adult rat are seen in A. In B, disintegrated myelin sheath (arrows) of an axon is seen at higher magnification. Scale bars, A = 0.5 μm; B = 0.2 μm.

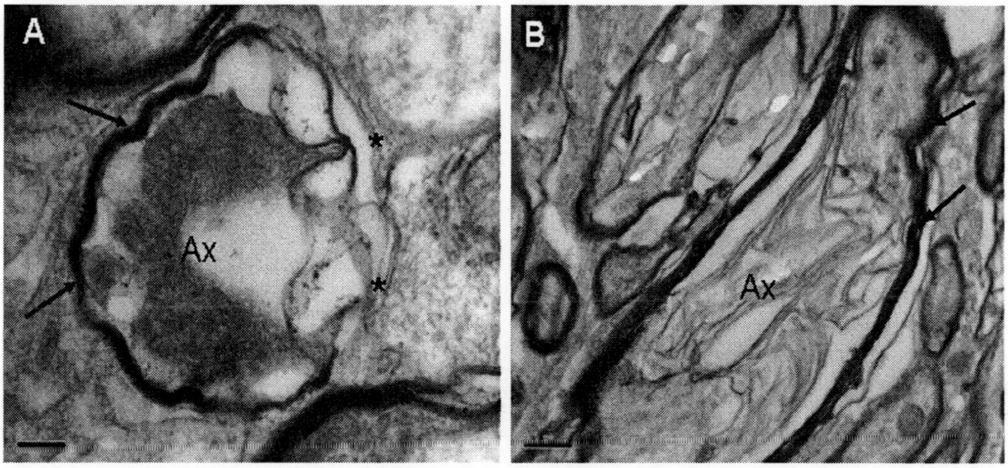

Figure 7. A and B show axons (Ax) undergoing degenerative changes following hypoxic injury. The myelin sheaths (arrows) are thinned out at many points (* in A). Scale bars, A and B = 0.5 μm.

NO derived from iNOS has also been implicated in damaging myelin-producing oligodendrocytes in neuropathological disorders such as multiple sclerosis and AD (Koprowski et al., 1993). Relatively high concentrations of NO have been reported to be produced at sites of inflammation in multiple sclerosis by macrophages and other cells (Bo et al., 1994; Cross et al., 1998; Hill et al., 2004). In vitro studies have shown that NO produced by the activated microglial cells is highly damaging to the oligodendrocytes (Merrill et al., 1993). In vivo studies have shown that axons degenerate when exposed to NO (Smith et al., 2001). Recent studies have suggested that peroxynitrite produced by iNOS in activated microglia may play an important role in the pathogenesis of white matter disorders (Jack et al., 2005; Li et al., 2005) as it is a highly reactive oxidant capable of inducing injury to a number of cell types. NO mediated impaired mitochondrial function is thought to be the most probable mechanism underlying cell damage (Bolanos et al., 1997).

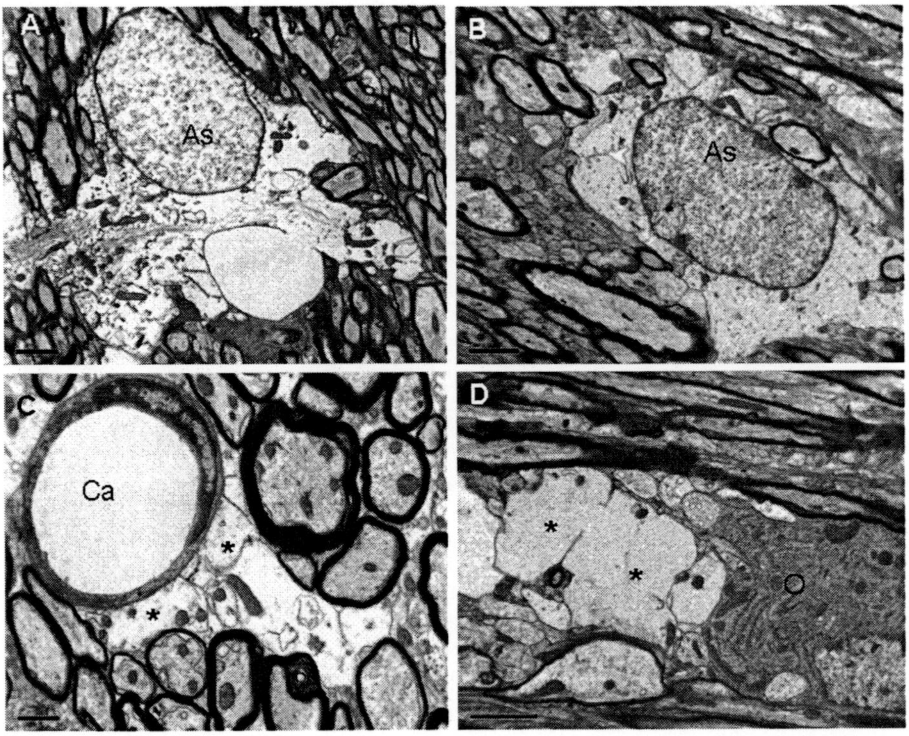

Figure 8. Swollen astrocytes (As) in the periventricular white matter of an adult rat subjected to hypoxia are seen in A and B. In C, swollen astrocyte end-feet (*) are in close association with a capillary (Ca). Edema of the white matter is evidenced in D by the presence of many swollen cell processes (*) in between myelinated axons and close to an oligodendrocyte (O). Scale bars, A and C = 0.2 μm; B = 0.5 μm; D = 1 μm.

Integrity of the BBB is important for the homeostatic regulation of the brain microenvironment (Abbott, 2002) and maintenance of the immune-privileged status of the brain by restricting the entry of T lymphocytes. BBB disruption is an early phenomenon in the formation of white matter lesions in many conditions such as multiple sclerosis (van Horssen et al., 2007) and AD (Bowman et al., 2007). Leakage of plasma proteins, which is enhanced in white matter lesions (van Horssen et al., 2007) due to endothelial tight junction abnormality, is widespread in the white matter in MS (Plumb et al., 2002; Kirk et al., 2003; Leech et al., 2007).

The transport of ions, nutrients, and other molecules across the BBB is regulated by NO (Janigro et al., 1994). Excess production of NO, however, increases the blood flow and permeability of the BBB allowing substances to enter into the brain passively (Shukla et al., 1996; Thiel and Audus, 2001). Other factors such as VEGF, which is potentiated in response to hypoxia, are also important for induction of vascular leakage in the brain (Schoch et al., 2002) leading to the formation of cerebral oedema. The increase in vascular permeability is thought to occur due to VEGF-mediated changes in the expression of zonula occludens-1 at the tight junctions of the endothelial cells (Fischer et al., 2002) and enhanced fenestration of the endothelial cells (Dobrogowska et al., 1998). Our studies have shown increased expression of VEGF in the astrocytes and leakage of tracers such as horseradish peroxidase in the periventricular white matter of adult rats following a hypoxic injury (Figure 9). VEGF has been described to contribute to inflammatory responses (Croll et al., 2004). It has been shown to enhance ICAM-1, VCAM-1 and major histocompatibility complex (MHC) class I and II expression in the brain (Proescholdt et al., 1999; Min et al., 2005). ICAM-1 is important for establishing adhesion of leukocytes to the vascular endothelium before their movement across the vessel walls into the tissue (Wang et al., 1994). Inflammatory demyelination in multiple sclerosis is thought to be initiated by T cells that recognize myelin antigens (Gran and Rostami, 2001).

Inflammation has also been suggested as another important factor for disruption of the BBB. Proinflammatory cytokines such as TNF- α, released in hypoxic-ischemic conditions, also induce expression of ICAM-1 and VCAM-1 on the endothelial cells which facilitate the leukocyte adhesion and infiltration into the areas of damage (Wong and Dorovini, 1992; Hess et al., 1994; McHale et al., 1999). VCAM-1 expression occuring on the macrophages/microglial cells has been related to the depletion of oligodendrocytes in brain lesions such as multiple sclerosis (Khadim et al., 2001; Peterson et al., 2002).

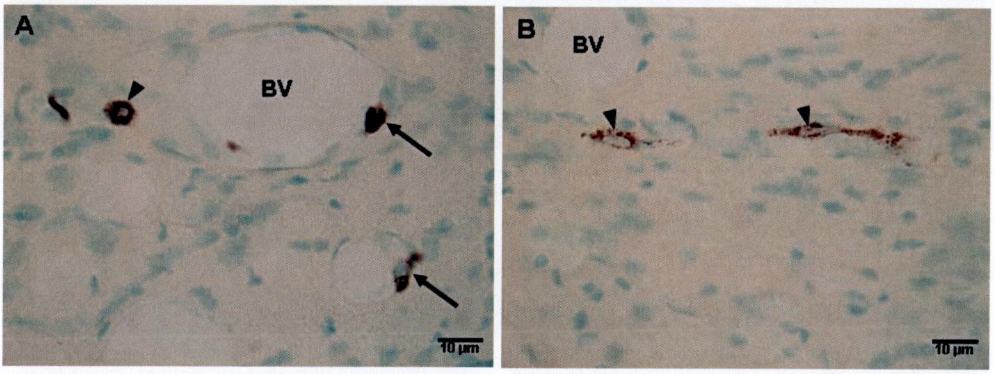

Figure 9. Leakage of horseradish peroxidase (HRP) in the periventricular white matter of an adult hypoxic rat is evidenced by HRP labeled perivascular cells (arrows) close to the blood vessels (BV) in A as well as other cells away from them (arrowheads) in A and B. Scale bar = 10µm.

Possible Therapeutic Strategies for Prevention of White Matter Damage in Adults

Prevention of glutamate toxicity with AMPA/kainate antagonists has been suggested as a possible effective therapy for preventing white matter injury in multiple sclerosis. The evidence in support of such a treatment comes from experimental studies in which specific antagonists of AMPA/kainate receptors ameliorated the neurological sequelae by increasing oligodendrocyte survival and reduced axonal damage in experimental autoimmune encephalomyelitis which mimics multiple sclerosis (Pitt et al., 2000; Smith et al., 2000; Werner et al., 2000; Groom et al., 2003). White matter injury has been reported to be reduced by AMPA/kainate antagonists in rodent models of transient focal cerebral ischemia (McCracken et al., 2002). In vitro studies have demonstrated that oligodendrocytes are protected from hypoxic injury by blockade of AMPA/kainate receptors thus preserving white matter function (McDonald et al., 1998; Fern and Moller, 2000; Yoshioka et al., 2000).

Strategies at reducing inflammation mediated damage have also been tested in multiple slerosis (Rieckmann and Smith, 2001; Rieckmann and Mäurer, 2002). A significant reduction in inflammatory white matter lesions has also been reported with interferon-beta-1b treatment (Barkhof et al., 2001). Strategies to reduce immune cell traffic across the blood-brain barrier (Tubridy et al., 1998; Liang et al., 2001) may afford protection to axons (Ghirnikar et al., 2001). NOS inhibitors, calcium antagonists, TNFRp55 blockers may be beneficial in preventing damage (Rieckmann and Mäurer, 2002).

Conclusion

Brain white matter is vulnerable to damage in the developing and adult brain. Damage to the axons and death of oligodendrocytes along with myelination disturbances have been widely reported in the immature as well as adult brain. Release of excess glutamate and enhanced production of NO, free radicals and inflammatory cytokines have been implicated in this damage. Although many neuroprotective agents have been reported to be of therapeutic value in ameliorating the damage to axons and oligodendrocytes, they seem to be more promising in animal studies than in actual clinical trials. Future studies aimed at in depth translational research to confirm the hypotheses underlying white matter damage in the developing and adult brain and their continued re-evaluation will lead to the development of effective therapies to minimize damage.

Acknowledgements

This work was supported by research grants (R181-000-0065-112 and R181-000-098-112) from the National University of Singapore. The technical help provided by Ms Y.G. Chan and Mr Y.Y. Deng is gratefully acknowledged.

References

Abbott, N. (2002). Astrocyte-endothelial interactions and blood-brain barrier permeability. *J. Anat., 200,* 527.

Adcock, L.M., Yamashita, Y., Goddard-Finegold, J. and Smith, C.V. (1996). Cerebral hypoxia-ischemia increases microsomal iron in newborn piglets. *Metab. Brain Dis., 11,* 359-367.

Akassoglou, K., Bauer, J., Kassiotis, G., Pasparakis, M., Lassmann, H., Kollias, G. and Probert, L. (1998). Oligodendrocyte apoptosis and primary demyelination induced by local TNF/p55TNF receptor signaling in the central nervous system of transgenic mice: models for multiple sclerosis with primary oligodendrogliopathy. *Am. J. Pathol., 153,* 801-813.

Akiguchi, I., Tomimoto, H., Suenaga, T., Wakita, H. and Budka, H. (1997). Alterations in glia and axons in the brains of Binswanger's disease patients. *Stroke, 28,* 1423-1429.

Amato, M. and Donati, F. (2000). Update on perinatal hypoxic insult: mechanism, diagnosis and interventions. *Eur. J. Paediatr. Neurol., 4,* 203-209.

Ballabh, P., Braun, A. and Nedergaard, M. (2004). Anatomic analysis of blood vessels in germinal matrix, cerebral cortex, and white matter in developing infants. *Pediatr Res., 56,* 117-124.

Baloh, R.W., Yue, Q., Socotch, T.M. and Jacobson, K.M. (1995a). White matter lesions and disequilibrium in older people. I. Case-control comparison. *Arch. Neurol., 52,* 970-974.

Baloh, R.W., Yue, Q. and Demer, J.L. (1995b). The linear vestibulo-ocular reflex in normal subjects and patients with vestibular and cerebellar lesions. *J. Vestib. Res.,* 5, 349-361.

Barkhof, F., van Waesberghe, J., Filippi, M., et al. (2001). T1 hypointense lesions in secondary progressive multiple sclerosis: effect of interferon beta-1b treatment. *Brain, 124,* 1396-1402.

Baud, O., Daire, J.L., Dalmaz, Y., Fontaine, R.H., Krueger, R.C., Sebag, G., Evrard, P., Gressens, P. and Verney, C. (2004). Gestational hypoxia induces white matter damage in neonatal rats: a new model of periventricular leukomalacia. *Brain Pathol., 14,* 1-10.

Bell, J.E. (2004). An update on the neuropathology of HIV in the HAART era. *Histopathology, 45,* 549-559.

Berger, R., Garnier, Y. and Jensen. A. (2002). Perinatal brain damage: underlying mechanisms and neuroprotective strategies. *J. Soc. Gynecol. Investig., 9,* 319-328.

Betz, A.L., Schielke, G.P. and Yang, G.Y. (1996). Interleukin-1 in cerebral ischemia. *Keio J Med., 45,* 230–237.

Beyer, M.K., Aarsland, D., Greve, O.J. and Larsen, J.P. (2006). Visual rating of white matter hyperintensities in Parkinson's disease. *Mov. Disord., 21,* 223-229.

Biran, V., Joly, L.M., Heron, A., Vernet, A., Vega, C., Mariani, J., Renolleau, S. and Charriaut-Marlangue, C. (2006). Glial activation in white matter following ischemia in the neonatal P7 rat brain. *Exp. Neurol., 199,* 103-112.

Bo, L., Mork, S., Kong, P.A., Nyland, H., Pardo, C.A. and Trapp, B.D. (1994). Detection of MHC class II-antigens on macrophages and microglia, but not on astrocytes and endothelia in active multiple sclerosis lesions. *J. Neuroimmunol., 51,*135-146.

Bolanos, J.P., Peuchen, S., Land, J.M., Clark, J.B. and Heales, S.J. (1997). Evaluation of the efficacy of potential therapeutic agents at protecting against nitric oxide synthase-mediated mitochondrial damage in activated astrocytes. *Brain Res. Brain Res. Protoc., 1,* 258-262.

Bolanos, J.P. and Almeida, A. (1999). Roles of nitric oxide in brain hypoxia-ischemia. *Biochim. Biophys. Acta., 1411,* 415-436.

Bondareff, W., Raval, J., Colletti, P.M. and Hauser, D.L. (1988). Quantitative magnetic resonance imaging and the severity of dementia in Alzheimer's disease. *Am. J. Psychiatry, 145,* 853-856.

Bonmann, E., Suschek, C., Spranger , M. and Kolb-Bachofen, V. (1997). The dominant role of exogenous or endogenous interleukin-1 beta on expression and activity of inducible nitric oxide synthase in rat microvascular brain endothelial cells. *Neurosci. Lett., 230,* 109-112.

Bots, M.L., van Swieten, J.C., Breteler, M.M., de Jong, P.T., van Gijn, J., Hofman, A. and Grobbee, D.E. (1993). Cerebral white matter lesions and atherosclerosis in the Rotterdam Study. *Lancet, 341,* 1232-1237.

Bowman, G.L., Kaye, J.A., Moore, M., Waichunas, D., Carlson, N.E. and Quinn, J.F. (2007). Blood-brain barrier impairment in Alzheimer disease: stability and functional significance. *Neurology, 68,* 1809-1814.

Bozzali, M., Falini, A., Franceschi, M., Cercignani, M., Zuffi, M., Scotti, G., Comi, G. and Filippi, M. (2002). White matter damage in Alzheimer's disease assessed in vivo using diffusion tensor magnetic resonance imaging. *J. Neurol. Neurosurg. Psychiatry, 72,* 742-746.

Brambilla, P., Cerini, R., Gasparini, A., Versace, A., Andreone, N., Vittorini, E., Barbui, C., Pelizza, L., Nose, M., Barlocco, L., Marrella, G., Gregis, M., Tournikioti, K., David, A.S., Keshavan, M.S, and Tansella, M. (2005). Investigation of corpus callosum in schizophrenia with diffusion imaging. *Schizophr. Res., 79,* 201-210.

Brambilla, P. and Tansella, M. (2007). The role of white matter for the pathophysiology of schizophrenia. *Int. Rev. Psychiatry, 19,* 459-468.

Breteler, M.M., van Swieten, J.C., Bots, M.L., Grobbee, D.E., Claus, J.J., van den Hout, J.H., van Harskamp, F., Tanghe, H.L., de Jong, P.T., van Gijn, J, et al. (1994). Cerebral white matter lesions, vascular risk factors, and cognitive function in a population-based study: the Rotterdam Study. *Neurology, 44,* 1246-1252.

Bruder, E.D., Lee, P.C. and Raff, H. (2004). Metabolic consequences of hypoxia from birth and dexamethasone treatment in the neonatal rat: comprehensive hepatic lipid and fatty acid profiling. *Endocrinology, 145,* 5364-5372.

Brun, A. and Englund, E. (1986). A white matter disorder in dementia of the Alzheimer type: a pathoanatomical study. *Ann. Neurol., 19,* 253-262.

Cai, Z., Sigrest, T., Hersey, K. and Rhodes, P.G. (1995). Intrauterine hypoxia-ischemia increases N-methyl-D-aspartate-induced cGMP formation and glutamate accumulation in cultured rat cerebellar granule cells. *Pediatr. Res., 38,* 107-112.

Cai, Z., Pang, Y., Xiao, F. and Rhodes, P.G. (2001). Chronic ischemia preferentially causes white matter injury in the neonatal rat brain. *Brain Res., 898,* 126-135.

Cammer, W. (2000). Effects of TNF-α on immature and mature oligodendrocytes and their progenitors in vitro. *Brain Res., 864,* 213–219.

Chew, L.J., Takanohashi, A. and Bell, M. (2006). Microglia and inflammation: impact on developmental brain injuries. *Ment. Retard. Dev. Disabil. Res. Rev., 12,* 105-112.

Chimowitz, M.I., Estes, M.L., Furlan, A.J. and Awad, I.A. (1992). Further observations on the pathology of subcortical lesions identified on magnetic resonance imaging. *Arch. Neurol., 49,*747-752.

Chui, H. (2001). Dementia due to subcortical ischemic vascular disease. *Clin. Cornerstone., 3,* 40-51.

Croll, S.D., Goodman, J.H. and Scharfman, H.E. (2004). Vascular endothelial growth factor (VEGF) in seizures: a double-edged sword. *Adv. Exp. Med. Biol., 548,* 57-68.

Cross, A., Manning, P., Keeling, R., et al. (1998). Peroxynitrite formation within the central nervous system in active multiple sclerosis. *J. Neuroimmunol., 88,* 45-56.

Dammann, O. and Leviton, A. (1997). Maternal intrauterine infection, cytokines, and brain damage in the preterm newborn. *Pediatr. Res., 42,* 1-8.

Debillon., T.C., Gras-Leguen, V., Ve′ rielle, N., Winer, J., Caillon, J., Roze′ , C. and Gressens, P. (2000). Intrauterine infection induces programmed cell death in rabbit periventricular white matter. *Pediatr. Res., 47,* 736–742.

Deguchi, K., Mizuguchi, M. and Takashima, S. (1996). Immunohistochemical expression of tumor necrosis factor alpha in neonatal leukomalacia. *Pediatr. Neurol., 14,* 13-16.

Deng, W., Rosenberg, P.A., Volpe, J.J. and Jensen, F.E. (2003). Calcium-permeable AMPA/kainate receptors mediate toxicity and preconditioning by oxygen-glucose deprivation in oligodendrocyte precursors. *Proc. Natl. Acad. Sci. USA, 100,* 6801-6806.

Deng, W., Wang, H., Rosenberg, P. A., Volpe, J. J. and Jensen, F. E. (2004). Role of metabotropic glutamate receptors in oligodendrocyte excitotoxicity and oxidative stress. *Proc. Natl. Acad. Sci. USA, 101,* 7751–7756.

De Stefano, N., Narayanan, S., Matthews, P.M., Francis, G.S., Antel, J.P. and Arnold, D.L. (1999). In vivo evidence for axonal dysfunction remote from focal cerebral demyelination of the type seen in multiple sclerosis. *Brain, 122,* 1933-1939.

Diaz, F.J., Merskey, H., Hachinski, V.C., et al. (1991). Improved recognition of leukoaraiosis and cognitive impairment in Alzheimer's disease. *Arch. Neurol., 48,* 1022-1025.

Dobrogowska, D.H., Lossinsky, A.S., Tarnawski, M. and Vorbrodt, A.W. (1998). Increased blood-brain barrier permeability and endothelial abnormalities induced by vascular endothelial growth factor. *J. Neurocytol., 27,* 163-173.

Dopp, J.M., Mackenzie-Graham, A., Otero, G.C. and Merrill, J.E. (1997). Differential expression, cytokine modulation, and specific functions of type-1 and type-2 tumor necrosis factor receptors in rat glia. *J. Neuroimmunol., 75,* 104-112.

Dougherty, R.F., Ben-Shachar, M., Deutsch, G., Potanina, P., Bammer, R. and Wandell, B.A. (2005). Occipital-callosal pathways in children: Validation and atlas development. *Ann. N. Y. Acad. Sci., 1064,* 98-112.

Edwards, S.G., Liu, C. and Blumhardt, L.D. (2001). Cognitive correlates of supratentorial atrophy on MRI in multiple sclerosis. *Acta. Neurol. Scand.,104,* 214–223.

El-Khoury, N., Braun, A., Hu, F., Pandey, M., Nedergaard, M., Lagamma, E.F. and Ballabh, P. (2006). Astrocyte end-feet in germinal matrix, cerebral cortex, and white matter in developing infants. *Pediatr. Res.*, 59, 673-679.

Englund, E. (1998). Neuropathology of white matter changes in Alzheimer's disease and vascular dementia. *Dement. Geriatr. Cogn. Disord., 9,* 6–12.

Fazekas, F., Chawluk, J.B., Alavi, A., Hurtig, H.I. and Zimmerman, R.A. (1987). MR signal abnormalities at 1.5 T in Alzheimer's dementia and normal aging. *AJR Am. J. Roentgenol., 149,* 351-356.

Fazekas, F., Kleinert, R., Offenbacher, H., Schmidt, R., Kleinert, G., Payer, F., Radner, H. and Lechner, H. (1993). Pathologic correlates of incidental MRI white matter signal hyperintensities. *Neurology, 43,* 1683-1689.

Fern, R. and Moller, T. (2000). Rapid ischemic cell death in immature oligodendrocytes: a fatal glutamate release feedback loop. *J. Neurosci., 20,* 34–42.

Fernando, M.S., Simpson, J.E., Matthews, F., Brayne, C., Lewis, C.E., Barber, R., Kalaria, R.N., Forste, G., Esteves, F., Wharton, S.B., Shaw, P.J., O'Brien, J.T. and Ince, P.G;

MRC Cognitive Function and Ageing Neuropathology Study Group. (2006). White matter lesions in an unselected cohort of the elderly: molecular pathology suggests origin from chronic hypoperfusion injury. *Stroke, 37,* 1391-1398.

Filley, C.M. (2005). White matter and behavioral neurology. *Ann. N. Y. Acad. Sci., 1064,* 162-183.

Fischer, S., Wobben, M., Marti, H.H., Renz, D. and Schaper, W. (2002). Hypoxia-induced hyperpermeability in brain microvessel endothelial cells involves VEGF-mediated changes in the expression of zonula occludens-1. *Microvasc. Res., 63,* 70-80.

Follett, P.L., Deng, W., Dai, W., Talos, D.M., Massillon, L.J., Rosenberg, P.A., Volpe, J.J. and Jensen, F.E. (2004). Glutamate receptor-mediated oligodendrocyte toxicity in periventricular leukomalacia: a protective role for topiramate. *J Neurosci, 24,* 4412-4420.

Folkerth, R.D. (2006). Periventricular leukomalacia: overview and recent findings. *Pediatr. Dev. Pathol., 9,* 3-13.

Fischer, S., Wobben, M., Marti, H.H., Renz, D. and Schaper, W. (2002). Hypoxia-induced hyperpermeability in brain microvessel endothelial cells involves VEGF-mediated changes in the expression of zonula occludens-1. *Microvasc. Res., 63,* 70-80.

Froen, J.F., Amerio, G., Stray-Pedersen, B. and Saugstad, O.D. (2002). Detrimental effects of nicotine and endotoxin in the newborn piglet brain during severe hypoxemia. *Biol. Neonate., 82*, 188-196.

Fu, L., Matthews, P.M., De Stefano, N., Worsley, K.J., Narayanan, S., Francis, G.S., Antel, J.P., Wolfson, C. and Arnold, D.L. (1998). Imaging axonal damage of normal-appearing white matter in multiple sclerosis. *Brain, 121,* 103-113.

Fukumura, D., Gohongi, T., Kadambi, A., Izumi, Y., Ang, J., Yun, C.O., Buerk , D.G., Huang, P.L. and Jain, R.K. (2001). Predominant role of endothelial nitric oxide synthase in vascular endothelial growth factor-induced angiogenesis and vascular permeability. *Proc. Natl. Acad. Sci. USA, 98,* 2604-2609.

Ghirnikar, R., Lee, Y. and Eng, L. (2001). Chemokine antagonist infusion promotes axonal sparing after spinal cord contusion injury in rat. *J. Neurosci. Res., 64,* 582-589.

Goldberg, M.P. and. Ransom, B.R. (2003). New Light on White Matter. *Stroke, 34,* 330-332.

Gootjes, L., Scheltens, P., Van Strien, J.W. and Bouma, A. (2007). Subcortical white matter pathology as a mediating factor for age-related decreased performance in dichotic listening. *Neuropsychologia, 45,* 2322-2332.

Gorman, D.F., Huang, Y.L. and Williams. C. (2001). Prolonged exposure to one percent carbon monoxide causes a leucoencephalopathy in un-anaesthetised sheep. *Toxicology, 165,* 97-107.

Grade, E., Mortensen, E.L., Krabbe, K., Rostrup, E. and Larsson, H.B. (2000). Relation between age-related decline in intelligence and cerebral white-matter hyperintensities in healthy octogenarians: a longitudinal study. *Lancet, 356,* 628-634.

Grafton, S.T., Sumi, S.M., Stimac, G.K., Alvord, E.C. Jr., Shaw, C.M. and Nochlin, D. (1991). Comparison of postmortem magnetic resonance imaging and neuropathologic findings in the cerebral white matter. *Arch. Neurol., 48,* 293-298.

Gran, B, and Rostami, A. (2001). T cells, cytokines, and autoantigens in multiple sclerosis. *Curr. Neurol. Neurosci. Rep., 1,* 263-270.

Grinker, R.R. (1925). Über einen Fall von Leuchtgasvergiftung mit doppelseitiger, Pallidumerweichung und schwerer degeneration des tieferen Großhirnmarklagers,*Z. ges. Neurol. Psychiat.*, *98,* 433-456.

Groom, A.J., Smith, T. and Turski, L. (2003). Multiple sclerosis and glutamate. *Ann. N. Y. Acad. Sci., 993,*229-275.

Hampel, H., Teipel, S.J., Alexander, G.E., Horwitz, B., Teichberg, D., Schapiro, M.B. and Rapoport, S.I. (1998). Corpus callosum atrophy is a possible indicator of region- and cell type-specific neuronal degeneration in Alzheimer disease: a magnetic resonance imaging analysis. *Arch. Neurol., 55,* 193-198.

Harrell, T.H., Honaker, L.M. and Davis, E. (1991). Cognitive and behavioral dimensions of dysfunction in alcohol and polydrug abusers. *J. Subst. Abuse., 3,* 415-426.

Harris, J.G., and Filley, C.M. (2001). CADASIL: neuropsychological findings in three generations of an affected family. *J. Int. Neuropsychol. Soc., 7,* 768-774.

Hess, D.C., Bhutwala, T., Sheppard, J. C., Zhao, W. and Smith. J. (1994). ICAM-1 expression on human brain microvascular endothelial cells. *Neurosci. Lett., 168*, 201.

Hill, K.E., Zollinger, L.V., Watt, H.E., Carlson, N.G. and Rose, J.W. (2004). Inducible nitric oxide synthase in chronic active multiple sclerosis plaques: distribution, cellular expression and association with myelin damage. *J. Neuroimmunol., 151,* 171-179.

Hortobagyi, T., Wise, S., Hunt, N., Cary, N., Djurovic, V., Fegan-Earl, A., Shorrock, K, Rouse, D. and Al-Sarraj, S. (2007). Traumatic axonal damage in the brain can be detected using beta-APP immunohistochemistry within 35 min after head injury to human adults. *Neuropathol. Appl. Neurobiol., 33,* 226-237.

Inder, T.E. and Volpe, J.J. (2000). Mechanisms of perinatal brain injury. *Semin. Neonatol., 5,* 3-16.

Jack, C., Ruffini, F., Bar-Or, A. and Antel, J.P. (2005). Microglia and multiple sclerosis. *J Neurosci. Res., 81,* 363-373.

Janigro, D., West, G.A., Nguyen, T.S., and Winn, H.R. (1994). Regulation of blood-brain barrier endothelial cells by nitric oxide. *Circ. Res., 75,* 528-538.

Jelinski, S.E., Yager, J.Y. and Juurlink, B.H. (1999). Preferential injury of oligodendroblasts by a short hypoxic-ischemic insult. *Brain Res., 81,* 150-153.

Jensen, F.E. (2002). The role of glutamate receptor maturation in perinatal seizures and brain injury. *Int. J. Dev. Neurosci., 20,* 339-347.

Jensen, A., Garnier, Y., Middelanis, J. and Berger, R. (2003). Perinatal brain damage--from pathophysiology to prevention. *Eur. J. Obstet. Gynecol. Reprod. Biol., 110, Suppl 1,* S70-79.

Jensen, F. E. (2005). Role of glutamate receptors in periventricular leukomalacia. *J. Child Neurol., 20,* 950–959.

Jiang, X., Mu, D., Sheldon, R.A., Glidden, D.V. and Ferriero, D.M. (2003). Neonatal hypoxia-ischemia differentially upregulates MAGUKs and associated proteins in PSD-93-deficient mouse brain. *Stroke, 34,* 2958-2963.

Johnston, M.V. (1997). Hypoxic and ischemic disorders of infants and children. Lecture for 38th meeting of Japanese Society of Child Neurology, Tokyo, Japan, July 1996. *Brain Dev., 19,* 235-239.

Johnston, M.V. (2001). Excitotoxicity in neonatal hypoxia. *Ment. Retard. Dev. Disabil. Res. Rev., 7,* 229-234.

Johnston, M.V. (2005). Excitotoxicity in perinatal brain injury. *Brain Pathol., 15,* 234-240.

Kadhim, H., Tabarki, B., Verellen, G., De Prez, C., Rona, A.M. and Sebire, G. (2001). Inflammatory cytokines in the pathogenesis of periventricular leukomalacia. *Neurology, 56,* 1278-1284.

Kadhim, H., Tabarki, B., De Prez, C., Rona, A.M. and Sebire, G. (2002). Interleukin-2 in the pathogenesis of perinatal white matter damage. *Neurology, 58,* 1125-1128.

Kadhim, H., Tabarki, B., De Prez, C. and Sebire, G. (2003). Cytokine immunoreactivity in cortical and subcortical neurons in periventricular leukomalacia: are cytokines implicated in neuronal dysfunction in cerebral palsy? *Acta Neuropathol. (Berl), 105,* 209-216.

Kadhim, H., Khalifa, M., Deltenre, P., Casimir, G. and Sebire, G. (2006). Molecular mechanisms of cell death in periventricular leukomalacia. *Neurology, 67,* 293-299.

Kalimo, H., Viitanen, M., Amberla, K., Juvonen, V., Marttila, R., Poyhonen, M,, Rinne, J.O, Savontaus, M., Tuisku, S. and Winblad, B. (1999). CADASIL: hereditary disease of arteries causing brain infarcts and dementia. *Neuropathol. Appl. Neurobiol., 25,* 257-265.

Karadottir, R., Cavelier, P., Bergersen, L. H. and Attwell, D. (2005). NMDA receptors are expressed in oligodendrocytes and activated in ischaemia. *Nature, 438,* 1162–1166.

Kaur, C. and You, Y. (2000). Ultrastructure and function of the amoeboid microglial cells in the periventricular white matter in postnatal rat brain following a hypoxic exposure. *Neurosci. Lett., 290,* 17-20.

Kaur, C., Sivakumar, V., Ang, L.S. and Sundaresan, A. (2006). Hypoxic damage to the periventricular white matter in neonatal brain: role of vascular endothelial growth factor, nitric oxide and excitotoxicity. *J. Neurochem., 98,* 1200-1216.

Kim, W.K. and Ko, K.H. (1998). Potentiation of *N*-methyl-D-aspartate-mediated neurotoxicity by immunostimulated murine microglia. *J. Neurosci. Res., 54,* 17-26.

Kim, J.H., Chang, K.H., Song, I.C., Kim, K.H., Kwon, B.J., Kim, H.C., Kim, J.H. and Han, M.H. (2003). Delayed encephalopathy of acute carbon monoxide intoxication: diffusivity of cerebral white matter lesions. *AJNR Am. J. Neuroradiol., 24,* 1592-1597.

Kirk, J., Plumb, J., Mirakhur, M. and McQuaid, S. (2003). Tight junctional abnormality in multiple sclerosis white matter affects all calibres of vessel and is associated with blood-brain barrier leakage and active demyelination. *J. Pathol., 201,* 319-327.

Koprowski, H., Zheng, Y.M., Heber-Katz E., Fraser, N., Rorke, L., Fu, Z.F., Hanlon, C. and Dietzschold, B. (1993). In vivo expression of inducible nitric oxide synthase in experimentally induced neurologic diseases. *Proc. Natl. Acad. Sci. USA, 90,* 3024-3027.

Kozachuk, W.E., DeCarli, C., Schapiro, M.B., Wagner, E.E., Rapoport, S.I. and Horwitz, B. (1990). White matter hyperintensities in dementia of Alzheimer's type and in healthy subjects without cerebrovascular risk factors. A magnetic resonance imaging study. *Arch. Neurol., 47,* 1306-1310.

Kurumatani, T., Kudo, T., Ikura, Y. and Takeda, M. (1998). White matter changes in the gerbil brain under chronic cerebral hypoperfusion. *Stroke, 29,* 1058-1062.

Ladiwala, U., Hewei, L., Antel, J. and Nalbantoglu, J. (1999). P53 induction by tumor necrosis factor-a and involvement of p53 in cell death of human oligodendrocytes. *J. Neurochem., 73,* 605–611.

Leech, S., Kirk, J., Plumb, J. and McQuaid, S. (2007). Persistent endothelial abnormalities and blood-brain barrier leak in primary and secondary progressive multiple sclerosis. *Neuropathol. Appl. Neurobiol., 33,* 86-98.

Leifer, D., Buonanno, F.S. and Richardson, E.P. Jr. (1990). Clinicopathologic correlations of cranial magnetic resonance imaging of periventricular white matter. *Neurology, 40,* 911-918.

Leys, D., Soetaert, G., Petit, H., Fauquette, A., Pruvo, J.P. and Steinling, M. (1990). Periventricular and white matter magnetic resonance imaging hyperintensities do not differ between Alzheimer's disease and normal aging. *Arch. Neurol., 47,* 524-527.

Li, J., Baud, O., Vartanian, T., Volpe, J.J. and Rosenberg, P.A. (2005). Peroxynitrite generated by inducible nitric oxide synthase and NADPH oxidase mediates microglial toxicity to oligodendrocytes. *Proc. Natl. Acad. Sci. USA, 102,* 9936-9941.

Liang, M., Mallari, C., Rosser, M., et al. (2000). Identification and characterization of a potent, selective, and orally active antagonist of the CC chemokine receptor-1. *J. Biol. Chem., 275,* 19000-19008.

Lindgren, A., Roijer, A., Rudling, O., Norrving, B., Larsson, E.M., Eskilsson, J., Wallin, L., Olsson, B. and Johansson, B.B. (1994). Cerebral lesions on magnetic resonance imaging, heart disease, and vascular risk factors in subjects without stroke. A population-based study. *Stroke, 25,* 929-934.

Ling, E.A. and Wong, W.C. (1993). The origin and nature of ramified and amoeboid microglia: a historical review and current concepts. *Glia, 7,* 9-18.

Liu, Y., Silverstein, F.S., Skoff, R. and Barks, J.D. (2002). Hypoxic-ischemic oligodendroglial injury in neonatal rat brain. *Pediatr. Res., 51,* 25-33.

Lopez, O.L., Becker, J.T., Rezek, D., Wess, J., Boller, F., Reynolds, C.F. 3rd. and Panisset, M. (1992). Neuropsychiatric correlates of cerebral white-matter radiolucencies in probable Alzheimer's disease. *Arch. Neurol., 49,* 828-834.

Lopez-Figueroa, M.O., Day, H.E., Lee, S., Rivier, C., Akil, H. and Watson, S.J. (2000). Temporal and anatomical distribution of nitric oxide synthase mRNA expression and nitric oxide production during central nervous system inflammation. *Brain Res., 852,* 239-246.

Louis, J.C., Magal, E., Takayama, S. and Varon, S. (1993). CNTF protection of oligodendrocytes against natural and tumor necrosis factor-induced death. *Science, 259,* 689-692.

Luyt, K., Varadi, A. and Molnar, E. (2003). Functional metabotropic glutamate receptors are expressed in oligodendrocyte progenitor cells. *J. Neurochem., 84,* 1452–1464.

Matute, C., Sanchez-Gomez, M.V., Martinez-Millan, L. and Miledi, R. (1997). Glutamate receptor-mediated toxicity in optic nerve oligodendrocytes. *Proc. Natl. Acad. Sci. USA, 94,* 8830–8835.

Matute, C., Alberdi, E., Domercq, M., Perez-Cerda, F., Perez-Samartin, A. Sanchez-Gomez, M.V. (2001). The link between excitotoxic oligodendroglial death and demyelinating diseases. *Trends Neurosci., 24,* 224–230.

McCracken, E., Fowler, J.H., Dewar, D., Morrison, S. and McCulloch, J. (2002). Grey matter and white matter ischemic damage is reduced by the competitive AMPA receptor antagonist, SPD 502. *J. Cereb. Blood Flow Metab., 22,* 1090–1097.

McDonald, J.W., Althomsons, S.P., Hyrc, K.L., Choi, D.W. and Goldberg. M.P. (1998). Oligodendrocytes from forebrain are highly vulnerable to AMPA/kainate receptor-mediated excitotoxicity. *Nat. Med., 4,* 291–297.

McHale, J.F., Harari, O.A., Marshall, D. and Haskard, D.O. (1999). TNF- α and IL-1 Sequentially Induce Endothelial ICAM-1 and VCAM-1 Expression in MRL/*lpr* Lupus-Prone Mice. *J. Immunol., 163,* 3993-4000.

Medana, I.M., Hunt, N.H. and Chaudhri, G. (1997). Tumor necrosis factor-alpha expression in the brain during fatal murine cerebral malaria: evidence for production by microglia and astrocytes. *Am. J. Pathol., 150,* 1473−1486.

Meng, S.Z., Arai, Y., Deguchi, K. and Takashima, S. (1997). Early detection of axonal and neuronal lesions in prenatal-onset periventricular leukomalacia. *Brain Dev., 19,* 480-484.

Ment, L.R., Stewart, W.B., Fronc, R., Seashore, C., Mahooti, S., Scaramuzzino, D. and Madri, J.A. (1997). Vascular endothelial growth factor mediates reactive angiogenesis in the postnatal developing brain. *Brain Res. Dev. Brain Res., 100,* 52-61.

Merrill, J.E. (1991). Effects of interleukin-1 and tumor necrosis factor-alpha on astrocytes, microglia, oligodendrocytes, and glial precursors in vitro. *Dev. Neurosci., 13,* 130-137.

Merrill, J.E., Ignarro, L.J., Sherman, M.P., Melinek, J. and Lane, T.E. (1993). Microglial cell cytotoxicity of oligodendrocytes is mediated through nitric oxide. *J. Immunol., 151,* 2132-2141.

Merrill, J.E. and Benveniste, E.N. (1996). Cytokines in inflammatory brain lesions: helpful and harmful. *Trends Neurosci., 19,* 331-338.

Micu, I., Jiang, Q., Coderre, E., *et al.* (2006). NMDA receptors mediate calcium accumulation in myelin during chemical ischaemia. *Nature,* 439, 988–992.

Min, J.K., Lee, Y.M., Kim, J.H., Kim, Y.M., Kim, S.W., Lee, S.Y., Gho, Y.S., Oh, G.T. and Kwon, Y.G. (2005). Hepatocyte growth factor suppresses vascular endothelial growth factor-induced expression of endothelial ICAM-1 and VCAM-1 by inhibiting the nuclear factor-kappaB pathway. *Circ. Res., 96,* 300-307.

Mirsen, T.R., Lee, D.H., Wong, C.J., Diaz, J.F., Fox, A.J., Hachinski, V.C. and Merskey H. (1991). Clinical correlates of white-matter changes on magnetic resonance imaging scans of the brain. *Arch. Neurol., 48,* 1015-1021.

Moody, D.M., Brown, W.R., Challa, V.R., Ghazi-Birry, H.S. and Reboussin, D.M. (1997). Cerebral microvascular alterations in aging, leukoaraiosis, and Alzheimer's disease. *Ann. N. Y. Acad. Sci., 826,*103-116.

Mulhern, R.K., Palmer, S.L., Reddick, W.E., Glass, J.O., Kun, L.E., Taylor, J., Langston, J. and Gajjar, A. (2001). Risks of young age for selected neurocognitive deficits in medulloblastoma are associated with white matter loss. *J. Clin. Oncol., 19,* 472-479.

Munoz, J.E., Vinolas, N., Castro, J., Graus, F., Conde, C., Castel, T., Biete, A., Tolosa, E. and Estape, J. (1993). [Cerebral metastasis of melanoma: study of 48 patients]. *Med. Clin. (Barc), 101,* 684-687.

Navia, B.A., Cho, E.S., Petito, C.K. and Price, R.W. (1986). The AIDS dementia complex: II. Neuropathology. *Ann. Neurol., 19,* 525-535.

Ness, J.K., Romanko, M.J., Rothstein, R.P., Wood, T.L. and Levison, S.W. (2001). Perinatal hypoxia-ischemia induces apoptotic and excitotoxic death of periventricular white matter oligodendrocyte progenitors. *Dev. Neurosci., 23,* 203-208.

Neubuerger, K.T. and Clarke, E.R. (1945). Subacute carbon monoxide poisoning with cerebral myelinopathy and multiple myocardial necroses. *Rocky Mountain. Med. J., 42,* 29-34.

Noda, M., Nakanishi, H., Nabekura, J. and Akaike, N. (2000). AMPA-kainate subtypes of glutamate receptor in rat cerebral microglia. *J. Neurosci., 20,* 251-258.

O'Sullivan, M., Jones, D.K., Summers, P.E., Morris, R.G., Williams, S.C. and Markus, H.S. (2001a). Evidence for cortical "disconnection" as a mechanism of age-related cognitive decline. *Neurology, 57,* 632–638.

O'Sullivan, M., Summers, P.E., Jones, D.K., Jarosz, J.M., Williams, S.C. and Markus, H.S. (2001b). Normal-appearing white matter in ischemic leukoaraiosis: a diffusion tensor MRI study. *Neurology, 57,* 2307–2310.

Oka, A., Belliveau, M.J., Rosenberg, P.A. and Volpe, J.J. (1993). Vulnerability of oligodendroglia to glutamate: pharmacology, mechanisms, and prevention. *J. Neurosci., 13,* 1441-1453.

Onaya, M. (2002). Neuropathological investigation of cerebral white matter lesions caused by closed head injury. *Neuropathology, 22,* 243-251.

Palmer, C., Menzies, S.L., Roberts, R.L., Pavlick, G. and Connor, J.R. (1999). Changes in iron histochemistry after hypoxic-ischemic brain injury in the neonatal rat. *J. Neurosci. Res., 56,* 60-71.

Pantoni, L. and Garcia, J.H. (1996). The significance of cerebral white matter abnormalities 100 years after Binswanger's report: a review. *Stroke, 26,* 1293-1301.

Pantoni, L., Garcia, J.H. and Gutierrez, J.A. (1996). Cerebral white matter is highly vulnerable to ischemia. *Stroke, 27,* 1641-1646.

Pantoni, L. and Garcia, J.H. (1997). Pathogenesis of leukoaraiosis. *Stroke,* 28, 652-659.

Patneau, D. K., Wright, P. W., Winters, C., Mayer, M. L. and Gallo, V. (1994). Glial cells of the oligodendrocyte lineage express both kainate- and AMPA-preferring subtypes of glutamate receptor. *Neuron, 12,* 357-371.

Pende, M., Holtzclaw, L. A., Curtis, J. L., Russel, J. T. and Gallo V. (1994) Glutamate regulates intracellular calcium and gene expression in oligodendrocyte progenitors through the activation of dl-α-amino-3-hydroxy-5-methyl-4-isoxazolepropionic acid receptors. *Proc. Natl. Acad. Sci. USA, 91,* 3215–3219.

Peterson, J.W., Bo, L., Mork, S., Chang, A., Ransohoff, R.M. and Trapp, B.D. (2002). VCAM-1-positive Microglia Target Oligodendrocytes at the Border of Multiple Sclerosis Lesions. *J. Neuropathol. Exp. Neurol., 61,* 539-546.

Piccini, P., Pavese, N., Canapicchi, R., Paoli, C., Del Dotto, P., Puglioli, M., Rossi, G. and Bonuccelli, U. (1995). White matter hyperintensities in Parkinson's disease. Clinical correlations. *Arch. Neurol.,* 52, 191-194.

Pitt, D., Werner, P. and Raine, C.S. (2000). Glutamate excitotoxicity in a model of multiple sclerosis. *Nat. Med., 6,* 67–70.

Pleasure, D., Soulika, A., Singh, S.K., Gallo, V. and Bannerman, P. (2006). Inflammation in white matter: clinical and pathophysiological aspects. *Ment. Retard. Dev. Disabil. Res. Rev., 12,* 141-146.

Plumb, J., McQuaid, S., Mirakhur, M. and Kirk, J. (2002). Abnormal endothelial tight junctions in active lesions and normal-appearing white matter in multiple sclerosis. *Brain Pathol., 12,* 154-169.

Proescholdt, M.A., Heiss, J.D., Walbridge, S., Muhlhauser, J., Capogrossi, M.C., Oldfield, E.H. and Merrill, M.J. (1999). Vascular endothelial growth factor (VEGF) modulates vascular permeability and inflammation in rat brain. *J. Neuropathol. Exp. Neurol., 58, 613*-627.

Qi, Y. and Dal Canto, M.C. (1996). Effect of Theiler's murine encephalomyelitis virus and cytokines on cultured oligodendrocytes and astrocytes. *J. Neurosci. Res., 45,* 364-374.

Revesz, T., Hawkins,C.P., du Boulay, E.P., Barnard, R.O. and McDonald, W.I. (1989). Pathological findings correlated with magnetic resonance imaging in subcortical arteriosclerotic encephalopathy (Binswanger's disease). *J. Neurol Neurosurg. Psychiatry., 52,* 1337-1344.

Rieckmann, P. and Smith, K.J. (2001). Multiple sclerosis: more than inflammation and demyelination. *Trends Neurosci., 24,* 435-437.

Rieckmann, P. and Mäurer, M. (2002). Anti-inflammatory strategies to prevent axonal injury in multiple sclerosis. *Curr. Opin. Neurol., 15,* 361-370.

Rose, S.E., Chen, F., Chalk, J.B., Zelaya, F.O., Strugnell, W.E., Benson, M., Semple, J. and Doddrell, D.M. (2000). Loss of connectivity in Alzheimer's disease: an evaluation of white matter tract integrity with colour coded MR diffusion tensor imaging. *J. Neurol. Neurosurg. Psychiatry., 69,* 528–530.

Rovaris, M., Iannucci, G., Falautano, M., Possa, F., Martinelli, V., Comi, G. and Filippi, M. (2002). Cognitive dysfunction in patients with mildly disabling relapsing-remitting multiple sclerosis: an exploratory study with diffusion tensor MR imaging. *J. Neurol. Sci., 195,*103–109.

Sawada, M., Suzumura, A. and Marunouchi, T. (1992). TNF alpha induces IL-6 production by astrocytes but not by microglia. *Brain Res., 583,* 296-299.

Sawada, M., Suzumura, A. and Marunouchi, T. (1995). Cytokine network in the central nervous system and its roles in growth and differentiation of glial and neuronal cells. *Int. J. Dev. Neurosci., 13,* 253-264.

Scarpelli, M., Salvolini, U., Diamanti, L., Montironi, R., Chiaromoni, L. and Maricotti M. (1994). MRI and pathological examination of post-mortem brains: the problem of white matter high signal areas. *Neuroradiology, 36,* 393-398.

Schmidt, R. (1992). Comparison of magnetic resonance imaging in Alzheimer's disease, vascular dementia and normal aging. *Eur. Neurol., 32,* 164-169.

Schoch, H.J., Fischer, S. and Marti, H.H. (2002). Hypoxia-induced vascular endothelial growth factor expression causes vascular leakage in the brain. *Brain, 125,* 2549-2557.

Selmaj, K. and Raine, C.S. (1988). Tumor necrosis factor mediates myelin and oligodendrocyte damage in vitro. *Ann. Neurol., 21,* 339-346.

Shukla, A., Dikshit, M. and Srimal, R.C. (1996). Nitric oxide-dependent blood-brain barrier permeability alteration in the rat brain. *Experientia, 52,* 136-140.

Silverstein, F.S., Naik, B. and Simpson, J. (1991). Hypoxia-ischemia stimulates hippocampal glutamate efflux in perinatal rat brain: an in vivo microdialysis study. *Pediatr. Res., 30,* 587-590.

Sjöbeck, M. and Englund, E. (2003). Glial levels determine severity of white matter disease in Alzheimer's disease: a neuropathological study of glial changes. *Neuropathol. Appl. Neurobiol., 29,*159-169.

Skoff, R.P. and Knapp, P.E. (1991). Division of astroblasts and oligodendroblasts in postnatal rodent brain: evidence for separate astrocyte and oligodendrocyte lineages. *Glia, 4,* 165-174.

Skoff, R.P., Bessert, D.A., Barks, J.D., Song, D., Cerghet, M. and Silverstein, F.S. (2001). Hypoxic-ischemic injury results in acute disruption of myelin gene expression and death of oligodendroglial precursors in neonatal mice. *Int. J. Dev. Neurosci., 19,* 197-208.

Skoog, I., Aevarsson, O., Beskow, J., Larsson, L., Palsson, S., Waern, M., Landahl, S. and Ostling, S. (1996). Suicidal feelings in a population sample of nondemented 85-year-olds. *Am. J. Psychiatry, 153,* 1015-1020.

Smith, T., Groom, A., Zhu, B. and Turski, L. (2000). Autoimmune encephalomyelitis ameliorated by AMPA antagonists. *Nat. Med., 6,* 62–66.

Smith, K., Kapoor, R., Hall, S. and Davies, M. (2001). Electrically active axons degenerate when exposed to nitric oxide. *Ann. Neurol., 49,* 470-476.

Sridhar, K., Kumar, P., Katariya, S. and Narang, A. (2001). Postasphyxial encephalopathy in preterm neonates. *Indian J. Pediatr, 68,* 1121-1125.

Steffens, D.C., Bosworth, H.B., Provenzale, J.M. and MacFall, J.R. (2002). Subcortical white matter lesions and functional impairment in geriatric depression. *Depress Anxiety, 15,* 23–28.

Stys, P.K. (1998). Anoxic and ischemic injury of myelinated axons in CNS white matter: from mechanistic concepts to therapeutics. *J. Cereb. Blood Flow Metab., 18,* 2-25.

Suhy, J., Rooney, W.D., Goodkin, D.E., Capizzano, A.A., Soher, B.J., Maudsley, A.A., Waubant, E., Andersson, P.B. and Weiner MW. (2000). 1H MRSI comparison of white matter and lesions in primary progressive and relapsing-remitting MS. *Mult. Scler.*, 6, 148-155.

Sze, G., De Armond, S.J., Brant-Zawadzki, M., Davis, R.L., Norman, D. and Newton, T.H. (1986). Foci of MRI signal (pseudo lesions) anterior to the frontal horns: histologic correlations of a normal finding. *AJR Am. J. Roentgenol., 147,* 331-337.

Tahraoui, S.L., Marret, S., Bodenant, C., Leroux, P., Dommergues, M.A, Evrard, P. and Gressens, P. (2001). Central role of microglia in neonatal excitotoxic lesions of the murine periventricular white matter. *Brain Pathol., 11,* 56-71.

Takashima, S.D., Armstrong, L. and Becker, L. E. (1978). Subcortical leukomalacia. *Arch. Neurol., 35,* 470–472.

Takashima, S., Iida, K. and Deguchi, K. (1995). Periventricular leukomalacia, glial development and myelination. *Early Hum. Dev., 43,* 177-184.

Thiel, V.E. and Audus, K.L. (2001). Nitric oxide and blood-brain barrier integrity. *Antioxid. Redox. Signal, 3,* 73-278.

Tekkok, S.B. and Goldberg, M.P. (2001). Ampa/kainate receptor activation mediates hypoxic oligodendrocyte death and axonal injury in cerebral white matter. *J. Neurosci., 21*, 4237-4248.

Tomimoto, H., Akiguchi, I., Wakita, H., Suenaga, T., Nakamura, S. and Kimura, J. (1997). Regressive changes of astroglia in white matter lesions in cerebrovascular disease and Alzheimer's disease patients. *Acta Neuropathol. (Berl), 94,* 146-152.

Tomimoto, H., Akiguchi, I., Ohtani, R., Yagi, H., Kanda, M., Shibasaki, H. and Yamamoto, Y. (2001). The coagulation-fibrinolysis system in patients with leukoaraiosis and Binswanger disease. *Arch. Neurol., 58,* 1620-1625.

Tsiskaridze, A., Shakarishvili, R., Janelidze, M., Vashadze, T. and Chikhladze, M. (1998). Cognitive correlates of leukoaraiosis in the early stages of Alzheimer's disease. *Funct. Neurol., 13,* 17-25.

Tsuji, M., Saul, J.P., du Plessis, A., Eichenwald, E., Sobh, J., Crocker, R. and Volpe. J.J. (2000). Cerebral intravascular oxygenation correlates with mean arterial pressure in critically ill premature infants. *Pediatrics, 106,* 625-632.

Tubridy, N., Behan, P., Capildeo, R., et al. (1998).The effect of anti-alpha4 integrin antibody on brain lesion activity in MS. The UK Antegren Study Group. *Neurology, 53,* 466-472.

van Horssen, J., Brink, B.P., de Vries, H.E., van der Valk, P. and Bø, L. (2007). The blood-brain barrier in cortical multiple sclerosis lesions. *J. Neuropathol. Exp. Neurol., 66,*321-328.

van Swieten, J.C., van den Hout, J.H., van Ketel, B.A., Hijdra, A., Wokke, J.H. and van Gijn, J. (1991). Periventricular lesions in the white matter on magnetic resonance imaging in the elderly: a morphometric correlation with arteriolosclerosis and dilated perivascular spaces. *Brain, 114,*761-774.

Vermersch, P., Roche, J., Hamon, M., Daems-Monpeurt, C., Pruvo, J.P., Dewailly, P. and Petit, H. (1996). White matter magnetic resonance imaging hyperintensity in Alzheimer's disease: correlations with corpus callosum atrophy. *J. Neurol., 243,* 231-234.

Volpe, J.J. (2001). Neurobiology of periventricular leukomalacia in the premature infant. *Pediatr Res., 50,* 553-562.

Volpe, J.J. (2003). Cerebral white matter injury of the premature infant—more common than you think. *Pediatrics ,112,*176-180.

Wang, X., Siren, A.L., Liu, Y., Yue, T.L., Barone, F.C. and Feuerstein, G.Z. (1994). Upregulation of intercellular adhesion molecule 1 (ICAM-1) on brain microvascular endothelial cells in rat ischemic cortex. *Brain Res. Mol. Brain Res., 26,* 61–68.

Werner, P., Pitt, D. and Raine, C.S. (2000). Glutamate excitotoxicity--a mechanism for axonal damage and oligodendrocyte death in Multiple Sclerosis? *J. Neural. Transm. Suppl., 60,* 375-385.

Whitford, T.J., Grieve, S.M., Farrow, T.F., Gomes, L., Brennan, J., Harris, A.W., Gordon, E. and Williams, L.M. (2007). Volumetric white matter abnormalities in first-episode schizophrenia: a longitudinal, tensor-based morphometry study. *Am. J. Psychiatry., 164,* 1082-1089.

Wong, D., Dorovini Zis, K. (1992). Upregulation of intercellular adhesion molecule-1 (ICAM-1) expression in primary cultures of human brain microvessel endothelial cells by cytokines and lipopolysaccharide. *J. Neuroimmunol., 39,* 11.

Wrathall, J.R., Choiniere, D. and Teng, Y.D. (1994). Dose-dependent reduction of tissue loss and functional impairment after spinal cord trauma with the AMPA/kainate antagonist NBQX. *J. Neurosci., 14,* 6598–6607.

Wu, C.Y., Lu, J., Cao, Q., Guo, C.H., Gao, Q. and Ling, E.A. (2006). Expression of 2',3'-cyclic nucleotide 3'-phosphodiesterase in the amoeboid microglial cells in the developing rat brain. *Neuroscience, 142,* 333-341.

Ye, P. and D'Ercole, A. (1999). Insulin-like growth factor I protects oligodendrocytes from tumor necrosis factor-a induced injury. *Endocrinology, 140,* 3063–3072.

Ylikoski, R., Ylikoski, A., Erkinjuntti, T., Sulkava, R., Raininko, R. and Tilvis, R. (1993). White matter changes in healthy elderly persons correlate with attention and speed of mental processing. *Arch. Neurol., 50,* 818-824.

Ylikoski, A., Erkinjuntti, T., Raininko, R., Sarna, S., Sulkava, R. and Tilvis, R. (1995). White matter hyperintensities on MRI in the neurologically nondiseased elderly. Analysis of cohorts of consecutive subjects aged 55 to 85 years living at home. *Stroke, 26,* 1171-1177.

Yoon, B.H., Kim, C.J., Romero, R., Jun, J.K., Park, K.H., Choi, S.T. and Chi, J.G. (1997). Experimentally induced intrauterine infection causes fetal brain white matter lesions in rabbits. *Am. J. Obstet. Gynecol., 177,* 797-802.

Yoshioka, H., Goma, H., Nioka, S, *et al.* (1995). Bilateral carotid artery occlusion causes periventricular leukomalacia in neonatal dogs. *Dev. Brain Res., 78,* 273-278.

Yoshioka, A., Yamaya, Y., Saiki, S., Kanemoto, M., Hirose, G., Beesley, J. and Pleasure, D. (2000). Non-N-methyl-D-aspartate glutamate receptors mediate oxygen–glucose deprivation-induced oligodendroglial injury. *Brain Res., 854,*207–215.

Zhang, K. and Sejnowski, T.J. (2000). A universal scaling law between gray matter and white matter of cerebral cortex. *Proc. Natl. Acad. Sci. USA, 97,* 5621–5626.

In: Neural Pathways Research
Editor: F. L. Pichler, pp. 157-174
ISBN: 978-1-60456-214-9

Chapter VII

Diffusion Tractography of the Human Memory System

Emi Takahashi*
Department of Radiology,
Harvard Medical School, Massachusetts General Hospital,
A. A. Martinos Cneter for Biomedical Imaging,
Building 149, 13th Street, Room 2301
Charlestown, MA 02129

Abstract

White matter neuronal fiber occupies most of the brain and is one of the bases of our cognitive functions. Recently, diffusion imaging has been applied to the study of neuronal fiber pathways, using tractography algorithms. With this technique, anatomical connections subserving higher cognitive functions in humans can be studied in a whole brain *in vivo*, in a relatively short amount of time. However, there are still a number of concerns that need to be addressed concerning the accuracy of current tractography techniques. Because of the limitations of today's methods, we need to be cautious in how we design tractography studies. Our recent experimental data demonstrates there are large-scale anatomical networks that serve as a basis of episodic memory in humans. In particular there are two anatomical pathways between frontal and temporal lobes. The existence of segregated lateral and medial parieto-temporal pathways in successful memory retrieval was also suggested. Here I review a selection of designs and models currently being used in tractography to study cognitive functions. I also discuss future applications of tractography.

* Email: emi@nmr.mgh.harvard.edu; Phone: +1-617-304-4100.

Detection of Whole Brain White Matter Connections

Recently, diffusion weighted imaging (DWI) was applied to study white matter neuronal connections. DWI is a technique that measures the diffusion properties of water molecules (Basser et al., 1994). Tractography algorithms are applied to DWI data to reconstruct connections between various brain regions (Mori et al., 1999; Jones et al., 1999; Conturo et al., 1999). Many studies have used DWI to show anatomical connections in the human brain (Conturo et al., 1999; Basser et al., 2000; Stieltjes et al., 2001; Xu et al., 2002; Behrens et al., 2003; Lehericy et al., 2004; Powell et al., 2004), and recent studies have used diffusion imaging to assess connectivity between functionally defined region of interests (e.g. Guye et al., 2003; Toosy et al., 2004; Dougherty et al., 2005; Kim et al., 2006; Takahashi et al., 2007, 2008).

DWI allows non-invasive mapping of anatomical connections and gives a comprehensive picture of connectivity throughout the entire brain by using various proposed tractography algorithms. One of the great advantages of diffusion tractography compared to tracer injection is that the entire fiber pathways in the brain can be studied at once in a relatively short amount of time. In general, a diffusion tensor is calculated and eigenvectors are computed from diffusion weighted images at every position in the brain. The eigenvector associated with the greatest eigenvalue indicates the principal direction of water diffusion. The fiber tract is propagated along this direction over a small distance to the next point where a new diffusion tensor is interpolated. We usually set the condition for terminating tractography fibers as the angle between two consecutive eigenvectors is greater than a given threshold, or as the diffusion fractional anisotropy (FA) value is smaller than a given threshold. The angle threshold contributes to reducing the error of the tractography in the initial analysis step, and the FA value threshold contributes to finding the gray/white matter border where fibers should be terminated.

We can study specific anatomical tracts and fibers from certain areas with diffusion tractography. We have been used functional magnetic resonance imaging (fMRI) to constrain the analysis of anatomical connectivity obtained by DWI, and studied fiber pathways underlying human memory functions using fMRI and diffusion tractography (Takahashi et al., 2007; Takahashi et al., 2008).

Network of Connections

We can study an anatomical network related to a region (e.g. activated area by a functional task) by performing diffusion tractography from the region. For example, we performed the following fMRI experiment, detected activated areas, and performed diffusion tractography from each activated area. The experiment consisted of two parts, an encoding phase and a retrieval phase (Figure 1). In the encoding phase, subjects were asked to perform 4 different “encoding” tasks: 1) make a living/nonliving judgment (deep encoding), 2) detect a given letter within a non-word letter sequence (shallow encoding), 3) press a random button, and 4) fixate on a central target. At the beginning of each block, an instruction was shown that specified the type of encoding task to be performed.

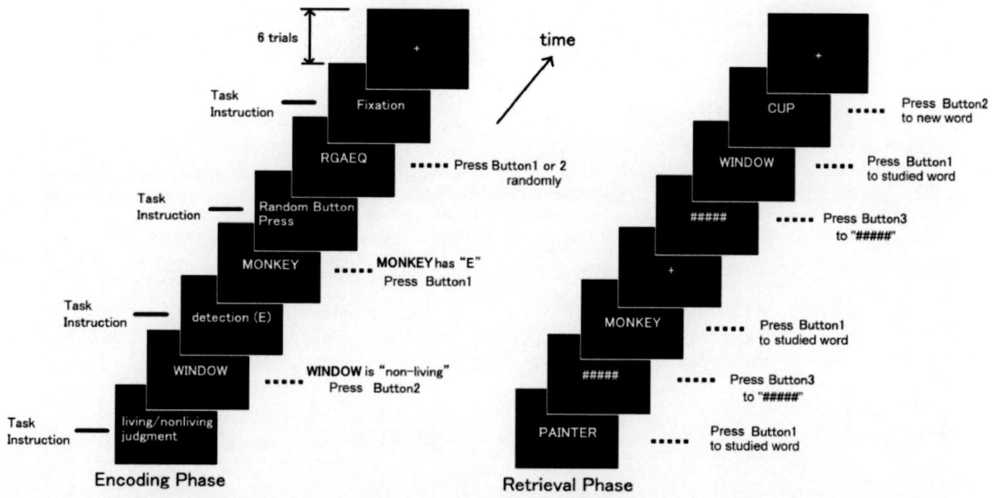

Figure 1. Experimental design for functional imaging in encoding (left) and retrieval (right). In encoding, there are four conditions: living/nonliving, detection, visuo-motor control, and fixation. In retrieval, there are retrieval trials and visuo-motor control trials.

In the "living/nonliving" blocks, subjects decided whether or not each word was animate. In the "detection" blocks, they decided whether each word contained "E". In both the "living/nonliving" and "detection" blocks, subjects reported their response by pressing one of two buttons held in the right hand. In the "random button press" (visuo-motor control) blocks, they looked at each non-word random letter sequence and pressed randomely one of the two buttons held in the right hand. In the fixation blocks, they looked at the fixation cross and did not press any buttons.

In the retrieval phase, subjects performed randomly intermixed retrieval trials, visuo-motor control trials, and fixation trials. In the retrieval trials, subjects made yes/no recognition memory judgments for previously studied and new stimuli. Subjects reported their response by pressing one of the two buttons held in the right hand. Half of the words from the encoding phase were presented again, along with new words. In the visuo-motor control trials, they looked at the number sign (####) and pressed the third button that was specifically assigned for this trial type. In the fixation trials, they looked at the fixation cross and did not press any buttons.

Diffusion tractography was performed from each activated cortical regions (but not from cerebellum and brain stem) in 20 healthy subjects. Results from single subject are shown in Figure 2. By studying the pathways among all activated regions, we can see an entire network that is involved in a specific cognitive task. We also examined connections between pairs of activation clusters in each subject. Fibers were obtained using any voxel in the starting cluster (as a seed) to any voxel within the end cluster, going in either direction. Examples of connections between activated regions are shown in Figure 3.

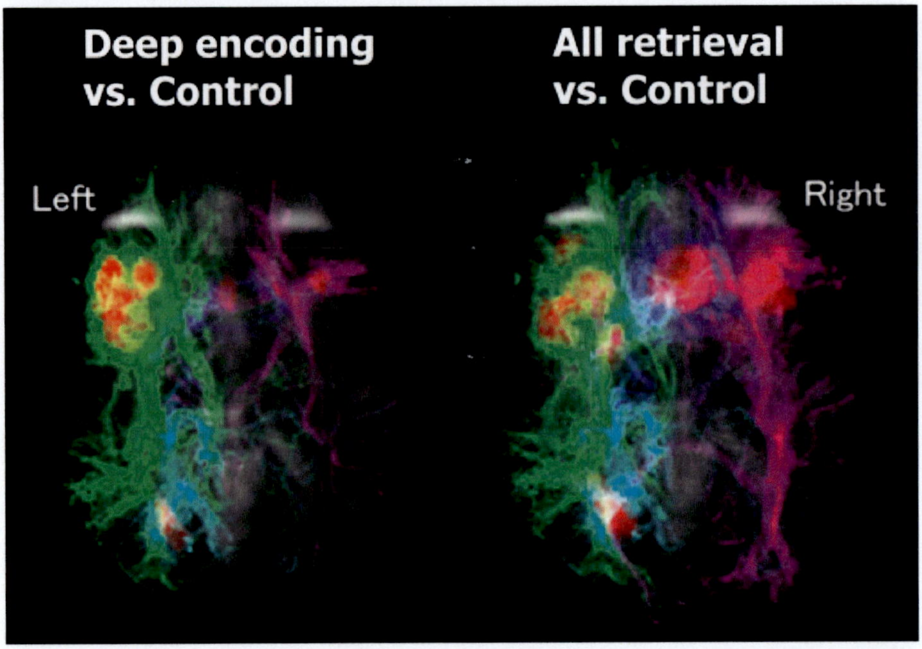

Figure 2. Activation and tractography from the activated areas (red) in the comparison "deep encoding versus visuo-motor control" (left) and "all retrieval versus visuo-motor control" (right). (Left) Fibers from left BA 9/46 (green), left BA 7/40 (light blue), medial BA 6 (purple), right BA 9/46 (pink). (Right) Fibers from left BA 45/44 and left BA 8/46 (green), left BA 7/40 (light blue). medial BA 6 (purple), right BA 45/44 (pink).

Selections of Cognitive Models

We need to select appropreate cognitive models to take advantages of diffusion tractography. We have used several models, but I first introduce one of them (Takahashi et al., 2007). Recent functional neuroimaging studies have shown that areas in the prefrontal, parietal and temporal cortices are involved in memory encoding and retrieval (Tulving et al., 1994; Demb et al., 1995; Buckner et al., 1998; Buckner and Koutstaal, 1998; Wagner et al., 1998; Buckner et al., 1999; Lepage et al., 2000). Among these memory-related areas, prefrontal cortex (PFC) is thought to subserve cognitive control processes (Goldman-Rakic, 1987; Schacter, 1987; Shimamura, 1995; Fuster, 1997; Takahashi and Miyashita, 2002), and send top-down signals to posterior cortices (Incisa della Rocchetta, 1993; Gazzaniga, 1995; Tomita et al., 1999; Takahashi and Miyashita, 2002). Recently, a two-stage model of PFC has been proposed (Petrides, 1994a, 1994b, 1996; Owen, Evans and Petrides, 1996). In this model, ventro-lateral prefrontal cortex (VLPFC) interacts with posterior cortices such as temporal cortex for active (or controlled) encoding and retrieval of information, whereas dorso-lateral prefrontal cortex (DLPFC) monitors and manipulates maintained information in VLPFC. Anatomical studies in nonhuman primates have shown direct corticocortical connections from VLPFC to temporal cortex (Petrides and Pandya, 2002), but direct connections from DLPFC to temporal cortex are still controversial (Seltzer and Pandya, 1989; Petrides and Pandya, 1994, 1999; Petrides, 2005).

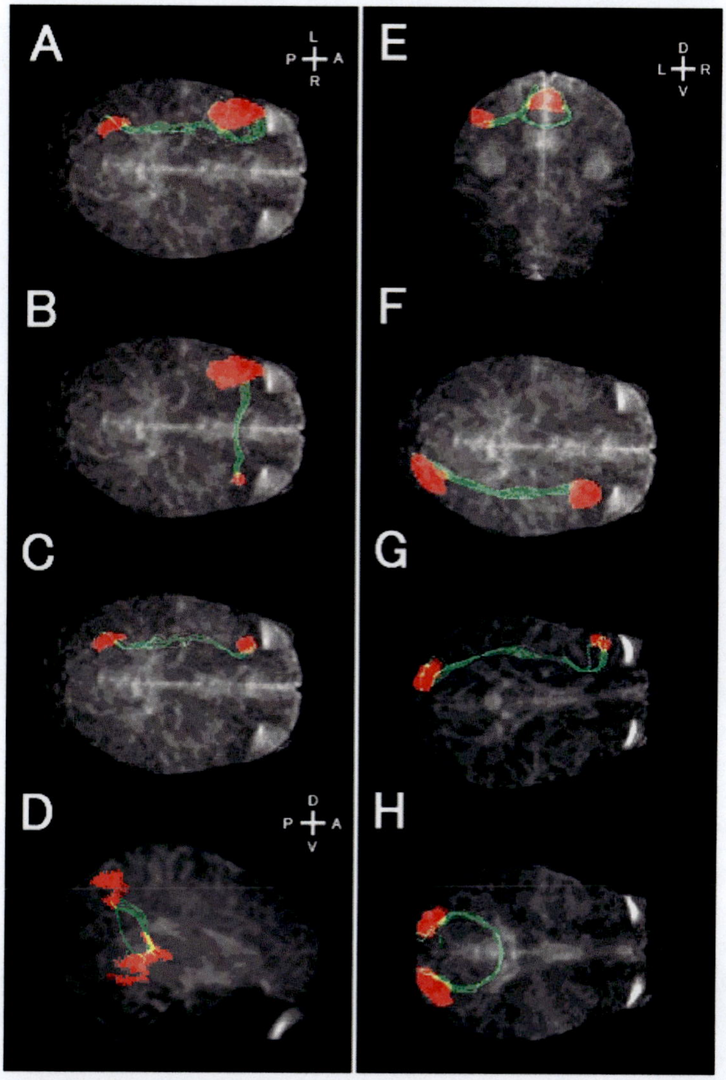

Figure 3. Examples of tractography between each pair of activated areas in single subject. (A-D) "deep encoding versus control" (A) left DLPFC (BA9/8/46/45) and left middle temporal to fusiform gyrus (BA22/37), (B) left and right operculum, (C) left inferior frontal gyrus (BA47) and left intraparietal sulcus, and (D) left intraparietal gyrus (BA7/40) and left middle temporal and fusiform gyri (BA37). (E-H) "all retrieval versus control" (E) left BA9/8 and medial BA6 (SMA and pre-SMA), (F) right frontal operculum (BA47/45) and right inferior occipital gyrus (BA19), (G) left frontal operculum (BA47) and left inferior occipital gyrus (BA19), and (H) left and right inferior occipital gyrus (BA19). (A: anterior, P: posterior, D: dorsal, V: ventral, L: left, R: right).

Based on these studies we hypothesized two possibilities for the anatomical connections between PFC and temporal cortex in humans; (i) Serial Pathway Model: DLPFC connects with VLPFC, and VLPFC connects with temporal cortex, but DLPFC does not connect with temporal cortex, (ii) Parallel Pathway Model: DLPFC and VLPFC both connect with temporal cortex. We found the connections between DLPFC and temporal cortex activation. I will explain about several ways that we applied to show the existence and specificity of the anatomical pathway between DLPFC and temporal cortex activation in several ways.

Specificity of Connections

If reconstructed fibers represented random connections between brain regions, then showing the existence of such fibers would be meaningless. Even though the connections are not entirely random, diffusely connected tracts between broad areas near the targets, are still less meaningful than they could otherwise be. To address this issue, we estimated the specificity in the connections between two activated clusters in the left DLPFC and the left temporal cortex in episodic encoding (Takahashi et al., 2007). For this purpose, we examined the degree of connectivity between the DLPFC activation and arbitrarily defined non-activating regions. If the connection was specific, the degree of connectivity between DLPFC and arbitrary regions should be significantly lower than the connectivity between DLPFC and temporal cortex activation. However, the degree of connectivity could be affected by many factors, such as distances between regions, and shapes of regions. Thus, we restricted the arbitrary regions to those satisfying the following conditions: (1) all voxels were inside the brain, (2) there was no overlap with the activated regions, (3) the volume and shape of the region were exactly the same as the temporal cortex activation, (4) the distance from the DLPFC was exactly the same as the distance between the DLPFC and the temporal cortex activation. Following these conditions, the arbitrary regions were selected 100 times for each subject, allowing their overlap. Then, the number of the fibers between the DLPFC and these arbitrarily defined regions were obtained. The fibers between the left DLPFC and the newly placed cluster were tracked from any voxel in one cluster to any voxel in the other cluster in both directions.

Arbitrary regions satisfying the selection criteria were used 100 times for each subject. Then, the number of fibers between the DLPFC and those arbitrarily defined regions were obtained. There was no connection between the DLPFC cluster and 55% of the random clusters. The median of the number of fibers between the DLPFC cluster and the random clusters was found to be 0, which is significantly different from the observed number of fibers (57 fibers). The random clusters that had more fibers than the actual number of fibers (57 fibers) were only 5%, and all of them were found in the precuneus in this particular subject.

We performed the same type of analysis on all 18 out of 20 subjects who had fibers between DLPFC and temporal cortex clusters. Eleven out of 18 subjects showed a significantly larger number of actual fibers relative to fibers generated from random clusters. At the population level, the number of actual fibers and the median number of fibers from random clusters for 18 subjects were significantly different. These analyses confirmed that the connections between DLPFC and temporal cortex activation were specific. Only some restricted areas (14.0 %) located in the right PFC (BA9/46) and precuneus (BA7) showed more connections with the left DLPFC cluster.

Estimation of Tractography Error

The streamline approach is based on the assumption that diffusion is locally uniform and can be accurately described by a single eigenvector. Unfortunately, this approach fails to describe crossing fibers (e.g. Alexander et al., 2001). To overcome this problem, tensorline approaches (e.g. tensor deflection) have been developed that use the entire tensor information instead of reducing it to a single eigenvector (Lazar et al., 2003). In our study (Takahashi et al., 2007), both algorithms, i.e. streamline (Basser et al., 2000) and tensor deflection (Lazar et al., 2003), were used to reconstruct tractography fibers.

The dispersion errors in white matter tractography were estimated by a statistical nonparametric bootstrap method (Lazar and Alexander, 2005). In each iteration, two samples were randomly selected with replacement from the pool of four independent DWI acquisitions, and were averaged. The procedure was repeated for each diffusion-encoding of the brain volume (15 directions) and for one non-diffusion weighted volume, resulting in one bootstrap DWI volume sample. The dispersion errors for the tractography were obtained by running the tractography algorithm over 100 iterations of the bootstrap DWI data set.

Thirty-three continuous seed points in the temporal cortex showed greater than 50 % probability of a connection to DLPFC. The seed cluster volume was 33 mm^3 (1 mm sampling of seed points), corresponding to 0.7 % of the entire temporal cortex activation (4848 mm^3). It is clear that these seed points did not distribute diffusively across the whole activated cluster, but were rather confined in a specific region. The Talairach coordinates of these 33 voxels were located in STS (x, y, z = -54 ± 5, -40 ± 2, -1 ± 2). These results indicate that the locations in temporal cortex that had connections with DLPFC were very specific.

We showed the results of boot-trac from a single seed point (Talairach: x, y, z = -49, -41, 0) in two ways. First, all the fibers for 100 bootstraps are shown in Figure 4A. Most of the fibers (78 %) went to the DLPFC, gradually diverging by the distance from the seed point in the temporal cortex activation. To assess uncertainty of this probability, we performed boot-trac analysis 1000 times. By shuffling the orders of the 1000 boot-trac samples 1000 times, an error for each iteration was obtained. At 100 iterations of boot-trac, the error was less than 5%. Second, the boot-trac from the same seed is shown as a probabilistic map (Figure 4B). This figure shows how many times, out of 100, the boot-trac fibers passed through each voxel. The probabilities were almost 100 % around the seed point in the temporal cortex activation, but soon went down less than 50 %. This is because of the divergence of the tracked fibers, which depends on the distance from the seed point, as displayed in Figure 4A.

We obtained a probabilistic map of connections from all 33 seed points with a connection probability greater than 50% (Figure 4C). These seed points are located in restricted locations, so they could compensate for the low probability of one seed point. The results indicate that the tractography from those 33 seed points reached the DLPFC activation with a probability of more than 80 %. The terminal location was the posterior end of the DLPFC cluster (around x, y, z = -49, 3, 36 in the Talairach coordinate), around the precentral sulcus to the middle frontal gyrus. This result indicates that not only the seed points, but also the terminal points were in very restricted locations within the activation. Figure 4C shows the probabilities of fibers from 33 voxels in the temporal cortex to anywhere in the whole brain. Although we did not selectively show the fibers that reached the DLPFC activation, high probabilities were found only along the pathway from the temporal cortex to DLPFC. Thus, the specificity of this pathway was remarkable.

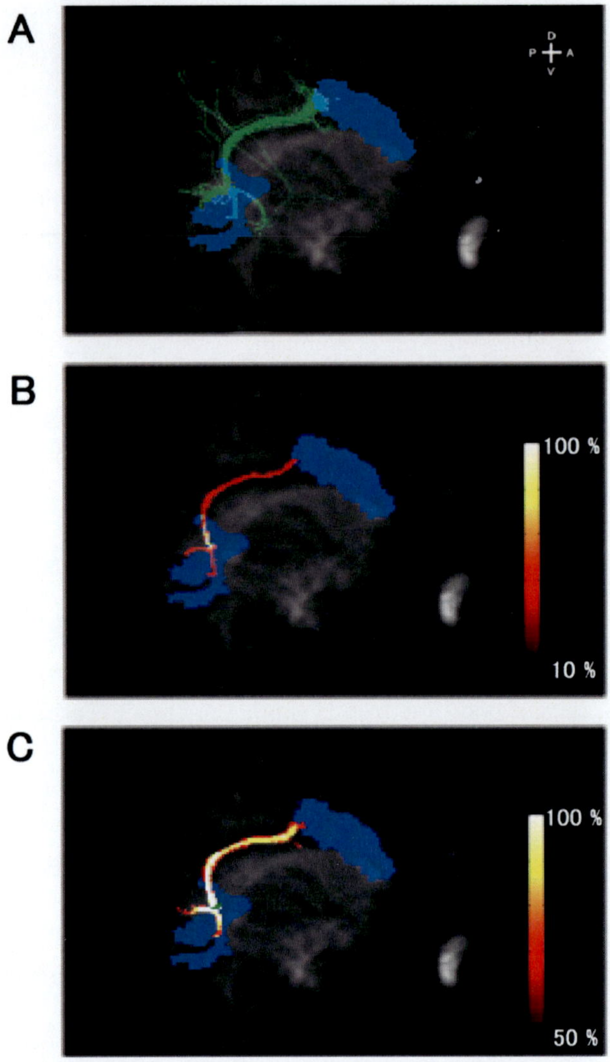

Figure 4. (A) All the fibers for 100 bootstraps from a single seed point at (-49, -41, 0). (B) A probabilitistic map of the connection from a single seed point (Talairach: x, y, z = -49, -41, 0). The voxels with more than 10 % probabilities were color-coded. (C) A probabilitistic map of the connection from the 33 seed points of high probabilities in the temporal cortex (green). Voxels that showed more than 50 % were color-coded. A: anterior, P: posterior, D: dorsal, V: ventral.

The boot-tracked fibers were diverging by the distance from the seed points, but in many cases, as shown in Figure 4A, most of the fibers reached the DLPFC activation. To see this observation quantitatively, we examined the probability of connections between any voxel in the temporal cortex and any voxel in the DLPFC. We found the connection with a probability of 100% in this subject. We performed this latter analysis for 18 subjects who showed connections between these regions. The existence of this pathway was crucial to distinguishing between the two models (the Serial and Parallel Pathway Models) mentioned above. Ten subjects out of the 18 subjects showed more than 50% probabilities for the connections between the left DLPFC and the left temporal cortex. The probability for the existence of this connection was very high in these 10 subjects (86.3 ± 18.0%, n=10). This result validates our major finding on the direct pathway between DLPFC and the temporal cortex activation.

Segregation of Multiple Pathways

Segregation is a different concept from specificity. For example, there is an unsolved relationship between the parietal cortex and temporal lobe in terms of episodic memory retrieval. The parietal cortex is thought to be involved in attention and eye-movement processes (Hyvarinen, 1982; Corbetta, 1998; Colby and Goldberg, 1999; Mesulam, 1999). In addition, recent functional neuroimaging studies have found activation in the parietal cortex during various memory tasks, especially those related to successful recognition memory (Henson et al. 1999; Konishi et al., 2000; Donaldson et al., 2001; Cansino et al., 2002; Dobbins et al., 2003; Rugg et al., 2003; Wheeler and Buckner, 2003; Herron et al, 2004). Most studies have identified several regions in the parietal cortex that respond to "Hits" (when subjects correctly recognize previously studied old items) more than to "Correct Rejections" (when they correctly identify new items). The regions in which this effect was identified consistently include intraparietal sulcus (IPS or Brodmann Areas (BA) 7), medial parietal cortex (the precuneus (PCu) or medial BA7, and posterior cingulate cortex or BA23/31), and several prefrontal regions. Although the roles of the prefrontal cortex in successful memory retrieval have been relatively well studied (Buckner et al., 1998a, b; Henson et al. 1999, followed by many other studies), those of the parietal cortex are still poorly understood (Shannon and Buckner, 2004; Naghavi and Nyberg, 2005; Wagner et al., 2005; Cavanna and Trimble, 2006). Since it is well known that declarative memory relies on the medial temporal lobe (MTL) and lateral temporal cortex, a key strategy for studying successful memory retrieval in the parietal cortex is to study its relationship with the temporal lobe. Given the accumulating knowledge on the functionally dissociated roles of the temporal lobe in terms of memory, studying connectivity between IPS/PCu and the temporal lobe should be fundamental to understand the roles of the parietal regions in memory.

Neuropsychological studies suggest that different types of memory depend on separated cortical structures (Squire, 1994, Takahashi and Miyashita, 2002). For example, patients with left lateral temporal lobe damage have impaired memory for semantic knowledge (De Renzi et al., 1987; Hart and Gordon, 1990; Snowden et al., 1989), but relatively preserved memory for episodic information (De Renzi et al., 1987; Snowden et al., 1994). Recent functional imaging studies also showed that lateral temporal cortex is activated by semantic memory retrieval or item-based memory, while the MTL is activated by episodic memory retrieval or relational memory (Wiggs et al., 1999; Lee et al., 2002; Konishi et al., 2006). Based on these functional dissociations between the MTL and lateral temporal cortex studies of the specific connections between these areas should provide insight into the process of memory retrieval in parietal cortex.

Group Study in Diffusion Tractography

Group analysis in the tractography study has been relatively less developed compared to that in the functional study. In our study (Takahashi et al., 2007; Takahashi et al., 2008), the coordinates of the activated clusters in Talairach coordinates were reverse normalized into each subject's coordinates, and used as the basis for tractography and determining the coordinates of the end points of the fibers. We then normalized the coordinates of the end points, averaged the data across all the subjects. This approach is better than normalizing diffusion weighted images (DWIs) directly because the resolution of DWIs was reduced when we resampled DWIs during normalization.

We reconstructed fibers originating from IPS and PCu where activation was found related to successful memory retrieval (Takahashi et al., 2008). The distribution of fiber terminals from 20 subjects are shown in Figure 5A. Probabilistic maps were obtained by a statistical nonparametric bootstrap method (Lazar and Alexander, 2005; Takahashi et al., 2007), and displayed in Figure 5B (note that only terminal points of the reconstructed fibers are displayed in Figure 5A, while the entire fibers are displayed in Fig. 5B). The IPS had connections with lateral temporal cortex (superior temporal gyrus (STG, BA 22), the middle temporal gyrus (MTG, BA 21) and the fusiform gyrus (FG, BA 37)) and bilateral superior colliculi (Figure 5A, upper row). PCu had connections with MTL (hippocampus, parahippocampal gyrus, the occipital cortex (BA17, 18, 19) and posterior cingulate cortex (Figure 5A, lower row).

Although DWI provides information about anatomical pathways *in vivo*, there is no functional interpretation in the reconstructed fibers themselves. We performed a whole-brain functional connectivity analysis, looking at the functional coupling between parietal regions of interest (Takahashi et al., 2008). This analysis based upon the fMRI data identified significant interactions with the temporal lobe that showed a regional specificity. We found that IPS was correlated with the lateral temporal cortex, and PCu was correlated with MTL. These results suggest that, with respect to successful memory retrieval, lateral and medial parieto-temporal pathways are anatomically and functionally dissociated.

Functionl Connectivity Analysis

Functional connectivity analysis was performed from the same regions of interests with diffusion tractography analyses. We obtained maps of correlation coefficients between residual time courses in the peak voxel (either IPS or PCu) and other voxels were calculated for individual subjects. To make inferences at the between-subject level about the functional connectivity encoded by these correlation coefficient maps, we transform them into summary statistics using Fisher's Z Transform. These within-subject maps were then passed to a second-level or between-subject one-sample t-test to provide SPMs in the usual way. To find areas that displayed a functional connectivity with both parietal regions, we show the voxels surviving an uncorrected threshold of $p = 0.001$ in both functional connectivity maps.

Regions with a significant functional correlation with an activated region in PCu (Talairach coordinates: x, y, z = 24, -62, 22) is shown in Figure 5C (upper row). Activity in bilateral hippocampi, parahippocampal gyri, occipital cortices (BA 17, 18, 19), posterior insula, anterior cingulate cortices (BA 32, 24), posterior cingulate cortices (BA 23, 31), precuneus (BA 7), inferior parietal lobes (BA39) were highly correlated with activity in PCu ($p < 0.05$, corrected for whole brain multiple comparisons at voxel-level).

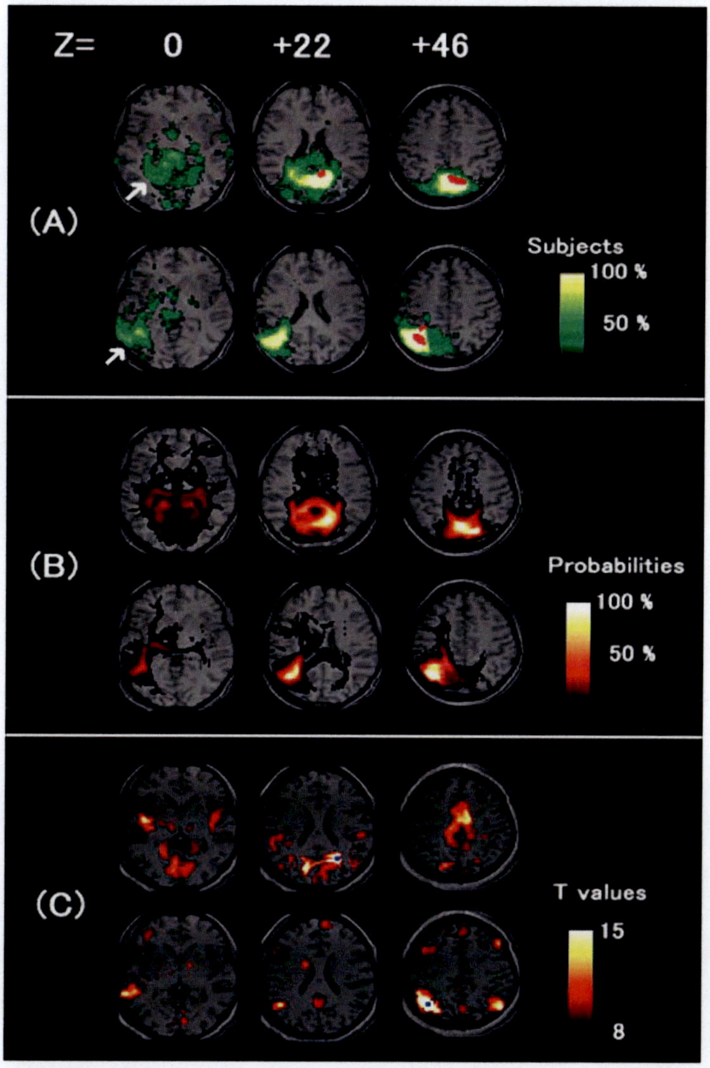

Figure 5. (A, B) Results of tractography analyses. (A) The average results of 20 subjects' terminal points of DTI fiber tracking. The left hemisphere of the brain corresponds to the left side of the image. Fiber tracking was performed from the PCu (upper row) and left IPS (lower row) activation clusters revealed by "Hits versus Correct rejections" condition in all the 20 subjects. The activation clusters are shown in red. The terminals of the reconstructed fibers are displayed in yellow/green. The color scale represents the proportion of the subjects. White arrows show MTL (upper row) and the lateral temporal cortex (lower row). (B) Probabilistic maps of fiber tracking based on the boot-trac analysis from the same activated clusters in (A) (upper row: PCu, lower row: left IPS). The color scale represents probabilities of fibers. (C) Results of functional connectivity analyses. The results of t-test on z transformed of correlation coefficients using activated regions in PCu (upper row) and left IPS (lower row) corrected for multiple comparisons for whole-brain at the voxel level ($p < 0.05$).

Figure 5C (lower row) shows maps of significant functional correlation with an activated region in left IPS (x, y, z = -44, -66, 46). The activity in bilateral frontal operculum (BA 47), IPS (BA 7), inferior parietal lobes (BA 39, 40), superior parietal lobes (BA 7), supplementary motor area (BA 8), precuneus (BA 7), middle frontal gyri (BA 8), left middle temporal gyrus (BA 21, 37), inferior frontal gyrus (BA 44), superior frontal gyrus (BA 6) were highly correlated with activity in the left IPS ($p < 0.05$, corrected for multiple comparisons at voxel-level).

Segregations of the Two Pathways in Anatomical and Functional Connectivity

The terminal points of reconstructed fibers from IPS and PCu exhibited very little overlap (11.0 % compared to the total terminal points from IPS, or 3.3 % compared to those from PCu; using a threshold of 5 subjects). Functional connectivity (at the threshold of $p < 0.05$, corrected for multiple comparisons for a whole brain at voxel level) also exhibited little overlap (5.68 % compared to all the correlated areas of IPS, and 3.58 % compared to those of PCu).

There are some possible limitations with this approach both in anatomical and functional connectivity analyses. One of the common concerns is that overlapping areas must depend on selections of thresholds. Using a very low threshold must provide us with highly overlapping pathways, *vice versa*. There is no consensus about which threshold represents proper suggestions on a degree of dissociation of multiple pathways so far.

Functional Segregations of the Two Pathways

Mean percent signal changes in PCu and IPS showed clearly different patterns in the "Hit" condition. In PCu, there is almost no signal change in the "Hit" compared to the control condition, while IPS showed almost the same amount of signal change with "Hit" and "Miss" conditions. In both areas, "False Alarm" and "Correct Rejection" conditions showed decreased signal changes compared to baseline.

Next, we showed that the pattern of activity in the clusters in the left middle temporal gyrus and parahippocampal gyrus depend on the memory performance of the subjects (Hits, Misses, False alarms, and Correct rejections). These clusters were defined by functional connectivity analysis with the IPS and the PCu, respectively. For the middle temporal gyrus, $F(4,95) = 4.66$, $p = 0.0018$, and there were significant differences between Misses and False alarms, and Misses and Correct rejections (Tukey's t-test $p < 0.05$). For the parahippocampal gyrus, $F(4,95) = 3.6$, $p = 0.009$, and there were significant differences between Misses and False alarms, and Misses and Correct rejections (Tukey's t-test $p < 0.05$). Correlations between the z-scores for functional connectivity and success rates (Buchel et al. 1999; Hampson et al., 2006) were also examined across all the subjects ($n = 20$). The functional connectivity between IPS and MTG was positively correlated to success rates ($r = 0.63$, $p = 0.0028$) and that between PCu and MTL was negatively correlated to success rates ($r = -0.7167$, $p = 0.004$). Thus, the middle temporal gyrus and parahippocampal gyrus show a similar pattern of activity to the IPS and PCu, which suggests that the connections between the IPS and middle temporal gyrus and that between PCu and parahippocampal gyrus are involved in successful memory retrieval.

Future Directions of Diffusion Tractography

One major challenge facing tractography is the crossing-fiber problem. Algorithms for tractography fail or produce errors when they encounter a voxel where neural fibers cross each other. Basically, at that point, a reconstructed fiber stops or goes in the wrong direction. There have been two approaches to solving this problem thus far: one is to assess tractography errors using a probabilistic approach, and the other is to improve the accuracy of tractography results.

A number of studies have assessed tractography results using the probabilistic approach (e.g. Behrens et al., 2003; Upadhyay et al., 2006; Takahashi et al., 2007). Although anatomical connections are conceptually incompatible with probability, this is one emerging trend in the study of connections *in vivo*. There are numbers of suggested methods to obtain probabilities of tracts (Lazar et al., 2003; Behrens et al., 2003; Chung et al., 2006; Jbabdi et al., 2007). However, we currently do not have a way to directly compare the results of different implementations of the probabilistic approach. This means that the same anatomical pathway can have different probabilities depending on the method used, even in the same brain. I believe that this problem needs to be addressed in the course of future tractography studies, in order to form a solid basis for comparative studies on large samples, similar to what has been done in tracer studies of connectivity of non-human primates. Another issue is that probability reflects both connections between the areas we are interested in, and the quality of the dataset (more precisely, various artifacts/noise on the pathway). The degree of artifact/noise could vary across different brain areas, so not addressing this problem may impair our ability to see the actual course of the pathway. These issues present challenging problems to be solved as the technology progresses.

On a different front, efforts to improve accuracy of tractography will also be very important. New techniques were proposed recently that apply multiple diffusion encoding directions (Tuch et al., 2002; Wedeen et al., 2005) together with different gradient strengths (Wedeen et al., 2005). In theory, these techniques provide us with more precise ways to uncover white matter fiber structures, by reconciling the different directions of reconstructed fibers in a single voxel. These techniques are powerful especially on post-mortem brains (Schmahmann et al., 2007). Applying them to human *in vivo* imaging is a promising future direction.

Acknowledgments

I greatefully thank Robert P. Levy for his very helpful editorial comments.

References

Alexander A.L., Hasan K.M., Lazar M., Tsuruda J.S. and Parker D.L. (2001). Analysis of partial volume effects in diffusion-tensor MRI. *Mgn. Reson, Med.* 45, 770-780.

Basser, P.J., Mattiello, J. and LeBihan, D. (1994). MR diffusion tensor spectroscopy and imaging. *Biophs. J.* 66, 259-267.

Basser P.J., Pajevic S., Pierpaoli C., Duda J. and Aldroubi A. (2000). In vivo fiber tractography using DT-MRI data. *Magn. Reson. Med.* 44, 625-632.

Behrens T.E., Johansen-Berg H., Woolrich M.W., Smith S.M., Wheeler-Kingshott C.A., Boulby P.A., Barker G.J., Sillery E.L., Sheehan K., Ciccarelli O., Thompson A.J., Brady J.M. and Matthews P.M. (2003). Non-invasive mapping of connections between human thalamus and cortex using diffusion imaging. *Nat. Neurosci.* 6, 750-757.

Buckner, R.L. and Koutstaal, W. (1998). Functional neuroimaging studies of encoding, priming, and explicit memory retrieval. *Proc. Natl. Acad. Sci. USA.* 95, 891-898.

Buckner R.L., Koutstaal W., Schacter D.L., Wagner A.D. and Rosen B.R. (1998). Functional-anatomic study of episodic retrieval using fMRI. I. Retrieval effort versus retrieval success. *Neuroimage* 7, 151-162.

Buckner, R.L., Kelley, W.M. and Petersen, S.E. (1999). Frontal cortex contributes to human memory formation. *Nat. Neurosci.* 2, 311-314.

Buckner RL, Koutstaal W, Schacter DL, Wagner AD, Rosen BR. 1998a. Functional-anatomic study of episodic retrieval using fMRI I. Retrieval effort versus retrieval success. *Neuroimage* 7:151-162.

Buckner RL, Koutstaal W, Schacter DL, Dale AM, Rotte M, Rosen BR. 1998b. Functional-anatomic study of episodic retrieval II. Selective averaging of event-related fMRI trials to test the retrieval success hypothesis. *NeuroImage* 7:163-175.

Cansino S, Maquet P, Dolan RJ, Rugg MD 2002. Brain activity underlying encoding and retrieval of source memory. *Cereb. Cortex* 12:1048-1056.

Cavanna AE, Trimble MR. 2006. The precuneus: a review of its functional anatomy and behavioural correlates. *Brain* 129:564-583.

Chung, S., Lu, Y., and Henry, R.G. (2006) Comparison of bootstrap approaches for estimation of uncertainties of DTI parameters. *Neuroimage*, 531-541.

Colby CL, and Goldberg ME. (1999). Space and attention in parietal cortex. *Ann. Rev. Neurosci.* 22:319-349.

Conturo T.E., Lori N.F., Cull T.S., Akbudak E., Snyder A.Z., Shimony J.S., McKinstry R.C., Burton H. and Raichle M.E. (1999). Tracking neuronal fiber pathways in the living human brain. *Proc. Natl. Acad. Sci. USA.* 96, 10422-10427.

Corbetta M. (1998). Frontoparietal cortical networks for directing attention and the eye to visual locations: identical, independent, or overlapping neural systems? *Proc. Natl. Acad. Sci. USA.* 95:831-838.

Demb J.B., Desmond J.E., Wagner A.D., Vaidya C.J., Glover G.H. and Gabrieli J.D. (1995). Semantic encoding and retrieval in the left inferior prefrontal cortex: a functional MRI study of task difficulty and process specificity. *J. Neurosci.* 15, 5870-5878.

De Renzi E, Liotti M, and Nichelli P. (1987). Semantic amnesia with preservation of autobiographical memory: A case report. *Cortex* 23:575-597.

Dobbins IG, Rice HJ, Wagner AD, and Schacter DL. (2003). Memory orientation and success: separable neurocognitive components underlying episodic recognition. *Neuropsychologia* 41:318-333.

Donaldson DI, Petersen SE, Ollinger JM, and Buckner RL. (2001). Dissociating state and item components of recognition memory using fMRI. *Neuroimage* 13:129-142.

Dougherty RF, Ben-Shachar M, Bammer R, Brewer AA, and Wandell BA. (2005). Functional organization of human occipito-callosal fiber tracts. *Proc. Natl. Acad. Sci. USA*. 102:7350-7355.

Fuster, J.M. (1997). The prefrontal cortex: Anatomy, physiology, and neuropsychology of the frontal lobe. (Philadelphia: Lippincott-Raven).

Gazzaniga, M.S. (1995). Principles of human brain organization derived from split-brain studies. *Neuron* 14, 217-228.

Goldman-Rakic, P.S. (1987). Circuitry of primate prefrontal cortex and regulation of behavior by representational memory. In Plum, F. and Mountcastle, V. eds. *Handbook of physiology: Section 1. The nervous system*. Vol. 5. Higher functions of the brain. Part 1, pp. 373-417. (Bethesda, MD: American Physiological Society).

Guye M, Parker GJ, Symms M, Boulby P, Wheeler-Kingshott CA, Salek-Haddadi A, Barker GJ, and Duncan JS. (2003). Combined functional MRI and tractography to demonstrate the connectivity of the human primary motor cortex in vivo. *Neuroimage* 19:1349-1360.

Hampson, M., Tokogly, F., Sun, Z., Schafer, R.J., Skudlarski, P., Gore, J.C., and Constable, R.T. (2006). Connectivity-behavior analysis reveals that functional connectivity between left BA39 and Broca's area varies with reading ability. *Neuroimage* 31, 513-519.

Hart J Jr, and Gordon B. (1990). Delineation of single-word semantic comprehension deficits in aphasia, with anatomical correlation. *Ann. Neurol.* 27:226-231.

Henson RNA, Rugg MD, Shallice T, Josephs O, and Dolan RJ. (1999). Recollection and familiarity in recognition memory: an event-related functional magnetic resonance imaging study. *J. Neurosci.* 19:3962-3972.

Herron JE, Henson RNA, and Rugg MD. (2004). Probability effects on the neural correlates of retrieval success: an fMRI study. *NeuroImage* 21:302-310.

Hyvarinen J. (1982). Posterior parietal lobe of the primate brain. *Physiol. Rev.* 62:1060-1129.

Incisa della Rocchetta, A. and Milner, B. (1993). Strategic search and retrieval inhibition: The role of the frontal lobes. *Neuropsychologia* 31, 503-524.

Konishi S, Wheeler ME, Donaldson DI, and Buckner RL. (2001). Neural correlates of episodic retrieval success. *NeuroImage* 12:276-286.

Konishi S, Asari T, Jimura K, Chikazoe J, and Miyashita Y. (2006). Activation shift from medial to lateral temporal cortex associated with recency judgements following impoverished encoding. *Cereb. Cortex*. 16:469-474.

Jbabdi S, Woolrich MW, Andersson JL, and Behrens TE. (2007). A Baysian framework for global tractography. *Neuroimage*. 37:116-129.

Jones, D.K., Horsfield, M.A. and Simmons, A. (1999). Optimal strategies for measuring diffusion in anisotropic systems by magnetic resonance imaging. *Magn. Reson. Med.* 42, 515-525.

Kim M, Ducros M, Carlson T, Ronen I, He S, Ugurbil K, and Kim D-S. (2006). Anatomical correlates of the functional organization in the human occipitotemporal cortex. *Magn. Reson. Imaging.* 24:583-590.

Lazar M., Weinstein D.M., Tsuruda J.S., Hasan K.M., Arfanakis K., Meyerand M.E., Badie B., Rowley H.A., Haughton V., Field A. and Alexander A.L. (2003). White matter tractography using diffusion tensor deflection. *Hum. Brain Mapp.* 18, 306-321.

Lazar, M. and Alexander, A.L. (2005). Bootstrap white matter tractography (BOOT-TRAC) *Neuroimage* 24, 524-532.

Lehericy S., Ducros M., Krainik A., Francois C., Van de Moortele P.F., Ugurbil K. and Kim D.S. (2004). 3-D diffusion tensor axonal tracking shows distinct SMA and pre-SMA projections to the human striatum. *Cereb. Cortex* 14, 1303-1309.

Lee AC, Robbins TW, Graham KS, and Owen AM. (2002). "Pray or Prey?" dissociation of semantic memory retrieval from episodic memory processes using positronemission tomography and a nocel homophone task. *Neuroimage* 16, 724-735.

Lepage, M., Ghaffar, O., Nyberg, L. and Tulving, E. (2000). Prefrontal cortex and episodic memory retrieval mode. *Proc. Natl. Acad. Sci. USA*. 97, 506-511.

Mesulam MM. (1999). Spatial attention and neglct: parietal, frontal and cingulated contributions to the mental representation and attentional targeting of salient extrapersonal events. *Philos. Trans R Soc. Lond B Biol. Sci.* 354, 1325-1346.

Mori, S., Crain, B.J., Chacko V.P. and van Zijl, P.C.M. (1999). Three-dimensional tracking of axonal projections in the brain by magnetic resonance imaging. *Annal. Neurol.* 45, 265-269.

Naghavi HR, and Nyberg L. (2005). Common fronto-parietal activity in attention, memory, and consciousness: Shared demands on integration? *Consci. Cogn.* 14, 390-425.

Owen, A.M., Evans, A.C., and Petrides, M. (1996). Evidence for a two-stage model of spatial working memory processing within the lateral frontal cortex: a positron emission tomography study. *Cereb. Cortex* 6, 31-38.

Petrides, M. (1994a). Frontal lobes and behaviour. *Curr. Opin. Neurobiol.* 4, 207-211.

Petrides, M. (1994b). Frontal lobes and working memory: evidence from investigations of the effects of cortical excisions in nonhuman primates. In: Boller, F., Grafman, J. (Eds.), *Handbook of Neuropsychology*, Vol.9. (Elsevier, Amsterdam), p.p. 59-82.

Petrides, M. (1996). Specialized systems for the processing of mnemonic information within the primate frontal cortex. *Philos. Trans. R. Soc. B.* 351, 1455-1462.

Petrides, M. (2005). Lateral prefrontal cortex: architectonic and functional organization. *Phil. Trans. R. Soc. B.,* 360, 781-795.

Petrides, M. and Pandya, D.N. (1999). Dorsolateral prefrontal cortex: comparative cytoarchitectonic analysis in the human and the macaque brain and corticocortical connection patterns. *Eur. J. Neurosci.* 11, 1011-1036.

Petrides, M. and Pandya, D.N. (2002). Comparative cytoarchitectonic analysis of the human and the macaque ventrolateral prefrontal cortex and corticocortical connection patterns in the monkey. *Eur. J. Neurosci.* 16, 291-310.

Powell H.W., Guye M., Parker G.J., Symms M.R., Boulby P., Koepp M.J., Barker G.J. and Duncan J.S. (2004). Noninvasive in vivo demonstration of the connections of the human parahippocampal gyrus. *Neuroimage* 22, 740-747.

Rugg, M.D., Henson, R.N.A., and Robb, W.G. (2003). Neural correlates of retrieval processing in the prefrontal cortex during recognition and exclusion tasks. *Neuropsychologia* 41:40-52.

Schacter DL. (1987). Memory, amnesia, and frontal lobe dysfunction. *Psychobiology* 15:21-36.

Schmahmann, J.D., Pandya, D.N., Wang, R., Dai, G., D'Arceuil, H.E., de Crespigny, A.J., and Wedeen, V.J. (2007). Association fibre pathways of the brain: parallel observations from diffusion spectrum imaging and autoradiography. *Brain* 130, 630-53.

Seltzer, B. and Pandya, D.N. (1989). Frontal lobe connections of the superior temporal sulcus in the rhesus monkey. *J. Comp. Neurol.* 281, 97-113.

Shannon, B.J., and Buckner, R.L. (2004). Functional-anatomic correlations of memory retriecal that suggest nontraditional processing roles for multiple distinct regions within posterior parietal cortex. *J. Neurosci.* 24:10084-10092.

Shimamura, A.P. (1995). Memory and frontal lobe function. In Gazzaniga, M.S. ed. *The cognitive neurosciences*. (Cambridge, MA: MIT press), pp. 803-813.

Snowden, J.S., Goulding, P.J., and Neary, D. (1989). Semantic dementia: A form of circumscribed cerebral atrophy. *Behav. Neurol.* 2:167-182.

Snowden, J.S., Griffiths, H., and Neary, D. (1994). Semantic dementia: Autobiographical contribution to preservation of meaning. *Cogn. Neuropsycol.* 11:265-288.

Squire L. (1994). Declarative and nondeclarative memory: multiple brain systems supporting learning and memory. In: Schacter DL, and Tulving E. (Eds.) *Memory systems*. (Cambridge MA: MIT Press.) p.p. 203-232.

Stieltjes B., Kaufmann W.E., van Zijl P.C., Fredericksen K., Pearlson G.D., Solaiyappan M. and Mori S. (2001). Diffusion tensor imaging and axonal tracking in the human brainstem. *Neuroimage* 14, 723-735.

Takahashi, E. and Miyashita, Y. (2002) In: Squire, L.R. and Schacter, D.L. (Eds.) *Neuropsychology of Memory* (The Guilford Press, New York)., p.p.301-310

Takahashi, E., Ohki, K., and Kim, D.S. (2007). Diffusion tensor studies dissociated two fronto-temporal pathways in the human memory system. *NeuroImage* 34:827-838.

Takahashi, E., Ohki, K., and Kim, D.S. (2008). Dissociated Pathways for Successful Memory Retrieval from the Human Parietal Cortex: Anatomical and Functional Connectivity Analyses. *Cerebral Cortex.* Advanced Access published January 8th, 2008, doi: 10.1093/cercor/bhm204.

Tomita, H., Ohbayashi, M., Nakahara, K., Hasegawa, I. and Miyashita, Y. (1999). Top-down signal from prefrontal cortex in executive control of memory retrieval. *Nature* 401, 699-703.

Toosy, A.T., Ciccarelli, O., Parker, G.J., Wheeler-Kingshott, C.A., Miller, D.H., and Thompson, A.J. (2004). Characterizing function-structure relationships in the human visual system with functional MRI and diffusion tensor imaging. *NeuroImage* 21:1452-1463.

Tuch, D.S., Rees, T.G., Wiegell, M.R., Makris, N., Belliveau, J.W., and Wedeen, V.J. (2002). High angular resolution diffusion imaging reveals intravoxel white matter fiber heterogeneity. *Magn. Reson. Med.* 4, 577-82.

Tulving, E., Kapur, S., Craik, F.I., Moscovitch, M., and Houle, S. (1994). Hemispheric encoding/retrieval asymmetry in episodic memory: positron emission tomography findings. *Proc. Natl. Acad. Sci. USA.* 91, 2016-2020.

Upadhyay, J., Ducros, M., Knaus, T.A., Lindgren, K.A., Silver, A., Tager-Flusberg, H., and Kim, D-S. (2006) Function and Connectivity in Human Primary Auditory Cortex: A Combined fMRI and DTI Study at 3 tesla. *Cereb. Cortex* 17, 2420-2432.

Wagner, A.D., Schacter, D.L., Rotte, M., Koutstaal, W., Maril, A., Dale, A.M., Rosen, B.R. and Buckner, R.L. (1998). Building memories: remembering and forgetting of verbal experiences as predicted by brain activity. *Science* 281, 1188-1191.

Wagner, A.D., Shannon, B.J., Kahn, I., and Buckner, R.L. (2005). Parietal lobe contributions to episodic memory retrieval. *Trends Cogn. Sci.* 9:445-453.

Wedeen, V.J., Hagmann, P., Tseng, W.Y., Rees, T.G., and Weisskoff, R.M. (2005) Mapping complex tissue architecture with diffusion spectrum magnetic resonance imaging. *Magn. Reson. Med.* 6, 1377-86.

Wheeler, M.E., and Buckner, R.L. (2003). Functional dissociation among components of remembering: control, perceived oldness, and content. *J. Neurosci.* 23:3869-3880.

Wiggs, C.L., Weisberg, J., and Martin, A. (1999). Neural correlates of semantic and episodic memory retrieval. *Neuropsycologia.* 37:103-118.

Xu D., Mori S., Solaiyappan M., van Zijl P.C. and Davatzikos C. (2002). A framework for callosal fiber distributon analysis. *Neuroimage* 17, 1131-1143.

In: Neural Pathways Research
Editor: F. L. Pichler, pp. 175-191
ISBN: 978-1-60456-214-9

Chapter VIII

Neuroplasticity in the Auditory Brainstem

Angelo Salami, Renzo Mora, Barbara Crippa and Massimo Dellepiane
ENT Department,
University of Genoa, Italy

Abstract

Neuroplasticity (brain plasticity or cortical plasticity) refers to the modifications that occur in the organization of the brain, and in particular changes that occur to the location of specific information processing functions, as a result of acquired brain injury. Neuroplasticity allows the neurons to compensate for injury and disease and to adjust their activities in response to new situations or to changes in their enviroment.

Although many plastic changes in the brain and in the receptor organs occur only in developing organisms as a continuation of embryological development, i.e., after birth, neuroplasticity has been also found in the adult.

An increasing interest in neuroplasticity and nerve regeneration within the auditory receptor and pathway has developed in recent years. In humans, the activation of the auditory system starts in the 26th-28th gestational week and continues for several years: Nerve regeneration and neuroplasticity have been demonstrated in the adult auditory receptors as well as throughout the auditory pathway. Many aspects of plasticity in the auditory system, both in the developmental period and during adulthood, are similar to the plasticity in other sensory systems and share similar mechanisms: the auditory system, like other sensory systems, is plastic, and as such is able to reorganize its structure and function after partial or total loss of receptor function.

The receptors of the auditory system are very vulnerable and several pathogenetic mechanisms may significantly alter their function, leading first to temporary threshold shifts and, in the case of prolonged or very intensive exposure, to permanent threshold shifts. Ultimately, the receptors die out, triggering a sequence of changes in the whole auditory system starting with the spiral ganglion cells and ending in the auditory cortex. Regardless of whether the loss of receptors (hearing loss) is based on mechanical lesions, ablations, chemical lesions by ototoxic drugs, noise-induced damage of hair cells, or programmed loss connected with aging, the sequence of basic changes in the central auditory system is similar and results in the reorganization of the cortical projection areas and in some cases the subcortical areas.

A better understanding of the neuroplasticity strategies regarding cortical reorganization and neuroplasticity could potentially be helpful in the designing of effective treatment and management strategies (e.g.,: cochlear implants or stem cells), but also in really understanding the pathogenesis of auditory or language diseases (e.g.,: deafness, tinnitus, dyslexia, etc.).

Introduction

Hearing is, probably, the most relevant sensory sensation for humans, because it is directly linked to language, which is the main characteristic that defines the human species. Thus, hearing in the human being is much more relevant than in other species, including other primates: the loss of the language perception is so relevant that it frequently results in depression or in the other behavioural changes.

The central auditory pathway starts with the cochlear nucleus: spiral ganglion neurons receive signals from the inner ear hair cells in the organ of Corti and conduct these signals through the eighth cranial nerve to the cochlear nucleus. From the cochlear nucleus, two ascending auditory systems arise:

- the lemniscal system comprises major nuclei of the superior olive in the rhombencephalon, the inferior colliculus in the midbrain and the medial geniculate body in the thalamus;
- a non-lemniscal system connects subnuclei of these regions where it interconnects with multimodal pathways.

Both systems conduct auditory signals to the cerebral cortex. The auditory cortical regions are linked to a variety of other cortical areas, including some in the temporal bone and frontal bone, but also in subcortical regions like the pons, cerebellum, and thalamus.

The auditory cortex also gives rise to a descending auditory system: it reaches the cochlea and the hair cells contained in it, allowing feedback interactions on all the levels of the central auditory system [1].

Neuroplasticity is classically accepted as the capacity of neurons to change the shape of their connection tree. Neuroplasticity appears during development and maturation of both the central and peripheral nervous systems. It can also, as a reactive process, be observed after nerve lesion. Neuroplasticity is involved in higher brain functions, as learning, memory or language. These processes are mainly conducted by trophic factors and by neurotransmitters, in particular glutamate (NMDA receptor - named after the selective agonist N-methyl-D-aspartate) and GABA (γ-aminobutyric acid) that participate in shape changes and functional adaptation of neurons. Glutamate is a major transmitter in central pain pathways. The excitatory effect of glutamate is mediated by two main types of ionotropic receptors: NMDA and AMPA (named after the selective agonist α-amino-3-hydroxy-5-methyl-isoxasole-4-

propionic acid). The NMDA receptors contribute to normal excitatory transmission and have special functions related to synaptic plasticity [2,3].

There are different degrees of plasticity on brain system:

a) some systems appear strongly determined and change very little even under extreme alterations of experience, e.g., complete deafness or blindness;
b) other neural systems do change considerably when experience is different but only during certain, limited "sensitive" time periods, and these times differ for different system;
c) a third of neural systems appear modifiable by experience throughout life.

Studies of non-human primate development have indicated that about 35% of axons/neurons are lost from the peak of development to adulthood, and that onset of this loss co-occurs with the period of synaptogenesis. Similar studies of the human brain also document periods of over-production of synapses [4].

The auditory receptor and its neural pathway towards the cortex are composed of highly complex neural circuits that reach their organization during development and maturation because of a high remodelling process [5].

Central auditory system neurons may be "born" at different times in ontogeny: the neurons are not generated in a temporal sequence from lower to higher centers along the auditory pathway [1, 6].

Recently, some authors have described development changes in the synaptic density for different cortical areas in language processing: primary auditory cortex, the angular gyrus and Broca's area. At birth, synaptic development (as measured by the time course of synaptic density) in the auditory cortex is more advanced than in the two language areas, but by 4 years of age the synaptic density is similar in this areas and is significantly greater in the adult (by about a factor of two) [7,8].

After birth, sensory input is mandatory to keep the course of development on a proper track: for instance, there are many more axons crossing from the cochlear nucleus to the inferior colliculus than remain uncrossed; if one cochlea is removed so that the contralateral inferior colliculus receives less crossed sensory activation, then more uncrossed fibers are encountered to innervate the inferior colliculus in the grown-up animal [9]. *The ability of neuronal networks to alter their structure and responsiveness using the changing patterns of incoming activity as instructor is called plasticity.*

The hearing improvement obtained from cochlear implants, and from some drug therapies, has stimulated further efforts to study neural plasticity, regeneration, or the use of stem cell grafting. However, these multidisciplinary studies require the study of research models during development, in vitro research, degeneration-regeneration analysis on adult animals, etc. The goal of this chapter is to focus on some main features related to this upcoming research field.

Neurobiology of the Auditory System

Current sensory neurobiology includes the search for factors involved in guiding nerve fibers toward their specific targets and, also, the influences that nerve fibers could exert on the target differentiation. Trophic interactions between development nerve fibers and their targets may influence their mutual differentiation: most of the sense organs fully differentiate only after they have been innervated and they degenerate after denervation. As a general rule, receptor cells start their differentiation after the influx and by contact by nervous fibers.

Therefore, during development, the arrangement of connections between neurons, or between neurons and other cells, contributes to further differentiation of the "interconnected partners".

Partial damage to the sensory receptor surfaces of the adult auditory system causes reorganization of the topographic cortical map that normally receive these inputs: for example, a small lesion within the cochlea does not permanently silence the cortical neurons that originally responded to that area of the auditory field; instead they acquired sensitivity to another part of the auditory field near the damaged region [10].

Inhibitory processes are involved in this plasticity, as receptor organ damage leads to anatomically defined loss of inhibitory inputs and physiologically defined unmasking of normally silent inputs in deprived sensory cortex: removal of the intracortical surround inhibition would unmask latent or suppressed inputs to cortical areas, deprived of their normal representations by the peripheral damage; then these latent input would become consolidated over time as the new representation in the deprived regions [11].

Neurotrophic factors can bring on a better survival and neurite outgrowth on neurons developed in culture but also after neural damage. Neurotrophic factors (a complex molecular group) act throughout cell membrane receptors, in particular through tyrosine kinase receptor (TrkA, TrkB or TrkCn) [12].

The nerve growth factor contributes to neurons survival and promotes nerve fiber sprouting [13].

Trophic actions involved in sensory differentiation may be mediated by either a diffusible trophic factor or by the interaction of membrane elements of nerve fibers and undifferentiated epithelial cells [14, 15].

Developing nerve fibers could participate in the differentiation of sensory cells by inducing molecular modifications on these cells. These actions may be mediated through cell-cell adhesion mechanisms that may be intimately involved in morphogenetic events: a lot of transcription molecules (trophic factors, retinoic acid, hormones, as thyroid hormone, etc.) have the core role in the normal development and maturation of auditory sensory cells and spiral ganglion neurons [16].

The auditory receptor and its neural pathway towards the cortex are composed of highly complex neural circuits that reach their organization during development and maturation because of a high remodelling process. Transitory nerve fibers, found during auditory receptor maturation, are very important because they indicate the early excess of sprounting that could be involved in cell maturation or on the remodelling of neural circuits. During the later stages of auditory maturation, transitory nerve fibers progressively retracted and degenerated. Trophic factors, such as NT3, add to cultures of auditory ganglion neurons promote neuronal survival and differentiation, while bFGF, NGF, transforming growth factors and other stimulate the regeneration process of auditory ganglion neurons [17-19].

Neurotransmitters are highly functional molecules that stimulate cell membrane receptors. It has been clearly stated that transitory expression of neurotransmitters, their enzymes and receptors, is really linked to their function as neurotrophic substances [20]: in the developing cochlea, some neurotransmitters such as dopamine, GABA or acetylcholine were early found in efferent fibers, and afferent neurons showed a very early maturation by specific reactivity to the administration of glutamate agonists [21,22]. In contrast, encephalins appearing at the end of maturation were found at the onset of function and auditory potentials [23].

The absence of neurotrophic factors or activity, oxydative stress, excitotoxicity linked to intracellular increase of Ca2+ concentration, and significant reduction of mitochondrial energetic process determine the degeneration of neurons. Glutamatergic neurons are highly sensitive to excessive release of glutamate: the glutamate receptors when markedly activated (e.g., high noise on the auditory receptor) allow the massive entrance of Ca2+ (which alternates the mitochondrial metabolism) and Na+ (that depolarised the neuron, allowing water to enter) resulting in neuronal edema [24].

In both physiological and neurotoxic conditions glutamate activates NMDA and no-NMDA (QUI-AMPA; KA-AMPA) receptors. Recent reports have involved NMDA receptors in the regenerating process of afferent nerve fibers [25, 26].

NMDA receptors are known to play indispensable roles in neural development and various forms of synaptic plasticity: the activation with different input patterns can induce distinct forms of synaptic plasticity, such as long-term potentiation and depression. It is well known that calyx-type auditory synapses are specialized for preserving the fidelity of synaptic transmission; NMDA receptors have a physiologically significant mechanism for synaptic fidelity, because they may potentially facilitate spike generation at synapse [27].

Because central synapses have various shapes and sizes, it is likely that the concentration of NMDA receptors may vary in different synapses. This concentration can be influenced by many factors, including the peak concentration of glutamate: the concentration of Ca2+ leads to various forms of synaptic plasticity in central synapses [28].

Recent study have highlighted that synaptic NMDA receptors are unsaturated; their activity can summate to different levels depending on the input patterns and are exposed to glutamate [29].

The mechanisms of cortical as well as subcortical plasticity in the auditory system are, similarly as in other systems, correlated with the pre- and postsynaptic activity: acetylcholine produces a postsynaptic excitation and increases the synaptic strength [30].

Other specific factors have been involved in the maintenance or protection of adult and developing auditory neurons or sensory cells, for example insulin or IGF-1, bFGF, EGF, mixed solutions of trophic factors including TGFa, retinoic acid, NT-3 and BDNF, FGF-2, or GDNF [31-34].

Plastic modification on the synaptic level are stimulated by genes: one of them is GAP-43, expressed in growing axons and presynaptic endings; it is present in the axons and their varicosities, one week after hearing loss [1].

Some authors have described the patterns of brain glucose utilization during human development: these studies show a rapid rise in cerebral metabolism during infancy, perhaps reflecting the burst of synaptogenesis described in the structural studies. This is followed by a decrease in brain glucose metabolism later in childhood, within a similar time frame as that observed for the loss of synapses [35].

These structural and physiological findings support the view of different maturational timetables for distinct brain structure mechanisms whereby environmental input may affect different brain systems to different degrees and at different times.

Plastic Changes in the Central Auditory System after Hearing Loss

1. Presbycusis

Presbycusis in human refers to a constellation of age-related auditory deficits that include a loss of hearing sensitivity and a decreased ability to understand speech, particularly in the presence of background noise.

The hearing loss tends to accelerate with age: age-related hearing deficits in humans have usually been attributed to changes in the cochlea, including a loss of a sensory cells, atrophy of stria vascularis, and a loss of spiral ganglion cells. However, recent studies have highlighted that important parts of presbycusis consist of changes in the central auditory system [36].

On animal models, many studies of plasticity mechanisms have suggested that:

- a diminished inhibition could be involved in changes observed during brain aging (age-related changes of the GABA receptors in the inferior colliculus):
- a synaptic loss is not an inevitable event in aging, but may be related to the preservation of peripheral auditory function and input to the neurons (similar results were observed in the superior olivary complex, which together with result from the inferior colliculus demonstrate that the mechanisms of binaural interaction might not be significantly influenced by aging) [37].

The signs of presbycusis in humans are similar to those observed in experimental animals: loss of inner hearing cells is typically present in the basal turn of the cochlea, whereas the loss of other hearing cells is more scattered, spreading up to the apical part.

Concomitantly with the loss of hair cells, presbycusis is accompanied by a loss of spiral ganglion cells, which usually comprises the loss of peripheral dendrites as well as central projections, i.e., auditory nerve fibers [38].

As a result of the cochlear pathology, neurons in the inferior colliculus that normally have low thresholds for high-frequency tones can no longer respond them; the best frequencies (the frequency for which a neuron has the lowest threshold) of ventral inferior colliculus neurons shift from high to middle frequencies so that the best frequencies tend to be similar throughout the inferior colliculus. Animal models show that, in principle, changes in the auditory cortex are similar to those observed in the inferior colliculus, i.e., with aging virtually the entire auditory cortex contained neurons with best frequencies in the middle-frequency range [36].

The efferent system seems to be more robust and does not degenerate to same extent as the afferent system [38].

2. Noise-Induced Hearing Loss

The effects of noise exposure on the central auditory system depend on the intensity, frequency content and duration of the exposure. Exposure to high intensities of noise even for a short period produces damage in the cochlea without recovery in hearing sensitivity: several studies indicate that intense acoustical stimulation changes the micromechanical properties of the sensory hair bundle at the apical pole of the hair cells; damage to the stereocilia correlates well with alterations in hearing sensitivity [39].

Hair cell loss in the cochlea after auditory overstimulation is connected with the degeneration of auditory nerve fibers: this process is consistent with the dystrophy resulting from the disappearance of inner hearing cells and auditory nerve fibers [39].

Calcium ions may play a potential role in the pathological processes following acoustical overstimulation: as significant elevations of Ca2+ is found in the cytoplasm of outer hearing cells after acoustic overstimulation, similarly, the increased permeability of the endolymph-perilymph barrier to potassium and sodium ions after noise exposure could be involved in the mechanism of hearing loss [40].

Acoustic trauma causes a disruption of dendrite endings below the inner hearing cells, leading to synaptic uncoupling; the dendrite damage might be due to an excessive release of the neurotransmitter glutamate from the inner hearing cells, which is toxic to the structure and function of the primary auditory neurons. Excess glutamate in the extracellular space below the inner hearing cells results in permeability changes in the postsynaptic membrane of the dendrite. The resulting osmotic imbalance causes water to move into the dendrites, resulting in dendrite swelling and ultimately membrane disruption. The application of the glutamate antagonist kynurenate protected dendrites of the primary auditory neurons from noise-induced damage [41].

Protection of the hair cells in the organ of Corti from noise-induced damage is associated with a decrease in concentration of calbindin, increases in heat shock proteins and changes in the cochlear antioxidant activity in the stria vascularis [42].

The role of cochlear efferents in acquired resistance to noise-induced hearing loss is less clear: conditioning-related protection may arise from a generalized stress response, an amplification of the olivocochlear reflex and a general enhancement of cochlear sensitivity [43].

In contrast to the situation in the auditory nerve and cochlear nucleus, the amplitudes of evoked responses in the inferior colliculus and in the auditory cortex are found to be enhanced, compared with the preexposure situation. It is possible to explain this enhanced amplitude by the lack of inhibition in these structures after exposure [44].

The discharge rate increased, particularly, in those neurons where the inhibitory influence is large at high sound intensities. These cortical modifications are related to change in the level of GABA and decreased levels of glutamate decarboxylase (GAD) [44].

These data strongly suggest that the function of the auditory cortex is seriously disturbed after noise exposure.

3. Unilateral Sensorineural Hearing Loss in Children

Children with unilateral sensorineural hearing loss (USNHL) show deficits in such areas as speech recognition, sound localization and academic performance. A better understanding of the compensatory strategies used in children with USNHL, especially regarding cortical reorganization and neuroplasticity, could potentially be helpful in the designing of effective treatment and management strategies [45].

Various studies have shown reorganization of auditory and language pathways in patients with USNHL: in subjects undergoing monaural stimulation, normal-hearing subjects displayed activation lateralized to the contralateral side, while unilaterally deaf subjects displayed bilateral activation patterns, indicating a functional reorganization of auditory pathways [46].

The ability of pediatric subjects with USNHL to compensate may depend greatly on the treatment and management approach; specifically, whether that approach fosters or hinders the optimal functional reorganization of auditory and language pathways. This reorganization, however, may differ in subjects with right and left USNHL, as evidenced by the "right-ear advantage" in early childhood [45].

In normal-hearing adult subjects, the presentation of monaural stimuli produces a contralaterality effect in the superior temporal gyrus. Crossing auditory pathways may have a greater number of fibers and faster transmission speed relative to ipsilateral pathways. In contrast, subjects with USNHL displayed greater activation on the side ipsilateral to the better-hearing ear. Thus, the hypothesized cortical reorganization may also be associated with changes in fiber organization, as white matter maturation has been observed to continue throughout the developmental period [47].

In normal adults, auditory stimuli elicit cortical responses that are large over temporal brain regions but small or absent over other regions. By contrast, in children the cortical responses are equally large over temporal and visual brain regions, consistent with the idea that there is a less specificity and more redundancy connections between the auditory and visual cortex at this time. These data suggest that early in human development there exists a redundancy of connections between auditory and visual areas and that this overlap decreases after birth. This loss of redundancy determines alterations in the organization of remaining sensory systems. The considerable variability in timing of sensitive periods may also be in part due to temporal differences in the occurrence of redundancy within the different systems [34].

Further research will be necessary to elucidate the precise mechanisms responsible for cortical reorganization in children with UNSHL.

Several works, on the plasticity of auditory brainstem nuclei in animal models, have shown that unilateral auditory deprivation produces the most pronounced effects on immature subcortical auditory pathways: changes range from neuronal loss in the cochlear nucleus to long term alterations of projections from the cochlear nucleus to the inferior colliculus [38,48].

4. Unilateral Sensorineural Hearing Loss in Adults

A fundamental characteristic of the central auditory pathways is the tonotopic organization of the main-line auditory nuclei and fields: the tonotopic organization of these structures reflects the organization of the cochlea [49].

In adult animals, after restricted monaural lesions of the cochlear epithelium, the tonotopic cortical map can reorganize such that cortical neurons deprived of their usual most sensitive afferent input now respond to tone frequencies adjacent to the frequency range damaged by the lesion (lesion edge frequencies). Further, the new responses occur with intensity thresholds similar to those recorder in the cortical regions normally representing the lesion edge frequencies [10].

In adult humans with USNHL, neurons in the deprived regions of the cortex appeared to respond to tone frequencies adjacent to the frequency range damage by the lesion: the increase of cortical strength represents an expansion of frequency representation of cortical areas adjacent to deprived area; the recruitment of a larger neuronal population may occur through a decrease in cortical lateral inhibition [49].

It also be noted that the effects observed in cortical maps may not have been due solely to cortical changes: past studies suggest that some of the changes in the cortical maps may be due to changes in the subcortical structures [10].

Studies on the plasticity of subcortical pathways in animal models have shown that unilateral auditory deprivation produces slower and less remarkable changes in the adult system, while the most pronounced effects can occur in younger children [48].

In adults with postchildhood onset of unilateral deafness, compared with normal-hearing subjects, the auditory evoked potentials show significant increases in interhemispheric waveform, cross-correlation coefficients and peak amplitude correlations: these increases provide evidence of substantial changes from the normal pattern of asymmetrical (controlateral > ispilateral amplitude) and asynchronous (controlateral earlier than ispilateral) central auditory system activation in normal subjects to a much more symmetrical and synchronous activation in the unilaterally deaf [39].

5. Bilateral Hearing Loss

Studies on cross-modal plasticity in animals models suggest that cortical areas can change their functional specificity depending on which inputs they receive. In humans there is converging evidence that secondary and association cortices can compensate for loss of one modality: this rewriting is believed to involve the stabilization of early multi-sensory cortico-cortical connections that exist in the newborn, but are eliminated in the presence of normal input: in the absence of competition from auditory inputs, the auditory cortex may become recruited for tactile or visual processing [35].

Deaf adults are better than hearing controls at detecting the onset or the direction of motion of a peripheral stimulus; these groups differences are most marked for peripheral stimuli [50].

These data are confirmed by functional magnetic resonance imaging studies and by animal studies: when the both cochlea are experimentally removed, the auditory cortex receives visually driven input [35]. The results are dramatic (especially in early brain development): neurons in the cortical region now receiving visual input respond to visual stimuli; axons, synapses and neurons of this region interact in this way to develop orderly local cell assemblies and modules typical for the visual cortex, and neurons are connected with other brain regions [1].

Animal studies highlight that as one ear of an adult rat has lost its function after a mono- or bilateral cochleotomy, the protein GAP-43 is expressed in the ipsilateral cochlear nucleus. This protein is known to be expressed in growing axons and in presynaptic endings under construction or reconstruction and highlights ongoing plastic modification on the synaptic level [51].

The complete absence of peripheral inputs causes in the central auditory system chemical modifications: physiological studies show a higher concentrations of GABA with a bilateral cortical inhibition [52].

This indicates that the plasticity response to hearing loss is not chaotic but specific with respect to the postsynaptic profile receiving altered innervation.

6. Cochlear Implant

Multiple-channel cochlear implants overcome basilar membrane or cochlear hair-cell damage by encoding acoustic information and providing electrical stimulation directly to the auditory nerve. The stimulation delivered to the auditory nerve is then conveyed much as in acoustic processing in normal-hearing subjects.

It is of neurophysiological interest to determine how auditory brainstem developmental plasticity can be promoted by chronic electrical stimulation.

It has been noted that local glucose utilization of the cerebral cortex changes in the course of human brain development. It starts out fairly low postnatally, increases dramatically to reach peak values around age 6, falls off afterwards, but does not settle on adults levels before the age of 16 years [53].

The maturation of cortical axons appears to be close to complete by about that age. In prelingual deaf patients, the auditory cortex remains in a hypometabolic state due to the lack of auditory input. These patients may well profit from the cochlear implant. If the cortical region shows normal metabolism before implantation, it appears to be occupied by non-auditory synapses due to cross-modal plasticity and will not be useful any more to receive activation along the ascending auditory pathway. For cochlear implant patients, hearing scores increase with the degree of hypometabolic auditory cortex present at the time of implantation [54].

However, recent research supports the theory that auditory pathways remain "frozen" during the period of sensory deprivation: in children with early-onset deafness, the "frozen" state is immature and stimulation with the implant causes the early stages of maturational processes (including the initial phase of rapid synaptogenesis and myelination) to resume. In children with late-onset deafness, these physiological processes may, at least in part, have begun before the onset of the deafness, and thus less additional maturation is required after the implant [55].

Positron emission tomography (PET) gives an objective measurement of changes in the brain of subjects with cochlear implants. The first PET imaging studies demonstrated a revival of brain metabolic activity in auditory cortex areas of deaf people after implantation [56,57].

PET studies, in users of cochlear implants, reveal that while noise activation in cochlear implants users does not significantly differ from that in normal subjects, hearing speech activates more cortical areas in cochlear implants users than in normal subjects: the PET imaging studies highlight a higher activation not only in the temporal cortices but also in Broca's areas and its right hemisphere homolog in the supplementary motor area and in the anterior cingulated gyrus [58].

7. Tinnitus

Subjective tinnitus is characterized by the perception of auditory signals experienced in the absence of any internal or external source of sound [59].

A frequent cause of tinnitus is noise exposure and presbycusis, but tinnitus may appear as a side effect of treatment with ototoxic drugs. In many cases, however, tinnitus originates in the central auditory system or in a brain structure related to the central auditory system (this statement is supported by the fact that tinnitus persists in patients with acoustic neurinoma after transaction of the auditory nerve) [60].

Magnetoencephalografic recordings and positron emission tomography demonstrate that tinnitus is related to plastic alterations in the auditory cortex which is a type of auditory phantom phenomenom: patients affected by tinnitus, compared with controls, show evidence of more widespread activation by tones at aberrant links between the limbic and auditory systems [59, 61].

These data confirm the results of experiments in animals and suggest that tinnitus in many cases is related to plastic changes in the auditory cortex and may be compared with phantom limb pain occurring after amputation.

It remains for future research to reveal whether tinnitus is maintained by cortical reorganization, whether cortical reorganization is a consequence of ongoing tinnitus, or whether both phenomena are triggered by a common source.

Neural Plasticity in Children with Learning Problems

Understanding speech during everyday listening places many demands on the auditory system: among these demands are the accurate representation of rapidly changing spectral information comprising the speech signal and the separation of speech from background noise. Recent studies are aimed at understanding the basic biologic processes underlying speech-sound perception in quiet and noise [62].

Many children with reading problems exhibit deficits in the perception of speech sounds: this has been attributed to inadequate phonological awareness skills and/or difficulties in perceiving brief sounds (including rapid spectral changes) which occur in speech. Of particular importance is that these perceptual deficits can be absent in quiet, but manifested in noise typical of everyday listening situations [63].

Commercially available auditory training software programs are widely used by learning-impaired children: the impact of these programs on neurobiological processes associated with auditory perceptual problems is not well known.

Neuronal plasticity resulting from training and perceptual learning has been shown to involve alteration in neural connections at multiple levels of the auditory pathway: when humans have been trained to discriminate speech sounds, changes in preattentive auditory neurophysiologic responses have been linked to perceptual changes [64].

Children with learning problems who practice with auditory training software exhibit plasticity of neural encoding of speech sounds at the cortical and subcortical levels of the auditory pathway: this plasticity is accompanied by improvement of behavioural performance [64].

Recent studies highlight that as the inferior colliculus may be the locus of the plastic activity, the cortex may modulate activity [65].

The corticofugal descending system is critical in manipulating signal encoding via positive feedback or lateral inhibition mechanisms; once trained or conditioned, egocentric selection allows the cortex to recognize the behavioural significance of an acoustic stimulus and then fine-tune its input by altering the sound representation at lower levels: specifically, the cortex modulates subcortical areas that encode basic stimulus features and thus improves subsequent cortical representation [66,67].

Children with delayed brainstem timing are particularly likely to profit from auditory training: a more precise brainstem response in noise may benefit the listener by providing a more accurate representation of the acoustic characteristics of the stimulus [65].

Neural Plasticity and Language in Deaf Humans

Different neural systems mediate the processing of lexical/semantic and grammar information in adults: recent studies suggest a greater role for more posterior temporal-parietal systems in lexical/semantic processing and for frontal-temporal systems within the left hemisphere in grammatical processing; in particular, the systems that mediate the processing of grammatical information are more modifiable and vulnerable in response to altered hearing than are those associated with lexical/semantic processing [35].

Following auditory deprivation, processes associated with the dorsal visual pathway are more altered than are functions associated with the ventral pathway, providing support for conceptions of visual system organization that distinguish function along these pathways [35].

These effects could be explained by the redundant connectivity of the human brain: in normal adults, auditory stimuli elicit responses that are large over temporal brain regions but small or absent over occipital regions; by contrast, in children (younger than 36 months old) the auditory responses are equally large over temporal and visual brain regions, consistent with the idea that there is less specificity and more redundancy of connection between the auditory and visual cortex at this time [34].

This loss of redundancy may be a boundary condition that determines when sensory deprivation can result in alterations in the organization of remaining sensory systems.

Conclusion

A knowledge of the plastic changes in the central auditory system occurring in humans may open new vistas in the treatment of the impairment.

It will be necessary to understand the influence of decreased inhibition in the central auditory system on the manifestation of hearing loss with the aim of starting rational treatment. Practical outcomes may also be expected from further studies of the long-lasting effect of wearing hearing aids. An understanding of the plastic changes in the auditory system during aging should serve to ameliorate the negative sequelae of this inevitable process or at least move the problems to an older age.

By far the most important function of the auditory system in humans is speech recognition. Speech understanding deteriorates with aging, especially in situations in which the voice of the speaker is accompanied by a noise background or by other voices. Improving speech recognition is an important goal in the treatment of hearing loss with hearing aids. Plastic changes in normally aided ears definitely do occur; research on this topic is just beginning.

Further research should be addressed to finally determine the whole complex world of substances that can help in further development of nerve and sensory epithelium regeneration of the inner ear.

Adequate knowledge of these substances, and their action on nerve fibers, could provide relevant information to develop culture tissues (e.g. from stem cells) for inner ear grafting. These advances in the research of auditory nerve neuroplasticity should open unexpected perspectives for deafness therapy in the near future.

References

[1] Illing, RB. Maturation and plasticity of the central auditory system. *Acta Otolaryngol.*, 2004, S552, 6-10.

[2] Wrang, L-Y. The dynamic range for gain control of NMDA receptor-mediated synaptic transmission at a single synapse. *The journal of neuroscience*, 2000, 20.

[3] Zheng W. and Knudsen E.I. (1999). Functional selection of adaptive auditory space map by GABAA – mediated inhibition. *Science,* 1999, 284, 962-65.

[4] Rakic 2000.

[5] Simmons DD. Development of the inner ear efferent system across vertebrate species. *J. Neurobiol*, 2002, 53, 228-250.

[6] Kandler K. and Friauf E. (1993). Pre- and postnatal development of efferent connections of the cochlear nucleus in the rat. *J. Comp. Neurol.*, 328, 161-84.

[7] Huttenlocher,1994.

[8] Dabholkar,1997.

[9] Moore D.R. and Kowalchuk N.E. (1988). Auditory brainstem of the ferret: effect of unilateral cochlear lesion on cochlear nucleus volume and projection to inferior colliculus. *J. Comp. Neurol.*, 272, 503-15.

[10] Rajan, R; Inrvine, DRF; Wise, LZ; Heil, P. Effect of unilateral partial cochlear lesions in adult cats on the representation of lesioned and unlesioned cochleas in primary auditory cortex. *J. Comp. Neurol.*, 1993, 338, 17-49.

[11] Rajan, R. Receptor organ damage causes loss of cortical surround inhibition without topographic map plasticity. *Nature Neuroscience*, 1998, 1, 138-143.

[12] Chao M.V. and Hempstead BL. (1995). P75 and Trk: a two-receptor system. *Tins*, 18, 321-326.

[13] Mbienne, JP; Favre, D; Sans, A. Early innervation and differentiation of hair cells in the vestibular epithelia of mouse embryos: SEM and TEM study. *Anat. Embryol*, 1998, 177, 331-340.

[14] Mbienne, JP; Dechesne, CJ; Schachner, M; Sans, A. Immunocytochemical characterization of the expression of cell adhesion molecule L1 during early innervation of mouse otocysts. *Cell. Tissue. Ress*, 1989, 255, 81-88.

[15] Eldman, GM. Cell adhesion molecules. *Science*, 1983, 219, 450-457.

[16] Rummler L.S. and Gupta R. (2004). Peripheral nerve repair: a review. *Current Op. Orthoped*, 15, 215-219.

[17] Rubel E.W. and Fritzsch B. (2002). Auditory system development: Primary auditory neurons and their targets. *Ann. Rev. Neurosci.*, 25, 51-101.

[18] Pirvola, U; Ylikoski, J; Palgi, J; Lehtonen, E; Arumae, U; Saarma, M. Brain-derived neurotrophic factor and neurotrophin 3 mRNAs in the peripheral target fields of developing inner ear ganglia. *Proc. Natl. Acad. Sci. USA*, 1992, 98, 9915-9919.

[19] Van de Water T.R. and Represa J. (1991). Tissue interactions and growth factors that control development of the inner ear. Neurol. Tube-Otic Anlage Interactio. *Ann. NY. Acad. SCI*, 630, 116-128.

[20] Emerit, MB; Riad, M; Hamon, M. Trophic effects of neurotransmitters during brain naturation. *Biol. Neonate*, 1992, 62, 193-201.

[21] Merchan-Perez, A; Gil-Loyzaga, P; Eybalin, M. Immunocytochemical detection of calcitonin gene related peptide in the post natal developing rat cochlea. *Int. J. Dev. Neurosci*, 1990, 8, 603-612.

[22] Gil Loysaga P. and Pujol R. (1990) Neurotoxicity of Kainic Acid in the rat cochlea during early developmental stages. *Europ. Arc. Otorhinolaryngol*, 248, 40-48.

[23] Gil-Loyzaga, P; Eybalin, M; Cupo, A. Met-Enkephalin and Met-Enkephalin-Arg6-Gly7-Leu8. Immunofluorecence in the developing guinea-pig organ of corti. *Dev. Brain Res*, 1988, 42, 142-145.

[24] Haberny, KA; Paule, MG; Scallet, AC; Sistare, FD; Leser, DS; Haning, JP; Slikker W. Ontogeny of the N-Metyl-D-Aspartate (NMDA) Receptor System and Susceptibility to Neurotoxitity. Toxicol. *SCI*, 2002, 68, 9-17.

[25] Puel, JL. Chemical synaptic transmission in the cochlea. *Progress Neurobiol*, 1995, 47, 449-476.

[26] D'Aldin, CG; Ruel, J; Assie, R; Pujol, R; Puel, JL. Implication of NMDA type glutamate receptors in neural regeneration and neoformation of synapses after excitotoxic injury in the guinea-pig cochlea. *Int. J. Devl. Neurosci*, 1997, 15, 619-629.

[27] Trussell, LO. Synaptic mechanisms for coding timing in auditory neurons. *Annu Rev Physiol*, 1999, 61, 477-96.

[28] Feldman, DE; Nicoll, RA; Malenka, RC. Synaptic plasticity at thalamocortical synapses in developing rat somatosensory cortex: LTP, LTD, and silent synapses. *J. Neurobiol.*, 1999, 41, 92-101.

[29] Wang, LY. The dynamic range for gain control of NMDA receptor-mediated synaptic transmission at a single synapse. *J. Neurosci.*, 2000, 20, 1-5.

[30] Weinberger, NM. Dynamic regulation of receptive fields and maps in the adults sensory cortex. *Annu Rev Neurosci*, 1995, 18, 129-158.

[31] Staeker H. and Van De Water T.R. (1998). Factors controlling hair cell regeneration/repair in the inner ear. *Curr. Opin. Neurobiol.*, 8, 480-487.

[32] Pirvola, U; Hallbook, F; Xing-Qun, L; Virkkala, J; Saarma, M; Ylikoski, J. Espression of neurotrophins and Trk receptor in the developing, adult, and regeneration avian cochlea. *J. Neurobiol.*, 1997, 33, 1019-1033.

[33] Zhou, X; Hossajn, WA; Rutledge, A; Bajer, C; Morest, DK. Basic fibroblast growth factor (FGF-2) affects development of acoustico-vestibular neurons in the chick embryo brain in vitro. *Hear. Res.*, 1996, 93, 147-166.

[34] Ylikoski, J; Pirvola, U; Virkkala, J; Suvanto, P; Liang, XQ; Magal, E; Altschuler, R; Miller, JM; Saarma, M. Guinea pig auditory neurons are protected by glial cell line-derived growth factor from degeneration after noise trauma. *Hear. Res.*, 1998, 124, 17-26.

[35] Neville H. and Bavelier D. (2002). Human brain plasticity: evidence from sensory deprivation and altered language experience. *Progress in brain research*, 138, 177-188.

[36] Willott, JF. Aging and the auditory system: anatomy, physiology, and psychophysics. San Diego: Singular; 1991.

[37] Caspary, DM; Milbrandt, JC; Helfert, RH. Central auditory aging: GABA changes in the inferior colliculus. *Exp. Gerontol.*, 1995, 30, 349-360.

[38] Moore, JK; Niparko, JK; Miller, MR; Linthicum FH. Effect of profound hearing loss on a central auditory nucleus. *Ann. J. Otol.*, 1994, 15, 588-595.

[39] Syka, J. Plastic changes in the central auditory system after hearing loss, restoration of function, and during learning. *Physiol. Rev.*, 2002, 82, 601-636.

[40] Konishi, T; Salt, AN; Hamrick, PE. Effects of exposure to noise on permeability to potassium of the endolymph-perilimph barrier in guinea pigs. *Acta Otolaryngol.*, 1982, 94, 395-401.

[41] Puel, JL; Ruel, J; d'Aldin, CG; Pujol, R. Excitotoxicity and repair of cochlear synapses after noise-trauma induced hearing loss. *Neuroreport*, 1998, 9, 2109-2114.

[42] Zheng, XY; Henderson, D; McFadden, SL; Hu, BH. The role of the cochlear efferent system in acquired resistance to noise-induced hearing loss. *Hear Res.*, 1997, 104, 191-203.

[43] Kujawa S.G. and Liberman C.M. (1997). Conditioning-related protection from acoustic injury: effects of chronic deeferentation and sham surgery. *J. Neurophysiol.*, 78: 3095-3106.

[44] Szczepaniak W.S. and Moller A.R. (1995). Evidence of decreased GABAergic influence on temporal integration in the inferior colliculus following acute noise exposure: a study of evoked potential in the rat. *Neurosci. Lett.*, 196, 77-80.

[45] Bess, FH; Tharpe, AM; Gibler, AM. Auditory performance of children with unilateral sensorineural hearing loss. *Ear Hear*, 1986, 7, 20–26.

[46] Schmithorst, VJ; Holland, SK; Ret, J; Duggins, A; Arjmand, E; Greinwald, J. Cortical Reorganization in Children with Unilateral Sensorineural Hearing Loss. *Neuroreport*, 2005, 16, 463–467.

[47] Schmithorst, VJ; Wilke, M; Dardzinski, BJ; Holland, SK. Correlation of White Matter Diffusivity and Anisotropy Changes with Age during Childhood: A Cross-Sectional Diffusion Tensor Imaging Study. *Radiology*, 2002, 222: 212–218.

[48] Moore, DR. Auditory brainstem of the ferret: long survival following cochlear removal progressively changes projections from the cochlear nucleus to the inferior colliculus. *J. Comp. Neurol.*, 1994, 339, 301-310.

[49] Dietrich, V; Nieschalk, M; Stoll, W; Rajan, R; Pantev, C. Cortical reorganization in patients with high frequency cochlear hearin loss. *Hearing Research*, 2001, 158, 95-101.

[50] Armstrong, B; Neville, H; Hillyard, S; Mitchell, T. Effects of auditory deprivation on color and motion processing, 2002.

[51] Illing, RB. Activity-dependent plasticity in the adult auditory brainstem. *Audiol. Neurootol.*, 2001, 6, 319-45.

[52] Kotak, VC; Fujisawa, S; Lee, FA; Karthikeyan, O; Aoki, C; Sanes, DH. Hearing Loss Raises Excitability in the Auditory Cortex. *The Journal of Neuroscience,* 2005, 25, 3908-3.

[53] Chugani HAT. A critical period of brain development: studies of cerebral glucose utilization with PET. *Prev. Med.*, 1998, 27, 184-8.

[54] Pujol, J; Vendrel, P; Junque, C; Marti-Vilalta, JL; Capdevila, A. When does human brain development end? Evidence of corpus callosum growth up to adulthood. *Ann. Neurol.*, 1993, 34, 71-5.

[55] Thai-Van, H; Cozma, S; Boutitie, F; Disant, F; Truy, E; Collet, L. The pattern of auditory brainstem response wave V maturation in cochlear-implanted children. *Clinical Neurophysiology*, 2007, 118, 676-689.

[56] Ito, J; Iwasaki, Y; Sakakibara, J; Yonekura, Y. Positron emission tomography study of auditory sensation in a patients with cochlear implant. *Ann. Otol. Rhinol. Larungol.*, 1993, 102, 797-801.

[57] Naito, Y; Okazawa, H; Honjo, I; Hirano, S; Takahashi, H; Shiomi, Y; Hoji, W; Kawano, M, Ishizu, K; Yonekura, Y. Cortical activation with sound stimulation in cochlear implant user demonstrated by positron emission tomography. *Cogn. Brain Res.*, 1995, 2, 207-214.

[58] Naito, Y; Tateya, I; Fujiki, N; Hirano, S; Ishizu, K; Nagahama, Y; Fukuyama, H; Kojima, H. Increased cortical activation during hearing of speech in cochlear implant users. *Hear Res.*, 2000, 143, 139-146.

[59] Mühlnickel, W; Elbert, T; Taub, E; Flor, H. Reorganization of auditory cortex in tinnitus. *Psycology*, 1998, 95, 10340-43.

[60] Matthies C. and Samii M. (1997). Management of vestibular schwannomas (acoustic neuromas): the value of neurophysiology for evaluation and prediction of auditory function in 420 cases. *Neurosurgery*, 40, 919-29.

[61] Lockwood, AH; Salvi, RJ; Coad, ML; Towsley, ML; Wack, DS; Murphy, BW. The functional neuroanatomy of tinnitus: evidence for limbic system links and neural plasticity. *Neurology*, 1998, 50, 114-120.

[62] Kraus, N. Auditory pathway encoding and neural plasticity in children with learning problems. *Audiol Neurootol*, 2001, 2, 221-227.

[63] Bradlow, AR; Kraus, N; Hayes, E. Speaking clearly for children with learning problems: sentence perception in noise. *J. Speech Lang Hear Reserch*, 2003.

[64] Hayes, ES; Warrier, CM; Nicol, TG; Zecker, SG; Kraus N. Neural plasticity following auditory training in children with learning problems. *Clinical Neurophysiology*, 2003, 114, 673-684.

[65] Russo, NM; Nicol, TG; Zecker, SG; Hayes EA; Kraus, N. Auditory training improves neural timing in the human brainstem. *Behavioural Brain Research*, 2005, 156, 95-103.

[66] Suga, N; Xiao, Z; Ma, X; Ji, W. Plasticità and corticofugal modulation for hearing in adult animals. *Neuron*, 2002, 36, 9-18.

[67] Zhang Y. and Suga N. (2000). Modulation of responses and frequency tuning of thalamic and collicular neurons by cortical activation in mustached bats. *J. Neurophysiol*, 2000, 84, 325-33.

In: Neural Pathways Research
Editor: F. L. Pichler, pp. 193-203
ISBN: 978-1-60456-214-9

Chapter IX

A Novel Approach to Understanding Neural Plasticity: Epigenetic Regulation of the *BDNF* Gene

Farah D. Lubin, Eric D. Roth, J. David Sweatt and Tania L. Roth
The Evelyn F. McKnight Brain Institute, Department of Neurobiology, University of Alabama at Birmingham, Birmingham, AL 35294

Abstract

Brain-derived neurotrophic factor (BDNF) plays a pivotal role in both central nervous system (CNS) development and neural plasticity throughout the lifespan. Indeed, abnormal expression and function of BDNF has been implicated in the pathogenesis of cognitive and mental disorders. Despite the known importance of BDNF function and gene expression in normal neural processes and CNS disorders, there has been little investigation into the molecular mechanisms responsible for complex BDNF transcriptional readout in the brain. In this commentary, we will shed light on a relatively new transcriptional regulatory mechanism, epigenetics, which may serve to regulate *bdnf* gene expression in the brain. We will highlight the epigenetic markings of the *bdnf* gene in adult learning and memory formation of both spatial and non-spatial tasks, by social experiences in early development, and in adult cognitive dysfunction following seizures. Such research offers novel concepts for understanding the transcriptional mechanisms subserving adult cognition and mental health, and furthermore promises avenues for therapeutic treatment of adult cognitive and mental dysfunction.

A Brief Introduction to BDNF

Brain-derived neurotrophic factor (BDNF) is a protein found throughout the brain, performing a multitude of actions pivotal for a healthy and functional CNS. BDNF not only helps to support the survival of existing neurons in the adult brain, but influences neuronal excitability, and promotes growth and differentiation of new neurons and synapses. BDNF-producing cells are found throughout the CNS, and BDNF signaling is mediated by at least two surface receptors that are its targets: neurotrophic receptors p75 and TrkB (reviewed in Hennigan et al., 2007).

The BDNF protein is synthesized from a gene that has a rather complex structure. Traditionally, the rat *bdnf* gene has been viewed as consisting of four 5' non-coding exons (I-IV) each linked to individual promoter regions (P1-4) and a 3' coding exon (V), which encodes for BDNF pre-protein. Recently, the *bdnf* gene has been re-characterized to now contain nine 5' non-coding exons (I-IX) linked to the common 3' coding exon (IX) (Aid et al., 2007; Liu et al., 2006). To minimize confusion, *bdnf* gene structure is presented in Figure 1 with old nomenclature depicted above new nomenclature. In addition, throughout this commentary reference to *bdnf* transcripts will be described using old *bdnf* nomenclature. Activity of specific promoter regions within the *bdnf* gene dictates spatial and temporal expression of specific *bdnf* transcripts.

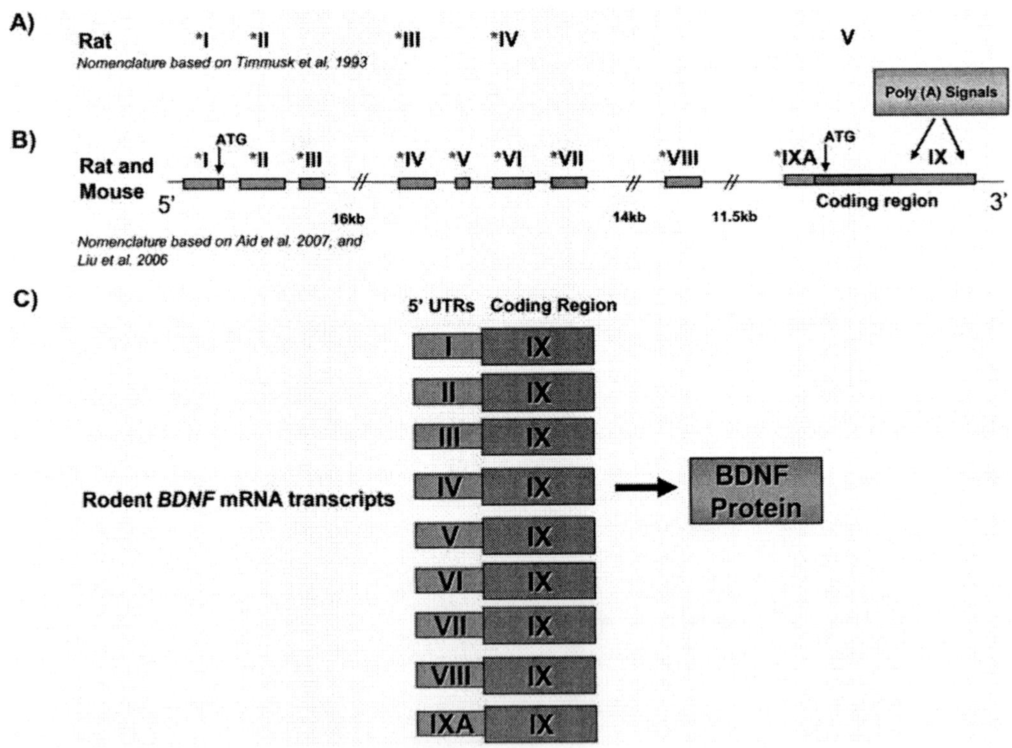

Figure 1. Schematic representation of the rodent brain-derived neurotrophic gene (BDNF). A) Previous analysis indicated that the rat bdnf gene contained four 5' non-coding exons, each with its own promoter (designated by an asterisk on the diagram), and a common 3' protein coding exon. B) The rodent bdnf gene now contains at least nine 5' non-coding exons, each with its own promoter, and a common coding exon. C) Transcription of the bdnf gene results in BDNF transcripts containing one of the 5' untranslated regions (5' UTRs) spliced to the common 3' protein coding region (exon IX).

Thus, known promoter regions for the *bdnf* gene are important regulatory sites for transcription factors and post-translationally modified histones that serve to direct differential expression and localization of *bdnf* exon-specific transcripts. To date, the biological significance of such a large number of transcripts is unknown.

Aberrant *BDNF* Gene Expression in Cognitive and Mental Dysfunction

The multitude of BDNF actions in the CNS supports the hypothesis that abnormal *bdnf* gene expression and function of the BDNF protein plays a prominent role in the development of cognitive disorders and mental illnesses. For example, one of the leading etiological hypotheses for Alzheimer's disease (AD) is that AD is due to a neurotrophic deficit. Indeed, *bdnf* gene activity and BDNF protein levels are greatly reduced with AD (reviewed in Elliott and Ginzburg, 2006). Since BDNF is pivotal to neuronal signaling and cell survival, such deficits in BDNF likely contribute to hippocampal dysfunction and neural degeneration prominent in AD. Furthermore, *in vitro* and *in vivo* studies demonstrate that BDNF plays a protective role against hippocampal and forebrain neuronal degeneration, thus further linking AD and BDNF expression (Tapia-Arancibia et al., 2004).

In relation to mental disorders, BDNF expression has been most characterized in schizophrenia and depression. For example, schizophrenic patients have decreased BDNF protein levels in both the prefrontal cortex and hippocampus, and in animal models of schizophrenia, there is also reduced total *bdnf* mRNA in these brain regions (reviewed in Angelucci et al., 2005). Similarly, patients with major depression have reduced plasma levels of BDNF. Animal models of depression also have parallel deficits in total *bdnf* mRNA and protein levels, and both the biochemical and behavioral deficits displayed by these animals can be rescued with antidepressants (reviewed in Duman and Monteggia, 2006; reviewed in Tsankova et al., 2006).

In addition to the role of BDNF in cognitive and mental disorders, social experiences have also been shown to engage *bdnf* gene activity and the BDNF protein throughout the CNS. For example, social isolation studies, such as the separation of infants from the mother and nest environment, provide us a unique perspective on the impact of social experiences on normal *bdnf* gene activity. If infant rodents are deprived of such experience, they grow up more anxious, have increased drug-seeking behavior, an array of cognitive deficits, and altered affiliative behaviors (reviewed in Lippmann et al., 2007). On the biochemical side, depriving an infant of such interaction yields a reduction in hippocampal and cortical *bdnf* mRNA and protein levels that persists well into adulthood (reviewed in Branchi et al., 2004; Fumagalli et al., 2007). Recently, Nair and colleagues (2007) have shown that maternal separation in the rat results in a differential upregulation of *bdnf* transcripts during postnatal development (exon II is upregulated at 14 days of age, while exons IV and V are upregulated at 21 days).

Overall, these data implicate aberrant *bdnf* gene expression in several disorders of cognition and mental health, and call for studies to begin to elucidate the molecular mechanisms controlling *bdnf* gene transcriptional readout in the brain. Future investigations into the transcriptional mechanism involved in *bdnf* gene expression will not only provide new appreciation of the molecular mechanisms controlling *bdnf* gene activity, but offer insight into viable avenues for therapeutic interventions to restore cognitive and mental health.

Evidence of Epigenetic Regulation of the *bdnf* Gene in Brain

Initial investigations into the transcriptional mechanisms underlying differential regulation of *bdnf* transcripts in the brain have focused on the role of transcription factors, such as cAMP response element binding protein and nuclear factor-kappa B, binding to *bdnf* promoter regions to initiate recruitment and activation of several coactivators of transcription. However, recent reports have highlighted another level of transcriptional regulation of the *bdnf* gene involving alterations in chromatin structure via post-translational modifications of histones such as acetylation, phosphorylation, and methylation, as well as by direct methylation of DNA (Figure 2). Collectively these transcriptional mechanisms are referred to as "epigenetic mechanisms." Historically, these forms of chromatin remodeling described above were thought to be involved exclusively in developmental processes; however, it has recently become clear that they are involved in a variety of long-term plastic changes in the adult brain.

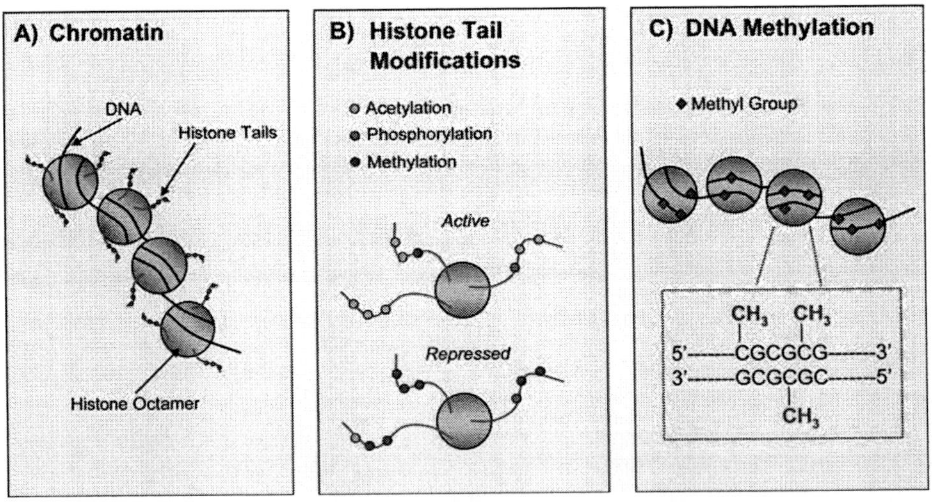

Figure 2. Schematic representation of epigenetic marking of the genome. A) DNA is compressed within the nucleus through interactions with histones, and this DNA-protein complex is referred to as chromatin. Two copies each of the histones H2A, H2B, H3 and H4 assemble, forming a histone octamer, around which DNA wraps and condenses. The N-terminal tail of a histone (B), contains many sites for epigenetic marking via histone acetylation, methylation, and phosphorylation. For example, acetylation of H3, shown as the addition of green circles to the tails, results in a relaxed chromatin state that promotes gene transcription, whereas methylation (shown via red circles) frequently results in a condensed chromatin state that represses gene transcription. C) Methylation of DNA is another method of epigenetic marking of the genome, where a methyl group (shown as red diamonds) is transferred to cytosines in what are referred to as CpG islands, which are regions in and around gene promoters that are rich in cytosine-guanine nucleotides. The addition of methyl groups in these regions represses gene transcription.

It has thus been proposed that epigenetic regulation of the *bdnf* gene is a transcriptional mechanism occurring in the adult brain that functions in learning and memory and in adult cognitive dysfunction related to pathological disorders.

In pursuit of this hypothesis, we and others have recently identified a potential role for chromatin structure regulation of the *bdnf* gene in either fear learning, spatial cognition, or following prolonged seizure activity and development of epilepsy (Bredy et al., 2007; Levenson et al., 2006; Taniura et al., 2006; Tsankova et al., 2004; Wood et al., 2006). Nestler's group has made seminal findings in chromatin remodeling of the *bdnf* gene in hippocampus after acute and chronic electroconvulsive seizures and in a mouse model of depression. Specifically, they found that chronic social defeat in adult mice produced a lasting downregulation of particular *bdnf* transcripts (III and IV) that were associated with increased histone methylation at the particular promoters (Tsankova et al., 2006). Huang and colleagues (2002) have shown increased histone H4 acetylation at the *bdnf* gene promoter in response to status epilepticus or prolonged seizure activity by pilocarpine in rat hippocampal neurons. In addition, we have discovered not only that DNA methylation is a dynamic process involved in controlling long-lasting changes in synaptic strength but that DNA Methyltransferase inhibitors applied acutely to hippocampal slices alter methylation of the *bdnf* gene at specific loci in upstream CpG islands, and block long-term potentiation (LTP) induction (Levenson et al., 2006). Together, these studies and our findings present a clearer understanding of the epigenetic regulation of the *bdnf* gene in several behaviors, and highlight a new activity-dependent transcriptional mechanism for gene expression changes in the adult brain.

Novel Research Approaches for Epigenetic BDNF Gene Regulation in Learning and Memory Formation

BDNF has been implicated in synaptic plasticity and memory processes in the CNS. As previously described, there are multiple *bdnf* mRNA transcripts that are all translated into the same protein. This versatility or complexity of *bdnf* transcript expression appears to play a key role in learning and memory processes. For example, one behavioral paradigm often used to study learning and memory is fear conditioning. This method trains organisms to learn to fear a neutral stimulus. This is accomplished by pairing an aversive stimulus (e.g, shock, unpleasant odor) with a neutral stimulus (e.g, context, tone). Using a fear conditioning paradigm in rats, Rattiner and colleagues (2004) found a selective increase in *bdnf* exons I and III mRNA in the amygdala, but observed no changes in exons II and IV.

The complex molecular interactions contributing to differential regulation of *bdnf* transcripts in learning and memory processes are unknown, but we hypothesize that epigenetic mechanisms are involved. Indeed, a recent study supports this hypothesis. Bredy and co-workers (2007) report that histone modifications around specific *bdnf* promoters influence memory formation. Specifically, they demonstrate that extinction of conditioned fear in mice is associated with an increase in histone H4 acetylation around the promoter of *bdnf* exon IV, as well as increases in *bdnf* exons I and IV mRNA in prefrontal cortex.

Preliminary studies from our work in the Sweatt lab suggest that in addition to differential hippocampal *bdnf* expression among exons, fear conditioning also results in DNA methylation differences across the *bdnf* gene. However, it remains to be determined if these patterns of exon specific expression and epigenetic markings are strictly related to fear conditioning or if they represent a more general phenomenon associated with learning and memory. Thus it is important to begin exploring these phenomena using other behavioral paradigms.

Another behavioral paradigm prevalent in the field of learning and memory is the Morris water maze. This method involves placing a rat or mouse in a pool of water. The animal is then highly motivated to explore or navigate the water maze to find an escape platform. In a Morris water maze paradigm, Silhol and colleagues (2007) demonstrated that spatial memory training modifies the expression of BDNF protein and BDNF tyrosine kinase receptors. Ressler's group has shown that hippocampal *bdnf* knockout mice have deficits in spatial learning and memory (Heldt et al., 2007). However, potential epigenetic mechanisms modulating *bdnf* transcription in spatial learning and memory have yet to be examined. Through our work, we hope to gain a better understanding of epigenetic contributions to spatial learning and memory by examining DNA methylation patterns of various *bdnf* transcripts following the performance of a spatial memory task. Interesting comparisons can then be made to the epigenetic patterns associated with fear conditioning and other behavioral paradigms. Although it is too early to further speculate on potential outcomes from these experiments, we strongly feel that future investigations into the epigenetic regulation of *bdnf* transcripts may provide fascinating insights into molecular mechanisms underlying learning and memory processes.

Discovering Epigenetic Marking of the *bdnf* Gene by Early Social Experiences

It has been suggested that the long-term behavioral consequences of early experiences may be mediated through epigenetic changes. To date, we understand very little of the impact of early experiences on the epigenome. To the best of our knowledge, there have been no published studies on any epigenetic changes to the *bdnf* gene resultant from early manipulations of the social environment. But notably, the Levitt's group has shown that methylation status of the *bdnf* gene is dynamic over the course of postnatal development in the mouse. Specifically, as de novo methylation of *bdnf* exon IV is correlated with suppression of the *bdnf* gene in the striatum, demethylation of this same exon is observed in the hippocampus, where *bdnf* mRNA expression substantially increases (Dennis and Levitt, 2005). Their work argues that dynamic DNA methylation is a crucial mechanism mediating *bdnf* gene transcription during normal development. Thus, it is reasonable to hypothesize that the quality of early experiences will have an impact on this mechanism.

What we do know in regard to the impact of early experiences on epigenetic mechanisms regulating gene activity is work from Meaney and colleagues. Adult offspring of mothers that frequently lick and groom their pups (LG-ABN) have a more moderate stress response that is attributable to increased hippocampal glucocorticoid receptor expression, which are receptors that help to moderate neural and behavioral responses to stress (reviewed in Szyf et al., 2007). In groundbreaking research, Meaney and co-workers (2007) demonstrated that patterns of DNA methylation of the glucocorticoid receptor gene in the hippocampus are influenced by maternal behavior. Specifically, adults derived from the high LG-ABN mothers have both less methylation at the glucocorticoid receptor gene promoter and greater overall histone acetylation within the hippocampus. They have also been able to demonstrate that these epigenetic markings can be reversed, either by cross-fostering of offspring between the two types of mothers or through drug treatment in adulthood. Thus, these studies have provided the first demonstration that maternal behavior influences the epigenome.

This leads us to suggest that child maltreatment likewise alters epigenetic mechanisms throughout the brain, resulting in a persistent cellular "memory" that not only canalizes brain development but behavior throughout the lifespan. We are currently exploring this hypothesis using a rodent model of infant abuse and neglect (Roth and Sullivan, 2005). Our results thus far demonstrate that caregiver abuse and neglect leaves persisting epigenetic marks on the *bdnf* gene in several brain regions, and that pharmacologic intervention in adulthood can reverse these marks and rescue *bdnf* gene activity. These observations leave us to speculate that increased susceptibility to cognitive and mental disorders in adults with a history of childhood maltreatment reflect the lasting imprint of early stress or on epigenetic mechanisms pivotal to normal cognition (learning and memory) and emotional health.

Exploring the Importance of Epigenetic *bdnf* Gene Regulation in Cognitive Dysfunction Associated with Epilepsy

Epilepsy is a neurological disorder wherein recurrent and unprovoked seizures occur. Epilepsy is a series of symptoms involving episodic abnormal electrical activity in the brain that lead to increased susceptibility to cognitive and mental impairments. Indeed, cognitive deficits are associated with epilepsy in both children and adults (reviewed in Helmstaedter et al., 2003; Leritz et al., 2006). However, to date, no effective treatments are available for cognitive and behavioral impairments in epilepsy. Thus, we need to begin to address how abnormal neuronal activity associated with epilepsy significantly impacts normal cognitive processes and behavior. In this section, we will briefly highlight an emerging research idea suggesting alterations in epigenetic transcriptional regulation of genes, such as *bdnf*, may underlie changes in hippocampal organization and function of mature neurons resulting in spontaneous seizures (epilepsy) and cognitive decline.

The *bdnf* gene is one of the numerous genes modulated in epilepsy models and in human temporal lobe epilepsy (Binder, 2004; Binder et al., 2001). Seizure activity increases the expression of *bdnf* mRNA and protein (Humpel et al., 1993; Vezzani et al., 1999). Molecular studies show that interfering with BDNF *in vivo* inhibits the development of the epileptic state (Binder et al., 1999). In contrast, other studies report that intrahippocampal BDNF infusion promoted the development of epilepsy (Koyama and Ikegaya, 2005; Scharfman et al., 2005). Additional studies are necessary to further elucidate the role of BDNF in the development of epilepsy and related cognitive impairments.

Given that *bdnf* gene expression is regulated in both epilepsy and memory formation, we propose that altered *bdnf* gene transcription via regulation of chromatin structure within mature hippocampal neurons may be responsible for cognitive decline in epileptic patients. The first experimental results linking chromatin structure regulation to seizure-induced cognitive decline were recently obtained in relation an antiepileptic drug, valproic acid (VPA), which has been described as a histone deacetylase inhibitor (Jessberger et al., 2007). In this study, VPA treatment protected epileptic animals from hippocampus-dependent cognitive impairment after kainate-induced seizures. This supports the notion that altered chromatin structure may play a role in the regulation of genes, such as *bdnf*, involved in seizure-induced cognitive decline. Interestingly, Bredy and colleagues (2007) demonstrate that VPA treatment enhanced long-term memory for extinction that was associated with an increase in histone H4 acetylation at *bdnf* promoter regions and *bdnf* exon IV mRNA expression in the prefrontal cortex. Further analysis of the molecular mechanisms by which gene expression of *bdnf* is induced by seizure activity could provide novel concepts and targets for cognitive regression therapy using antiepileptic drugs such as VPA.

Although much of the epigenetics literature has focused on histone modifications across the *bdnf* gene, current studies in our laboratory have identified a role for DNA methylation of the *bdnf* gene as a direct transcriptional mechanism for differential expression of *bdnf* transcripts in the adult hippocampus in response to prolonged seizure activity. Thus, we found that seizure activity results in the demethylation of the *bdnf* gene at specific promoter regions that correspond to increased *bdnf* gene activity in the adult hippocampus. Therefore, studies aimed at understanding *bdnf* gene expression changes during prolonged seizure activity and in epilepsy should involve analysis of transcription-regulating mechanisms such as DNA methylation.

Future Direction

Epigenetic mechanisms are increasingly being recognized for their importance in exerting control over gene expression and behavior, and offer a plausible molecular mechanism responsible for stable changes in brain function and behavior. In relation to *bdnf* gene expression, what is needed now are correlative studies that begin to link the functional significance between differential expression of *bdnf* transcripts to changes in BDNF protein levels and how this contributes to neural plasticity or cognitive disorders. The study of epigenetic *bdnf* gene regulation as a mechanism for gene transcription adds yet another level of regulation of BDNF expression that will have to be incorporated in future molecular studies.

References

Aid, T., Kazantseva, A., Piirsoo, M., Palm, K., and Timmusk, T. (2007). Mouse and rat BDNF gene structure and expression revisited. *J. Neurosci. Res.* 85, 525-535.

Angelucci, F., Brene, S., and Mathe, A. A. (2005). BDNF in schizophrenia, depression and corresponding animal models. *Mol. Psychiatry*. 10, 345-352.

Binder, D. K. (2004). The role of BDNF in epilepsy and other diseases of the mature nervous system. *Adv. Exp. Med. Biol.* 548, 34-56.

Binder, D. K., Croll, S. D., Gall, C. M., and Scharfman, H. E. (2001). BDNF and epilepsy: too much of a good thing? *Trends Neurosci.* 24, 47-53.

Binder, D. K., Routbort, M. J., Ryan, T. E., Yancopoulos, G. D., and McNamara, J. O. (1999). Selective inhibition of kindling development by intraventricular administration of TrkB receptor body. *J. Neurosci.* 19, 1424-1436.

Branchi, I., Francia, N., and Alleva, E. (2004). Epigenetic control of neurobehavioural plasticity: the role of neurotrophins. *Behav. Pharmacol.* 15, 353-362.

Bredy, T. W., Wu, H., Crego, C., Zellhoefer, J., Sun, Y. E., and Barad, M. (2007). Histone modifications around individual BDNF gene promoters in prefrontal cortex are associated with extinction of conditioned fear. *Learn Mem.* 14, 268-276.

Dennis, K. E., and Levitt, P. (2005). Regional expression of brain derived neurotrophic factor (BDNF) is correlated with dynamic patterns of promoter methylation in the developing mouse forebrain. *Brain Res. Mol. Brain Res.* 140, 1-9.

Duman, R. S., and Monteggia, L. M. (2006). A neurotrophic model for stress-related mood disorders. *Biol. Psychiatry* 59, 1116-1127.

Elliott, E., and Ginzburg, I. (2006). The role of neurotrophins and insulin on tau pathology in Alzheimer's disease. *Rev. Neurosci.* 17, 635-642.

Fumagalli, F., Molteni, R., Racagni, G., and Riva, M. A. (2007). Stress during development: Impact on neuroplasticity and relevance to psychopathology. *Prog. Neurobiol.* 81, 197-217.

Heldt, S. A., Stanek, L., Chhatwal, J. P., and Ressler, K. J. (2007). Hippocampus-specific deletion of BDNF in adult mice impairs spatial memory and extinction of aversive memories. *Mol. Psychiatry,* 12, 656-670.

Helmstaedter, C., Kurthen, M., Lux, S., Reuber, M., and Elger, C. E. (2003). Chronic epilepsy and cognition: a longitudinal study in temporal lobe epilepsy. *Ann. Neurol.* 54, 425-432.

Hennigan, A., Trotter, C., and Kelly, A. M. (2007). Lipopolysaccharide impairs long-term potentiation and recognition memory and increases p75NTR expression in the rat dentate gyrus. *Brain. Res.* 1130, 158-166.

Huang, Y., Doherty, J. J., and Dingledine, R. (2002). Altered histone acetylation at glutamate receptor 2 and brain-derived neurotrophic factor genes is an early event triggered by status epilepticus. *J. Neurosci.* 22, 8422-8428.

Humpel, C., Wetmore, C., and Olson, L. (1993). Regulation of brain-derived neurotrophic factor messenger RNA and protein at the cellular level in pentylenetetrazol-induced epileptic seizures. *Neuroscience* 53, 909-918.

Jessberger, S., Nakashima, K., Clemenson, G. D., Jr., Mejia, E., Mathews, E., Ure, K., Ogawa, S., Sinton, C. M., Gage, F. H., and Hsieh, J. (2007). Epigenetic modulation of seizure-induced neurogenesis and cognitive decline. *J. Neurosci.* 27, 5967-5975.

Koyama, R., and Ikegaya, Y. (2005). To BDNF or not to BDNF: that is the epileptic hippocampus. *Neuroscientist* 11, 282-287.

Leritz, E. C., Grande, L. J., and Bauer, R. M. (2006). Temporal lobe epilepsy as a model to understand human memory: the distinction between explicit and implicit memory. *Epilepsy. Behav.* 9, 1-13.

Levenson, J. M., Roth, T. L., Lubin, F. D., Miller, C. A., Huang, I. C., Desai, P., Malone, L. M., and Sweatt, J. D. (2006). Evidence that DNA (cytosine-5) methyltransferase regulates synaptic plasticity in the hippocampus. *J. Biol. Chem.* 281, 15763-15773.

Lippmann, M., Bress, A., Nemeroff, C. B., Plotsky, P. M., and Monteggia, L. M. (2007). Long-term behavioural and molecular alterations associated with maternal separation in rats. *Eur. J. Neurosci.* 25, 3091-3098.

Liu, Q. R., Lu, L., Zhu, X. G., Gong, J. P., Shaham, Y., and Uhl, G. R. (2006). Rodent BDNF genes, novel promoters, novel splice variants, and regulation by cocaine. *Brain Res.* 1067, 1-12.

Nair, A., Vadodaria, K. C., Banerjee, S. B., Benekareddy, M., Dias, B. G., Duman, R. S., and Vaidya, V. A. (2007). Stressor-specific regulation of distinct brain-derived neurotrophic factor transcripts and cyclic AMP response element-binding protein expression in the postnatal and adult rat hippocampus. *Neuropsychopharmacology* 32, 1504-1519.

Rattiner, L. M., Davis, M., and Ressler, K. J. (2004). Differential regulation of brain-derived neurotrophic factor transcripts during the consolidation of fear learning. *Learn Mem.* 11, 727-731.

Roth, T. L., and Sullivan, R. M. (2005). Memory of early maltreatment: neonatal behavioral and neural correlates of maternal maltreatment within the context of classical conditioning. *Biol. Psychiatry* 57, 823-831.

Scharfman, H., Goodman, J., Macleod, A., Phani, S., Antonelli, C., and Croll, S. (2005). Increased neurogenesis and the ectopic granule cells after intrahippocampal BDNF infusion in adult rats. *Exp. Neurol.* 192, 348-356.

Silhol, M., Arancibia, S., Maurice, T., and Tapia-Arancibia, L. (2007). Spatial memory training modifies the expression of brain-derived neurotrophic factor tyrosine kinase receptors in young and aged rats. *Neuroscience* 146, 962-973.

Szyf, M., Weaver, I., and Meaney, M. (2007). Maternal care, the epigenome and phenotypic differences in behavior. *Reprod. Toxicol.* 24, 9-19.

Taniura, H., Sng, J. C., and Yoneda, Y. (2006). Histone modifications in status epilepticus induced by kainate. *Histol. Histopathol.* 21, 785-791.

Tapia-Arancibia, L., Rage, F., Givalois, L., and Arancibia, S. (2004). Physiology of BDNF: focus on hypothalamic function. *Front Neuroendocrinol.* 25, 77-107.

Tsankova, N. M., Berton, O., Renthal, W., Kumar, A., Neve, R. L., and Nestler, E. J. (2006). Sustained hippocampal chromatin regulation in a mouse model of depression and antidepressant action. *Nat. Neurosci.* 9, 519-525.

Tsankova, N. M., Kumar, A., and Nestler, E. J. (2004). Histone modifications at gene promoter regions in rat hippocampus after acute and chronic electroconvulsive seizures. *J. Neurosci.* 24, 5603-5610.

Vezzani, A., Ravizza, T., Moneta, D., Conti, M., Borroni, A., Rizzi, M., Samanin, R., and Maj, R. (1999). Brain-derived neurotrophic factor immunoreactivity in the limbic system of rats after acute seizures and during spontaneous convulsions: temporal evolution of changes as compared to neuropeptide Y. *Neuroscience 90*, 1445-1461.

Wood, M. A., Attner, M. A., Oliveira, A. M., Brindle, P. K., and Abel, T. (2006). A transcription factor-binding domain of the coactivator CBP is essential for long-term memory and the expression of specific target genes. *Learn Mem.* 13, 609-617.

In: Neural Pathways Research
Editor: F. L. Pichler, pp. 205-220
ISBN: 978-1-60456-214-9

Chapter X

From Affective Blindsight to Affective Blindness: When Cortical Processing Suppresses Subcortical Information

Jacob Jolij*
School of Psychology
University of Exeter

Abstract

Some patients with a lesion to their primary visual cortex (V1) show a phenomenon called 'affective blindsight': they can correctly guess the emotional expression of a face presented to their blind field. This ability has been attributed to an evolutionary old subcortical route for processing of emotional information. Neuroimaging studies show that this route is also functioning in normal observers: emotional faces, made invisible using backward masking, still elicit responses of the amygdala via a subcortical pathway. However, contrary to blindsight patients, normal observers cannot overtly identify the expression of unseen emotional faces. In this chapter, I will introduce and discuss a model describing the relation between conscious and unconscious processing that explains this discrepancy.

Introduction

Visual processing and visual awareness are not the same: recent advances in the cognitive neurosciences have made it clear that what we consciously perceive is in fact only a tip of the iceberg of the vast amount of information that is processed by the brain. This seems especially the case for emotional stimuli, such as emotional facial expressions: every time we

* E-mail: j.jolij@exeter.ac.uk

perceive a face, an extensive network of interacting brain areas processes different aspects of that face, such as identity, gender, age, and emotional expression (Vuilleumier and Pourtois, 2007). We use this information to guide our behaviour, for example to initiate an approach reaction towards a happy person, or avoid someone when there is a clear expression of anger on her or his face. Even more basis responses, such as fight or flight reactions can be triggered by facial expressions: an expression of fear, for example, may signal that there is a potential threat imminent, and it may be a good idea to prepare to run away (Ohman and Soares, 1993; LeDoux, 1996).

Generally, we believe that we are in control of our actions, and that we guide our behaviour by what we consciously perceive. In the example above, we may think we base our decision to avoid a friend who looks extremely angry on the fact that we have consciously perceived his facial expression. But is this true? Studies in patients with a lesion to the primary visual cortex (V1) show that these patients have the ability to correctly identify the emotional expressions of faces that are shown to them, even though these patients are blind. This is called 'affective blindsight'. Affective blindsight is thought to be mediated by an evolutionary old subcortical pathway that processes visual information independently of the cortical face processing areas that are necessary of conscious awareness (Morris, Ohman, and Dolan, 1999). This pathway activates the amygdala, a structure responsible for the regulation of emotional behaviour. Apparently, blindsight patients are able to use the information processed in this pathway in order to discriminate between different emotional expressions. Since even patients with fairly recent lesions show affective blindsight, it is not likely that this ability is the result of structural changes in the brains of these patients (Pegna, Khateb, Lazeyras, and Seghier, 2005). The subcortical pathway mediating affective blindsight has projections to the parietal somatosensory association cortex. Activations in this area have been demonstrated to correlate with subjective feelings of emotional experience in absence of visual awareness in patients showing affective blindsight (Anders, Birbaumer, Sadowski, Erb, Mader, Grodd, and Lotze, 2004). According to some researchers, affective blindsight shows that we in fact are 'blindly led by our emotions' when responding to emotional expressions: our reactions are guided by unconsciously processed information; conscious perception is just a slow epiphenomenon (DeGelder, Vroomen, Pourtois, and Weiskrantz, 1999; DeGelder, Vroomen, Pourtois, and Weiskrantz, 2000; Heywood and Kentridge, 2000).

Indeed there is ample evidence that the emotional significance of facial expressions is processed by a subcortical pathway involving the amygdala in normal observers (Morris, Ohman, and Dolan, 1999). Even when facial expressions cannot be consciously identified, the amygdala shows differential responses to emotional expressions of fear, anger, sadness, and happiness (Whalen, Rauch, Etcoff, McInerney, Lee, and Jenike, 1998; Kilgore and Yurgelun-Todd, 2004). Apparently, the same pathway that mediates blindsight is doing exactly the same in normal observers: processing information about emotional expressions of faces, and activate the amygdala, suggesting that evolution has equipped us with a backup route for processing of emotions (LeDoux, 1996). However, there is an interesting discrepancy between the work on affective blindsight and the work on unconscious emotion processing in normal observers: the very definition of 'unconscious emotion processing' seems to be conflicting with the notion of 'affective blindsight'. In virtually all studies on the neural correlates of unconscious emotion processing in normal observers, unawareness of the faces used is assessed by asking an observer to report the emotional expression of the faces used in the experiment. If the observer is unable to do so, this is taken as evidence of unawareness, and indeed, even though the observer cannot tell the difference between a neutral and a fearful expression, his amygdala can (e.g. Whalen *et al.*, 1998). However, in affective blindsight, we see the opposite: these patients are not able to consciously perceive facial expression shown to them, yet they can discriminate between different emotional expressions, an ability shown to be associated with differential activations of the amygdala (Morris, De Gelder, Weiskrantz, and Dolan, 2000; Pegna *et al.*, 2005). How can we explain this paradoxical situation, where the same pathway seems to support successful emotion discrimination in patients, but not in normal observers?

In this chapter I will lay out a hypothesis explaining this discrepancy, based on my work on transcranial magnetic stimulation induced blindsight in normal observers (Jolij and Lamme, 2005). In short, I propose that the information in the fast but crude subcortical pathways is often ignored or even actively inhibited by the executive areas of the brain: the executive systems show a strong preference for cortically processed information when generating a response. Only when the quality of cortically processed information is compromised over a longer period of time, the brain's executive systems take subcortically processed information into account.

Subcortical Processing of Emotional Expressions: Fast but Crude

The idea that emotional content of visual stimuli is processed in a subcortical pathway is not new. In his famous two route model of emotion processing, LeDoux (1996) proposes that potentially threatening elements in a visual scene (for example, a snake, or a fearful face) can be detected in two ways: first, it could be that the object sensitive areas in cortical visual system identify the object and activate the amygdala. This route, however, can be quite slow: although latencies of object sensitive neurons in the temporal cortex have been shown to be in the range of 100 – 110 ms (e.g. Kiani, Esteky, and Tanaka, 2005), latencies of 200 – 300 ms have been reported for basic visual processes, such as scene segmentation, and visuomotor transformations (e.g. Heinen, Jolij, and Lamme, 2005; Jolij, Van Gaal, Scholte, and Lamme, *in press*), especially under non-optimal viewing conditions. That may be too slow if a tiger is looming in the jungle: by the time the cortical visual system has identified the blur coming at you as a tiger, it may be too late. LeDoux theorizes that evolution has equipped the brain with a shortcut to rapidly analyze potentially threatening elements in a visual scene. This so-called subcortical route involves the superior culliculi, the pulvinar, and the amygdala, and can detect stimuli that potentially require a fight or flight response, such as snakes, spiders, and fearful faces. Because of the tuning properties of the neurons in this pathway, it is less sensitive to fine details (Sahraie *et al.*, 2003), but it is faster, and can activate the amygdala without requiring the extensive processing of the cortical system. This allows an organism to prepare or even initiate a fight or flight reaction more rapidly. It is believed that this very direct perception-action system in fact is a very old evolutionary system, and may have been a forerunner of the more advanced cortical visual system.

LeDoux's (1996) model is quite generally accepted, but direct anatomical evidence of a subcortical processing route for emotional information processing in humans is rare. The recent neuroimaging studies on unconscious processing of emotional expressions do seem to confirm the existence of this route, however. In most studies on the neural correlates of unconscious processing of facial expressions, a backward visual masking paradigm is used to prevent conscious identification of pictures of faces with an emotional expression. There is evidence that masking derives its effectiveness from interfering with cortical processing of the masked stimulus: either the mask 'catches up' with the masked stimulus (Breitmeyer and Ogmen, 2000; Macknick, 2006), or interferes with cortico-cortical interactions (Lamme, Zipser and Spekreijse, 2002; Fahrenfort, Scholte, and Lamme, 2007), and prevents the masked stimulus from being processed up to a level of conscious awareness. It is therefore assumed that any residual processing of masked emotional faces must be restricted to subcortical areas. A consistent finding in neuroimaging studies of unconscious emotion processing is that unseen fearful faces activate the amygdala, or more specifically, the right amygdala. Left amygdala activations are reported when faces are consciously perceived by the observers (Pessoa, 2005). Studies in blindsight patients seem to confirm the role of particularly the right amygdala in unconscious emotion detection (Morris *et al.*, 2000; Pegna *et al.*, 2005). Unconscious detection of fear is not just based on configural processing of a face. Apparently, presentation of a pair of fearful eyes is enough to evoke an amygdala response even the eyes are masked (Whalen *et al.*, 2004). Finally, the amygdala is not just capable of detecting fearful stimuli, but also shows differential responses to happy and sad expressions, suggesting that unconscious processing of facial expressions is not just limited to fear (Kilgore and Yurgelun-Todd, 2004).

More direct evidence of a subcortical pathway to the amygdala has been provided by Morris, Ohman, and Dolan (2000): they observed increased connectivity in this pathway when the amygdala showed an increased response to fearful expression, but for unseen faces only, a finding corroborated by Lidell *et al.* (2005), using completely undetectable stimuli, and Morris *et al.* (2000) in a blindsight patient. As is to be expected from the neurophysiological characteristics of the neurons in this pathway, the subcortical pathway meditating unseen emotions is sensitive to specifically the lower spatial frequency components of faces, containing the coarse details of a face, mainly the eyes and the mouth – elements that play a critical role in signalling emotional expression. Finer details, such as wrinkles are processed by areas in the neocortex (Vuilleumier, Armony, Driver, and Dolan, 2003; Winston, Vuilleumier, and Dolan, 2003).

Summarizing, even when visual stimuli such as faces with an emotional expression are masked, and not processed by the cortical visual system, the amygdala receives input via a network of subcortical structures signalling the coarse details of the unseen stimulus. Apparently, this information is sufficient to allow the amygdala to differentiate between different emotional expressions.

Affective Blindness or Affective Blindsight?

Even though normal observers cannot overtly discriminate between masked facial expression while the information necessary to make such discriminations is present in the brain, unseen emotional expressions do affect behaviour. For example, unseen fearful expressions can increase the galvanic skin response, a measure of autonomic nervous activity associated with acute stress, indicating that even an unseen fearful expression may trigger a fight/flight reaction (Ohman and Soares, 1993). An even more compelling demonstration of how unseen facial expressions can subtly influence behaviour has been provided in a classical study by Murphy and Zajonc (1993). In this study, observers had to rate Chinese characters they were unfamiliar with as either positive or negative. Critically, pictures of either positive or negative facial emotional expressions were presented very briefly, preventing conscious discrimination of the facial expression, right before presentation of the Chinese character. It turned out that unseen negative emotional expressions resulted in a more negative evaluation of Chinese characters, while positive emotions resulted in a more positive evaluation. It is not unlikely that these subtle effects are mediated by the same subcortical pathways mediating affective blindsight.

Apparently, observers are not as blind to unseen facial expressions as they seem. Could it be that the assessment of unawareness of emotional expressions in studies of unconscious neural processing may have been not sensitive enough to pick up residual emotion processing? A recent study by Pessoa, Japee, Sturman and Ungerleider (2005) seems to suggest so: as in other studies on unconscious emotion processing, Pessoa *et al.* measured neural activity evoked by masked expressions of fear. However, they found that some of the masked faces were perceived by their participants, particularly by a sub-population of 'over-achievers'. When this was factored in the analysis of the fMRI data, it turned out that the amygdala only showed a response to these 'seen' trials. Since earlier studies did not use a trial-by-trial analysis, Pessoa *et al.* suggest that the 'unconscious' processing of emotional expression may not be so unconscious after all. Apparently, there are large inter- and intra-individual differences in the detection of the emotion at near-threshold presentation conditions (Pessoa, Japee, and Ungerleider, 2005; Pessoa and Padmala, 2005; Szczepanowski and Pessoa, 2007). It may be that the alleged unconscious perception of emotions may in fact be an example of 'weak conscious processing', and that the remarkable inability of normal observers to discriminate between masked facial expressions may be due to the fact that unconscious effects can only be revealed if consciously processed information is completely absent in stead of reduced (Snodgrass, Bernat, and Shevrin, 2004; Snodgrass and Shevrin, 2006).

Especially this latter point is important. It may potentially explain the discrepancy between affective blindsight in patients and affective blindness in normal observers: if normal observers perform at a level of 'weak conscious processing', this residual processing may override unconsciously processed information – a phenomenon that has been demonstrated in unconscious semantic priming (Snodgrass, Bernat, and Shevrin, 2004; Snodgrass and Shevrin, 2006). To really make sure that any residual processing can only be attributed to purely unconscious processing, and can actually be measured, one needs to make sure that stimuli are presented below the threshold of objective awareness, that is, participants must be at chance level in a stimulus *detection* task in stead of on a *discrimination* task in order to properly assess unawareness: participants should not only be unable to tell the emotion of a masked face (that is, they are subjectively unaware of a facial expression); they should even be unable to report that a face has been presented *at all* – something called objective unawareness (Snodgrass, Bernat, and Shevrin, 2004).

In most of the work on unconscious processing of emotions, masking of stimuli has prevented conscious discrimination of stimuli in stead of conscious detection of stimuli. It might be that this has allowed some conscious effects on emotion processing in subcortical structures, and potentially interfered with subtle effects of unconscious processing on over behaviour. However, even in blindsight there are some reports of residual visual awareness, which, though unlikely, may mediate weak conscious processing of unseen signals of fear (Sahraie, Weiskrantz, Barbur, Simmons, Williams, and Brammer, 1997; Silvanto, Cowey, Lavie, and Walsh, 2007). In the first study on affective blindsight, indeed patient GY, who has unilateral damage to his primary visual cortex, did report to be aware of the on- and offset of stimuli, even though he reported not to consciously perceive the emotional expressions (DeGelder *et al.*, 1999). This finding shows that it is hard to meet the criteria of objective unawareness, even in patients with cortical damage, though these patients still show effects of unconsciously processed information on overt behaviour.

However, Lidell *et al.* (2005) have demonstrated that even with undetectable stimuli a brainstem-amygdala-cortical network is activated in normal observers, despite the objective unawareness of the stimuli; furthermore affective blindsight has been reported in patients with bilateral destruction of the primary visual cortex, excluding the possibility that contralesional cortical visual pathways contribute to their residual visual processing, as has been hypothesized in the case of unilateral patients such as G.Y. (Hamm, Weike, Schupp, Treig, Dressel, and Kessler, 2003; Pegna *et* al., 20051 Sahraie *et al.*, 1997). These studies demonstrate that even in cases of objective unawareness, emotion processing still occurs in a subcortical pathway, and, more importantly, still show the same paradoxical pattern where normal observers cannot discriminate between unseen facial expressions (Liddell *et al.*, 2005), but patients can (Pegna *et al.*, 2005).

Recently, we have published a study in which we report both affective blindsight and affective blindness in normal observers (Jolij and Lamme, 2005). In this study, we let observers discriminate between happy and sad schematic faces, while we stimulated the early visual cortex on different moments after stimulus presentation, ranging from 50 to 300 ms after stimulus onset. We replicated the basic finding that TMS of the occipital pole suppresses perception of visual stimuli when applied around 100 ms after stimulus onset (Amassian, Cracco, Maccabee, Cracco, Rudell, and Eberle, 1989; Corthout, Uttl, Ziemann, Cowey, and Hallett, 1999), as assessed with both localization and detection tasks, but despite objective unawareness of the stimuli, observers were still able to guess the emotion of schematic faces above chance level in those trials in which TMS successfully suppressed the conscious percept of the stimuli. However, when we increased the presentation time of our stimuli from 17 ms to 33 ms, we still were able to successfully suppress conscious awareness of the stimuli when TMS was applied around 100 ms after stimulus onset, but in this condition the observers' ability to successfully report the emotional expression of the schematic faces was suppressed as well.

This seems rather counterintuitive. With longer presentation times, one would expect that the signal in the subcortical pahtways would be stronger, leading to even better unconscious recognition of affect, in stead of the reverse effect we report. A further analysis revealed that a similar effect was present in the later trials of the short presentation time condition: after prolonged exposure to the task, unconscious recognition of emotional expression diminished. Disappearance of the blindsight effect seemed to covary with how well the observers performed in the approximately five out of every six trials in which TMS was not effective in suppressing the visual stimulus because of the stimulus-TMS interval: when overall performance on the emotion detection and stimulus localisation tasks was around 70% in these unsuppressed trials, TMS around 100 ms resulted in suppression of conscious awareness of the stimulus, but not in suppression of emotion recognition. However, when performance on both tasks in unsuppressed trials was 90% or higher, TMS around 100 ms did not only suppress conscious awareness, but also emotion recognition (Jolij and Lamme, 2005). We replicated this finding in two observers by manipulating stimulus luminance, with the same effect (Jolij, unpublished data).

Apparently, observers can only respond unconsciously processed emotions in quite specific circumstances. There seems to be an interaction between overall state of certainty about a given task, and the ability to respond to unconsciously processed information: the better one performs, the more likely one is to respond to what one consciously perceives, and ignore unconsciously processed information. However, if a task is particularly hard, unconsciously processed information might be taken into account when making a response (Jolij and Lamme, 2005). But how does this interaction work? How does the brain decide which type of information to use? And why would the brain's executive systems ignore unconsciously processed information?

Repression of Unconscious Information by Conscious Processing: A Model

Our study (Jolij and Lamme, 2005) suggests an interaction between conscious perception, or better, task accuracy, and unconscious processing in emotion recognition: only in certain circumstances observers take into account unconsciously processed emotional information when responding to unseen stimuli. This may potentially explain the paradoxical finding that blindsight patients can report the emotional expression of unseen emotional faces, while normal observers cannot. In the remainder of this chapter, I will lay out a basic model of this interaction between conscious and unconscious processing.

I will take the dual route model of emotional face processing as a starting point (Johnson, 2005; Ledoux, 1996; Vuilleumier, Armony, Driver, and Dolan, 2003; Vuilleumier and Pourtois, 2007). According to this model, the emotional expression of a face is processed in parallel by both cortical areas, and subcortical areas. Processing in the cortical areas is necessary for conscious perception of the face (though not all cortical processing will result in conscious awareness; cf. Lamme, 2003), while on the other hand processing in the subcortical areas cannot support visual awareness of the face.

The accuracy of the cortical processing route under normal viewing conditions is near perfect: performance on emotion discrimination tasks of unmasked facial expressions is near 100% (cf. Jolij and Lamme, 2005). Obviously, there is a large number of factors that can strongly impair the accuracy of cortically processed information: in our experiment, shortening presentation times yielded a performance drop of almost 25% to 75%, but masking or lowering luminance of stimuli will also degrade a conscious percept, leading to a lower accuracy of the cortical pathways. The subcortical pathways are less accurate even under optimal viewing conditions. If we assume that the performance of blindsight patient GY is based on subcortically processed information only, accuracy of subcortically processed information should be around 60% - 80% for optimally presented stimuli such as videos of about 3s (De Gelder *et al.* 1999). Interestingly, our results seem to indicate a similar performance under far less optimal viewing conditions: accuracy on emotion discrimination of briefly presented, low contrast stimuli in fully suppressed trials in our experiments was around 65% - 70%. A safe estimate of the accuracy of the subcortical processing route for facial expressions would therefore be about 70%, but apparently accuracy of this route is not as sensitive to decreases in stimulus visibility as the cortical processing route, as neuroimaging studies indeed do suggest (cf. Killgore and Yurgelun-Todd, 2004; Lidell *et* al., 2005; Morris, Ohman, and Dolan, 1998, 2000; Whalen *et al.*, 1998).

Both the cortical and subcortical processing routes project to the executive areas in the prefrontal cortex (Morris, Ohman, and Dolan, 2000; Anders *et al.*, 2004; Haxby, Gobbini, Furrey, Ishai, Schouten, and Pietrini, 2001), and we may safely assume that both cortically and subcortically contribute to visually guided behaviour. Which of both signals is used in order to initiate a response is most likely to be determined by the nature of the response required: interpreting someone's emotional expression in a social context requires accuracy (misinterpreting someone's sad expression for a happy expression could be very awkward in social interactions), while initiating a flight reaction in response to a fearful face may require a quick response – a false alarm in this scenario may not be as harmful as waiting until it is too late to respond to a potential threat. So, how does the brain select which information to use in order to initiate a response? One potential mechanism may be that unconsciously processed information is only used in response to signals indicating threat or fear. However, the literature does not seem to support this hypothesis: normal observers cannot respond to unconsciously processed fear, indicating that simple presence of a signal of fear in the subcortical pathways is not sufficient to initiate an overt response, even though some physological responses are initiated (Morris *et* al., 2000; Ohman and Soares, 1993; Whalen *et al.*, 1998).

Our study on TMS induced blindsight suggest that overall viewing conditions play a role in determining which type of information processing (cortical or subcortical) drives a visuall guided response, something that probably relates to the difference in sensitivity for visibility reduction in the cortical and subcortical pathways. As stated above, under optimal viewing conditions, there is a large discrepancy between the accuracy of cortically (accuracy around 100%) and subcortically processed information (accuracy around 70%). However, when visibility decreases, this discrepancy is reduced, and may even reverse: in blindsight patients, the subcortical pathways are clearly more accurate than the damaged cortical pathways.

Assuming that the brain uses the most accurate source of information to generate a response could potentially explain the discrepancy between affective blindness in normal observers and affective blindsight in patients, or the paradoxical disappearance of the affective blindsight experiment in our experiment: if viewing conditions *in general* are good, cortically processed information provides the most accurate information, and is therefore used to generate a response. However, determining whether source of information is the most accurate is not a matter of trial-by-trial evaluation. Our work suggests that general visibility (that is, visibility of stimuli over a longer period of time) is used to determine which source of information to use; even if that means that in some instances performance will suffer. It is important here to realize that 'visibility' should be interpreted in a real-world sense: in the work on unconscious face processing, the emotional faces were masked using neutral faces. The visibility of these neutral faces was good – subjects could clearly see them. It is not unlikely that this good visibility of neutral faces may have triggered the executive systems of the brain to use cortically processed information. After all, the brain has not evolved for laboratory experiments with masked stimuli. In order to explain why normal observers cannot respond to unconsciously processed information, we need to assume that generating a visually guided response is not simply a matter of accumulating evidence from the cortical and subcortical pathways: if that were the case, observers should be able to respond to subcortically processed information in all cases. In stead, it appears that subcortically processed information is actively ignored or repressed.

A possible reason for this may be that this avoids response conflict: if we assume that cortically processed information has a near-perfect accuracy under normal viewing conditions, but subcortical have an accuracy of around 70%, these different pathways come up with conflicting outputs on average one out of every three to four occasions. Repressing the least accurate source of information in stead of integrating when preparing a response may potentially speed up processing because the conflict does not need to be resolved. However, when overall visibility decreases, subcortically processed information becomes relatively more accurate, and may be used to generate responses as well. In our TMS experiment, even unsuppressed stimuli in the one frame condition were relatively hard to see, simulating a more realistic situation of poor visibility of emotional facial expressions, possibly explaining why we did find effects of unconscious processing in our study.

One key assumption of this model is that the brain is somehow capable of evaluating its own accuracy. Research on error monitoring and awareness indicates that that is indeed the case: when an error is made, specific error related signals are generated in the anterior cingulate cortex, and in the temporal areas of the brain. These signals have been shown to play a role in adapting behaviour and learning (Ridderinkhof, Van Den Wildenberg, Segalowitz, and Carter, 2004; Wills, Lavric, Croft, and Hodgson, 2007). Moreover, in a recent study by Lau and Passingham (2006), areas in the dorsolateral prefrontal cortex have been identified which seem to signal the subjective quality of visual information regardless of task performance: participants were engaged in a visual discrimination task, and had to indicate whether they had clearly seen the stimulus or whether they had guessed when making a response. Lau and Passingham identified two stimulus-mask intervals at which performance on the discrimination task was equal, but the subjective experience of the stimuli (i.e. the number of 'seen' versus 'guessed' responses) was different. Functional imaging revealed that the dorsolateral prefrontal cortex signalled this difference in subjective quality. Summarizing, it appears that there indeed are neural circuits capable of evaluating performance (Ridderinkhof *et al.*, 2004; Wills *et al.*, 2007), and even areas that signal the subjective quality of visually presented information (Lau and Passingham, 2006): it is quite likely that the brain *is* capable of evaluating its own accuracy.

Taking all the above into account, a formal model of how consciously and unconsciously processed information interact in the processing of emotional facial expressions is represented in figure 1. Information about a face (here a schematic face) is processed in parallel in both cortical and subcortical areas. Under normal viewing conditions, the cortical areas have near-perfect accuracy, while the accuracy of the subcortical pathways is capped at 70%. Both pathways project to the executive areas of the brain, responsible for response generation. Information in the subcortical pathways is actively ignored, or repressed, if the accuracy of cortically processed information exceeds that of subcortically processed information to avoid response conflict (figure 1b). However, if in this situation on cortical information processing is disrupted, for example by masking or by a TMS pulse, the executive systems of the brain do not have access to any information about a stimulus: subcortically processed information is ignored, and cortical information is compromised. This situation leads to the affective blindness observed in normal subjects (figure 1c). However, when the accuracy of the cortical pathways is deteriorated over a longer period of time, and approaches that of the subcortical pathways (such as in low visibility conditions, or cortical damage), subcortically processed information may contribute to visually guided behaviour (figure 1d). If in this case cortically processed information is disrupted by masking or TMS there is information available to base a response on, resulting in affective blindsight. Areas involved in error monitoring such as the anterior cingulate cortex or the dorsolateral prefrontal cortex may play a role in the hypothesized repression of subcortically processed information.

A mechanism like this clearly has its benefits: it assures that we are *not* blindly led by emotions. Misinterpreting an emotional expression may have severe consequences for social beings like humans, and apparently the need for accurate identification outweighs the need for fast fight or flight responses. Still, this mechanism allows for such responses if circumstances change. Moreover, covert responses based on subcortically information are not repressed. These covert responses may speed up an overt response when needed, thus sacrificing relatively little on speed, while gaining in accuracy. The model described above fits the data published thus far about unconscious emotion processing and is able to explain the paradoxical discrepancy between affective blindsight observed in patients and the affective blindness of normal observers.

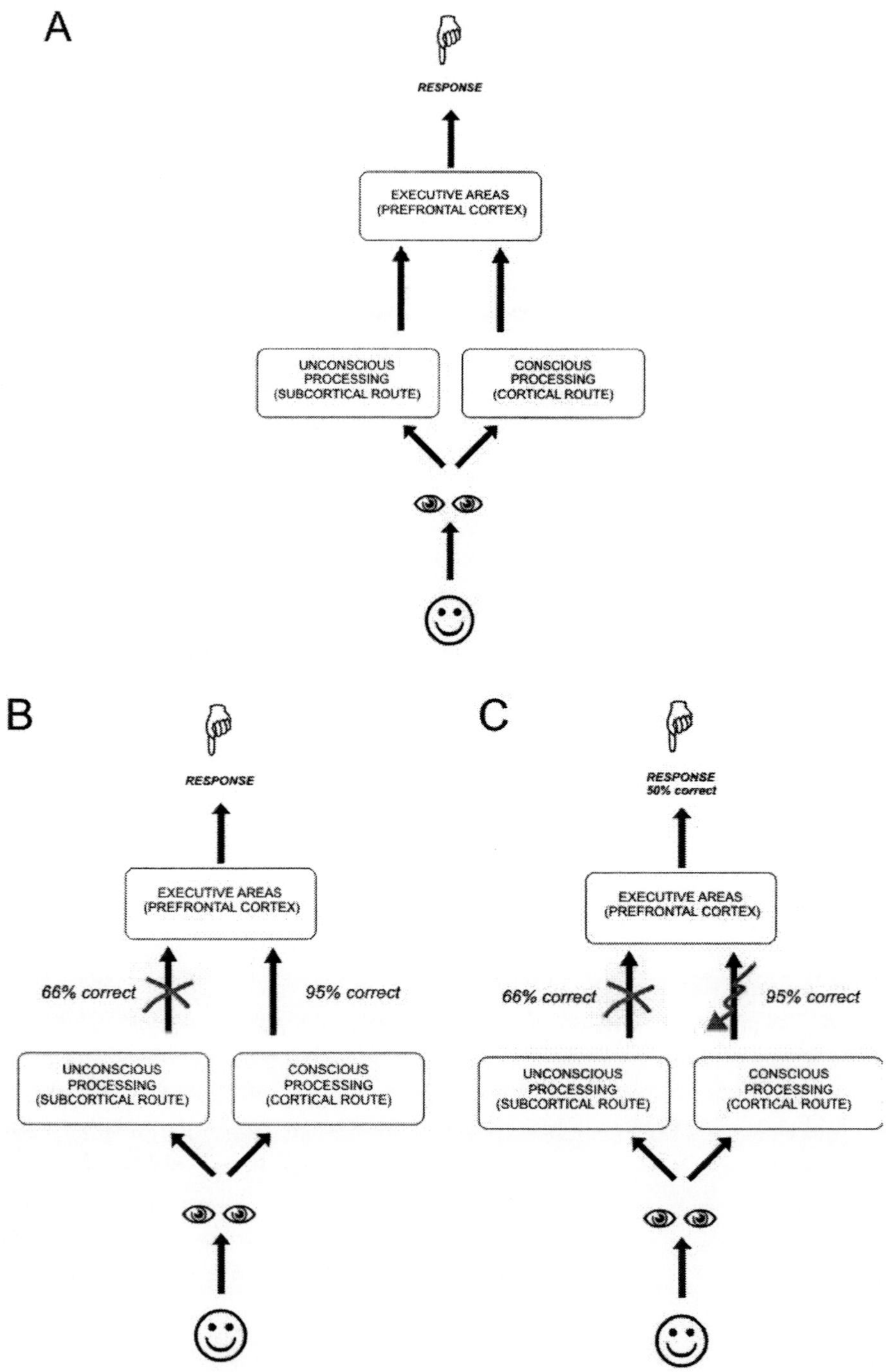

Figure 1. Continued on next page.

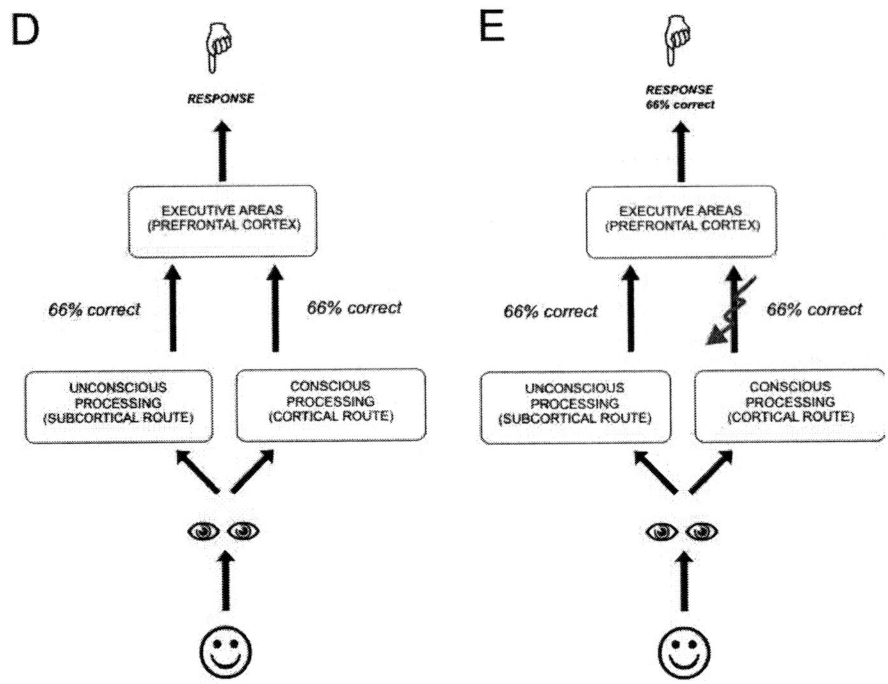

Figure 1.

Conclusion

There is no doubt that emotional facial expressions can be processed in total absence of awareness, and that these unconsciously processed emotions can modulate behaviour and even result in emotional feelings (Anders *et al.*, 2004; DeGelder *et al.*, 1999; Murphy and Zajonc, 1993; Ohman and Soares, 1993; Pegna *et al.*, 2005; Whalen *et al.*, 1998). However, this does not mean that we are blindly led by emotions (Heywood and Kentridge). The accuracy of unconsciously processed information is limited (DeGelder *et al.*, 1999; Jolij and Lamme, 2005), and guiding responses purely based on this information may have serious disadvantages for social beings like humans: an inappropriate response to a wrongly interpreted facial expression may lead to awkward social situations. Therefore, in stead of guiding responses by unconsciously processed information, the brain guides its responses by consciously processed information when possible, and represses unconsciously processed information to avoid response conflicts. Only when the accuracy of consciously processed information is compromised over a longer period of time, the brain changes its strategy and takes unconsciously processed information into account when generating overt responses.

Obviously, this model remains to be tested empirically, and a lot of questions remain unanswered. Exactly when does the brain switch from 'conscious processing' to 'unconscious processing'? Are there interpersonal differences in strategy? And, what does 'conscious' exactly mean? Recent empirical and theoretical work on visual awareness suggests that there may be a distinction between 'phenomenal awareness' (the actual qualitative experience) and 'access awareness' (the contents of awareness we can report upon) (Block, 2005; Lamme,

2003; Snodgrass, Bernat, and Shevrin, 2004). Are 'consciously' guided responses based on phenomenal awareness or on access awareness? And, finally, can this model be generalized to other types of visually guided behaviour? Not only emotional stimuli are processed unconsciously – in fact, unconscious processing has been demonstrated for a vast array of different stimuli (e.g. Dehaene *et al.*, 1998; Lau and Passingham, 2007). Yet, direct demonstrations of overt responses to unconsciously processed stimuli in normal observers are rare and not entirely undisputed (Boyer, Harrison, and Ro, 2005; Jolij and Lamme, 2005; Kolb and Braun, 1995; Meeres and Graves, 1990; Morgan, Mason, and Solomon, 1998), while blindsight in patients has been demonstrated for a wide variety of stimulus attributes, such as colour, form, and motion (Weiskrantz, 2003). Could a same model explain this discrepancy as well? We are currently working on behavioural and neuroimaging experiments in our lab to test this model, and find an answer to the questions posed above.

References

Amassian, V.E., Cracco, R.Q., Maccabee, P.J., Cracco, J.B., Rudell, A., and Eberle. L. (1989). Suppression of visual perception by magnetic coil stimulation of human occipital cortex. *Electroencephalography and Clinical Neurophysiology*, 74, 458-462.

Anders, S., Birbaumer, N., Sadowski, B., Erb, M., Mader, I., Grodd, W., and Lotze, M. (2004). Parietal somatosensory association cortex mediates affective blindsight. *Nature Neuroscience*, 7, 339-340.

Block, N. (2005). Two neural correlates of consciousness. *Trends in Cognitive Sciences*, 9, 46-52.

Boyer, J.L., Harrison, S., and Ro, T. (2005). Unconscious processing of orientation and color without primary visual cortex. *Proceedings of the National Academy of Sciences of the USA*, 102, 16875-16879.

Breitmeyer, B.G., and Ogmen, H. (2000). Recent models and findings in visual backward masking: a comparison, review, and update. *Perception and Psychophysics*, 62, 1572-1595.

Corthout, E., Uttl, B., Ziemann, U., Cowey, A., and Hallett, M. (1999). Two periods of processing in the (circum)striate visual cortex as revealed by transcranial magnetic stimulation. *Neuropsychologia*, 37, 137-145.

De Gelder B., Vroomen, J., Pourtois, G., and Weiskrantz, L. (1999). Non-conscious recognition of affect in the absence of striate cortex. *Neuroreport*, 10, 3759-3763.

De Gelder, B., Vroomen, J., Pourtois, G., and Weiskrantz, L. (2000) Affective blindsight: are we blindly led by emotions? Response to Heywood and Kentridge. *Trends in Cognitive Sciences*, 4, 125-126.

Dehaene, S., Naccache, L., Le Clec, H.G., Koechlin, E., Mueller, M., Dehaene-Lambertz, G., van de Moortele ,P.F., and Le Bihan, D. (1998). Imaging unconscious semantic priming. *Nature*, 395, 597-600.

Fahrenfort, J.J., Scholte, H.S., and Lamme, V.A. (2007). Masking disrupts reentrant processing in human visual cortex. *Journal of Cognitive Neuroscience*, 19, 1488-1497.

Hamm, A.O., Weike, A.I., Schupp, H.T., Treig, T., Dressel, A., and Kessler, C. (2003). Affective blindsight: intact fear conditioning to a visual cue in a cortically blind patient. *Brain*, 126, 267-275.

Haxby, J.V., Gobbini, M.I., Furey, M.L., Ishai, A., Schouten, J.L., and Pietrini, P. (2001). Distributed and overlapping representations of faces and objects in ventral temporal cortex. *Science*, 293, 2425-2430.

Heinen, K., Jolij, J., and Lamme, V.A.F. (2005). Two temporally distinct periods of activity in V1 are required for figure-ground segregation: a TMS study. *Neuroreport*, 16, 1483-1487.

Heywood, C.A., and Kentridge, R.W. (2000). Affective blindsight? *Trends in Cognitive Science*, 4, 125-126.

Johnson, M.H. (2005). Subcortical face processing. *Nature Reviews Neuroscience*, 6, 766-774.

Jolij, J., and Lamme, V.A.F. (2005). Repression of unconscious information by conscious processing: Evidence from affective blindsight induced by transcranial magnetic stimulation. *Proceedings of the National Academy of Sciences of the USA*, 102, 10747-10751.

Kanwisher, N., and Yovel, G. (2006). The fusiform face area: a cortical region specialized for the perception of faces. *Philosophical Transactions of the Royal Society of London, Series B: Biological Sciences*, 361, 2109-2028.

Kiani, R., Esteky, H., and Tanaka, K. (2005). Differences in onset latency of macaque inferotemporal neural responses to primate and non-primate faces. *Journal of Neurophysiology*, 94, 1587-1596.

Killgore, W.D., and Yurgelun-Todd, D.A. (2004). Activation of the amygdala and anterior cingulate during nonconscious processing of sad versus happy faces. *Neuroimage*, 21, 1215-1223.

Kolb, F.C., and Braun J. (1995). Blindsight in normal observers. *Nature*, 377, 336-338.

Lamme, V.A. (2003). Why visual attention and awareness are different. *Trends in Cognitive Sciences*, 7, 12-18.

Lamme, V.A. (2006). Towards a true neural stance on consciousness. *Trends in Cognitive Sciences*, 10, 494-501.

Lamme, V.A., Zipser K., and Spekreijse, H. (2002). Masking interrupts figure-ground signals in V1. *Journal of Cognitive Neuroscience*, 14, 1044-1053.

Lau, H.C., and Passingham, RE. (2006). Relative blindsight in normal observers and the neural correlate of visual consciousness. *Proceedings of the National Academy of Sciences of the USA*, 103, 18763-18768..

Lau, H.C., and Passingham, R.E. (2007). Unconscious activation of the cognitive control system in the human prefrontal cortex. *Journal of Neuroscience*, 27, 5805-5811.

LeDoux, J. (1996). *The Emotional Brain*. Simon and Schuster: New York.

Liddell, B.J., Brown, K.J., Kemp, A.H., Barton, M.J., Das, P., Peduto, A., Gordon, E., and Williams L.M. (2005). A direct brainstem-amygdala-cortical 'alarm' system for subliminal signals of fear. *Neuroimage*, 24, 235-243.

Macknik, S.L. (2006). Visual masking approaches to visual awareness. *Progress in Brain Research*, 155, 177-215.

Meeres, S.L., and Graves, R.E. (1990). Localization of unseen visual stimuli by humans with normal vision. *Neuropsychologia*, 28, 1231-1237.

Morgan, M.J., Mason, A.J., and Solomon J.A. (1997). Blindsight in normal subjects? *Nature*, 385, 401-402.

Morris, J.S., De Gelder, B., Weiskrantz, L., and Dolan, R.J. (2001). Differential extrageniculostriate and amygdala responses to presentation of emotional faces in a cortically blind field. *Brain*, 124, 1241-1252.

Morris, J.S., Ohman, A., and Dolan, R.J. (1998). Conscious and unconscious emotional learning in the human amygdala. *Nature*, 393, 467-470.

Morris, J.S., Ohman, A., and Dolan R.J. (1999). A subcortical pathway to the right amygdala mediating "unseen" fear. *Proceedings of the National Academy of Sciences of the USA*, 96, 1680-1685.

Murphy, S.T., and Zajonc, R.B. (1993). Affect, cognition, and awareness: affective priming with optimal and suboptimal stimulus exposures. *Journal of Personality and Social Psychology*, 64, 723-739.

Ohman, A., and Soares, J.J. (1993). On the automatic nature of phobic fear: conditioned electrodermal responses to masked fear-relevant stimuli. *Journal of Abnormal Psychology*, 102,121-132.

Pegna, A.J., Khateb, A., Lazeyras, F., and Seghier, M.L. (2005). Discriminating emotional faces without primary visual cortices involves the right amygdala. *Nature Neuroscience*, 8, 24-25.

Pessoa, L. (2005). To what extent are emotional visual stimuli processed without attention and awareness? *Current Opinion in Neurobiology*, 15, 188-196.

Pessoa, L., Japee, S., Sturman, D., and Ungerleider L.G. (2006). Target visibility and visual awareness modulate amygdala responses to fearful faces. *Cerebral Cortex*, 16, 366-375.

Pessoa, L., Japee, S., and Ungerleider L.G. (2005). Visual awareness and the detection of fearful faces. *Emotion*, 5, 243-247.

Pessoa, L., and Padmala, S. (2005). Quantitative prediction of perceptual decisions during near-threshold fear detection. *Proceedings of the National Academy of Sciences of the USA*, 102, 5612-5617.

Ridderinkhof, K.R., Van Den Wildenberg, W.P., Segalowitz, S.J., Carter, C.S. (2004). Neurocognitive mechanisms of cognitive control: the role of prefrontal cortex in action selection, response inhibition, performance monitoring, and reward-based learning. *Brain and Cognition*, 56, 129-140.

Sahraie, A., Trevethan, C.T., Weiskrantz, L., Olson, J., MacLeod, M.J., Murray, A.D., Dijkhuizen, R.S., Counsell, C., and Coleman, R. (2003). Spatial channels of visual processing in cortical blindness. *European Journal of Neuroscience*, 18, 1189-1196.

Sahraie, A., Weiskrantz, L., Barbur, J.L., Simmons, A., Williams, S.C., and Brammer, M.J. (1997). Pattern of neuronal activity associated with conscious and unconscious processing of visual signals. *Proceedings of the National Academy of Sciences of the USA*, 94, 9406-9411.

Silvanto, J., Cowey, A., Lavie, N., and Walsh, V. (2007). Making the blindsighted see. *Neuropsychologia, in press.*

Snodgrass, M., Bernat, E., and Shevrin, H. (2004). Unconscious perception: a model-based approach to method and evidence. *Perception and Psychophysics*, 66, 846-867.

Snodgrass M., and Shevrin, H. (2006). Unconscious inhibition and facilitation at the objective detection threshold: replicable and qualitatively different unconscious perceptual effects. *Cognition*, 101, 43-79.

Szczepanowski, R., and Pessoa, L. (2007). Fear perception: can objective and subjective awareness measures be dissociated? *Journal of Vision*, 7, 10.

Vuilleumier, P., Armony, J.L., Driver, J., and Dolan, R.J. (2003). Distinct spatial frequency sensitivities for processing faces and emotional expressions. *Nature Neuroscience*, 6, 624-631.

Vuilleumier, P., and Pourtouis, G. (2007). Distributed and interactive brain mechanisms during emotion face perception: evidence from functional neuroimaging. *Neuropsychologia*, 45, 174-194.

Whalen, P.J., Kagan, J., Cook, R.G., Davis, F.C., Kim, H., Polis, S., McLaren, D.G., Somerville, L.H., McLean, A.A., Maxwell, J.S., and Johnstone, T. (2004). Human amygdala responsivity to masked fearful eye whites. *Science*, 17, 306, 2061.

Whalen, P.J., Rauch, S.L., Etcoff, N.L., Mclnerney, S.C., Lee, M.B., and Jenike, M.A. (1998). Masked presentations of emotional facial expressions modulate amygdala activity without explicit knowledge. *Journal of Neuroscience*, 18, 411-418.

Wills, A.J., Lavric, A., Croft, G.S., and Hodgson, T.L. (2007). Predictive learning, prediction errors, and attention: evidence from event-related potentials and eye tracking. *Journal of Cognitive Neuroscience*, 19, 843-854.

Winston, J.S., Vuilleumier, P., and Dolan, R.J. (2003). Effects of low-spatial frequency components of fearful faces on fusiform cortex activity. *Current Biology*, 13, 1824-1829.

Reviewed by Hakwan Lau, Columbia University

In: Neural Pathways Research
Editor: F. L. Pichler, pp. 221-230
ISBN: 978-1-60456-214-9

Chapter XI

Neostriatal Connections and Their Relation to Behavior and Memory Processing

Konstantinos I. Tsamis
Laboratory of Neuropathology and Neuroimmunology,
1st Department of Neurology, Aristotle University,
Thessaloniki, Greece

Abstract

The classic view of neostriatum placed it among the structures that modulate motor functions. Under that view, research in the field of pathways shed light in the established connections with other nuclei of basal ganglia and motor cortex. However, converging data of the last decades emphasize more and more the role of the neostriatum in non-motor functions, such as behavior and memory. This evidence came from studies on animals with experimental lesions on the neostriatum and clinical studies on the non-motor symptoms of the diseases affecting primarily the neostriatum (Huntington's disease, Parkinson's disease, neuroacanthocytosis). Furthermore, recent data on psychiatric disorders underline the central role of the striatal pathology in the appearance of mood and behavioral manifestations. The anatomical substrate of these neostriatal functions is not yet very clear. The striatum is known to be closely connected to the cerebral cortex, especially the frontal cortex. Besides, the ventral striatum is connected with the limbic system and there are also indications about direct and indirect connections of the neostriatum with the hippocampus, thalamus and amygdala. Future research will possibly reveal the role of these connections in the modulation of behavior and in memory processing.

The neostriatum is part of the basal ganglia and it consists of the caudate nucleus and the putamen. The caudate nucleus is anatomically located deep in each hemisphere, being part of the outer wall of the lateral ventricle, sitting astride the thalamus. In the rostrocaudal axis it is divided into three parts: the head, the body and the tail. The putamen together with the more centrally located globus pallidus comprises what was formerly known as the lenticular nucleus. It is divided by the caudate nucleus and thalamus with the internal capsule and by the claustrum with the external capsule. The ventral part of both the caudate nucleus and putamen comprises the ventral striatum, which is composed by the nucleus accumbens and the olfactory tubercle.

As far as histology is concerned, the neostriatum (and ventral striatum) is principally composed of the so-called medium sized spiny neurons, which are projecting GABAergic neurons [7, 8, 21, 66]. The size of their cell body is about 20μm. They protrude three to five primary dendrites which give rise to several secondary and tertiary dendritic branches, covering an area of about 200μm in diameter [44]. All the dendrites are densely covered by dendritic spines and this is the reason why these neurons are called "spiny". There are two subpopulations of spiny neurons that contain different neuroactive peptides (substance P and dynorphin or enkephalin) and express different types of dopamine receptors (D1 and D2) [40]. The neostriatum also contains a small number of interneurons that use mainly acetylcholine and GABA as neurotransmitters. Cholinergic interneurons are large, with diameter of the cell body 25 to 40μm and dendrites almost smooth with very sparse dendritic spines [13, 14]. GABAergic interneurons are smaller (15μm – 20μm) and their dendrites are smooth, too [21, 59, 61].

The striatal neurons are not uniformly scattered throughout the neostriatum, but they are organized in striosomes and matrix. This compartmentalization was primarily observed in acetylcholinesterase staining, and later studies confirmed its existence also in neurotransmitters, receptors and neuropeptides [23, 25]. In addition, neurons of striosomes and matrix were found to establish afferent and efferent connections with different structures [17, 18, 19, 45, 46].

From the functional point of view, it is known (even from the beginning of the previous century) that the neostriatum has a central role in organizing and controlling movements. However, studies of the last decades emphasize its involvement in cognitive and behavioral functions. The interest in these functions developed from studies on animals with experimental lesions on the neostriatum (rodents and primates) and from observations on patients suffering from diseases, which affect primarily the neostriatum. Furthermore, recent data on psychiatric disorders underline the central role of the striatal pathology in the appearance of mood and behavioral manifestations [48, 49].

In the second half of the previous century, more and more studies on animals revealed cognitive deficits after experimental manipulations of the neostriatum. The fact that similar deficits were obtained after lesions in the prefrontal cortex suggested a functional link between these structures. Later studies specified better the neostriatal cognitive functions and related them with memory processes, association learning, response to reward, analysis of spatial information and ability to switch behavioral strategies. Initially, the involvement of neostriatum in memory was attributed to habitual actions, but recent data support the idea of a more general role in memory formation and in mediating sensorimotor associations. Furthermore, the rewarding process – being a crucial factor for the modulation of one's behavior through motivation – is mediated by the ventral striatum together with the neostriatum [64, 67].

The functional consequences of neostriatal alterations in humans were mainly observed on patients suffering from Huntington's disease and Parkinson's disease. In the first disease, mood disorders, antisocial personality disorders and memory deficits are observed even in the early stages, when there are almost no pathologic alterations of the striatum. In Parkinson's disease, patients suffer also from mood disorders, cognitive slowing, decreased attention resources and difficulties in learning. Even if some of these disorders can be attributed to cortical dysfunction, the involvement of the neostriatum, as the primarily affected area, can't be excluded. In psychiatric disorders – i.e., schizophrenia and depression – the central role of the neostriatum, together with the prefrontal neocortex, in the pathogenesis of these disorders, is well documented.

The anatomical substrate of the cognitive functions of the neostriatum is not yet very clear, despite the fact that the connections have been extensively studied, mainly in mammals and primates. The connections that are better described and characterized are those established with other nuclei of the basal ganglia, thalamus, motor and premotor cortex [6,55].

The neostriatum receives massive excitatory glutamatergic inputs deriving from the neocortex. These fibers originate mainly from pyramidal neurons of the fifth layer and terminate on the head of the dendritic spines of striatal spiny neurons [28]. The transmission of information in these synapses is mediated by NMDA and AMPA receptors (ionotropic receptors) and by metabotropic glutamate receptors [15, 30, 31]. These receptors (apart from NMDAR) have been proved to be essential for long term depression (LTD), which is thought to be the process underlying learning and memory in the neostriatum [5,56]. The second main input source for the neostriatum is the substantia nigra pars compacta, with dopaminergic fibers. These fibers terminate on the shafts of the dendritic spines where dopamine receptors D1 and D2 are mainly used. Dopamine exerts a regulatory role on the neostriatum [2, 37]. The neostriatum also receives input from the thalamus and subthalamic nucleus, which are glutamatergic, too [29].

The efferent connections are GABAergic, originating from medium sized spiny neurons, and they target mostly the substantia nigra pars reticulata, the internal and the external segment of the globus pallidus. The external segment of the globus pallidus is thence connected to the internal through the subthalamic nucleus. In this way efferent connections of the neostriatum are organized in two circuits, the direct and the indirect [6, 54]. The final recipient subcortical structure of this information processing is the thalamus (mainly nuclei VA, VL and MD) which is subsequently linked back to the cortex.

The above described connections were found to be segregated in five partly "closed" circuits named "basal ganglia – thalamocortical loops". Each loop initiates from a number of interconnected cortical areas and through the neostriatum, globus pallidus, substantia nigra and thalamus terminates back in one of the areas of the cerebral cortex from where it derived. In the course through these structures each loop remains segregated from the others. This concept had been proposed in 1986 by Alexander et al. and it is until today valid in its general rudiments [1, 24]. The five circuits described are the motor circuit, the oculomotor and three prefrontal circuits (dorsolateral prefrontal, lateral orbitofrontal and anterior cingulate). The two first circuits are better characterized, but in the light of recent physiological, pharmacological and pathological data prefrontal circuits have drawn new attention.

Among the cortical areas that are connected to the neostriatum, the frontal cortex has a central position, and this was initially attributed to its involvement in motor functions (motor – premotor cortex). Later studies in experimental animals, especially primates, revealed that cortical areas related to motor functions are connected predominantly with the putamen. Besides, there are abundant connections established between the prefrontal cortex and neostriatum, mainly with the head of the caudate nucleus [45, 65, 68]. These connections may be involved in the modulatory role of the caudate nucleus in behavior, since it is known that the prefrontal cortex is essential for the configuration of one's character, social behavior, memory and learning [10, 11, 22, 52].

The function of the prefrontal cortex is related to attention, memory and spatial processing. It is also considered to be involved in the ability to differentiate among conflicting thoughts, to predict future consequences of current activities, to adjust the behavior with the social environment, to determine the relations with others and even to determine self-estimation. All these functions are crucial for the modulation of one's behavior [50]. It is thus important to note that the ultimate connections between the prefrontal cortex and caudate nucleus are not "one-way", but the neostriatum is also affecting the neocortex through the thalamus. This has been proven to apply to motor and oculomotor circuits and, in case the same is true for prefrontal circuits, it is a possible explication for the regulatory role of the neostriatum in behavior [33, 47].

However, the exact localization of fronto-striatal connections hasn't been yet fully described. Physiological and anatomical studies on primates revealed connectivity of some prefrontal areas with different regions of the caudate nucleus [42]. Furthermore, behavioral studies on primates suggest that the prefrontal cortex doesn't have a unitary function, but specific prefrontal regions serve differential processes [50]. Thus, the previously described prefronto-striatal circuits have been ascribed with specific functions. Connections from the dorsolateral prefrontal cortex to the dorsolateral portion of the head of the caudate nucleus are thought to be engaged in spatial memory. Connections from the lateral orbitofrontal regions to the ventromedial sector of the head of the caudate nucleus are presumed to be involved in the capacity of making appropriate switches in behaviour [1]. Connections from the medial prefrontal region to the dorsomedial portion of the head and to the medial sector of the body of the caudate nucleus may have a role in motivational processes. Finally, connections from the superior prefrontal area to the dorsal-intermediate portion of the head of the caudate nucleus are supposed to subserve kinaesthetic, spatial as well as mnemonic (working or representational memory) functions [34].

These functional implications of the prefronto-striatal circuits have been mostly revealed by studies on primates [62] and they cannot be used to draw direct conclusions for respective connections in humans. The prefrontal cortex in human is very voluminous, even in comparison to the primates' prefrontal cortex, and there is not an exact correspondence between their anatomical subdivisions. This is possibly because of evolutionary differences, and it reflects the behavioral differences between the species. So, it appears necessary to acquire anatomical data from studies on human brain tissue in order to make certain assumptions on fronto-striatal connections in humans [63]. In a recent study by S.E. Leh et al., an attempt towards this direction was made, confirming some of the connections previously described in animals [32].

The efferent connections of the neostriatum that are involved in behavioral functions and memory are complex and so less understood [38]. The general concept of neostriatal connectivity through the globus pallidus, substantia nigra and thalamus back to the cerebral cortex is considered to be valid in this case, too. Despite this, the idea of segregated loops in prefrontal circuits, after the level of the caudate nucleus, has yet to be proved [24, 39]. In particular, the dorsolateral part of the head of the caudate nucleus is presumed to be connected to the lateral portion of the globus pallidus and the substantia nigra, which project to the ventral anterior thalamic nucleus (VA) and to the medial dorsal thalamic nucleus (MD), respectively. The ventromedial sector of the head of the caudate nucleus is connected to the medial part of the globus pallidus and the substantia nigra, which also project to the ventral anterior thalamic nucleus and to the medial dorsal thalamic nucleus. The data about these connections derives also from studies on experimental animals (primates) and their existence and function in humans haven't yet been proved [1, 34].

Furthermore, other connections that have been reported may have a role in the modulation of cognitive functions. The thalamus was found to project to the caudate nucleus and the putamen transferring sensory information in the neostriatum without the mediation of the cerebral cortex [29, 53]. In this way, the neostriatum may be involved in sensory perception and sensorimotor associations [20, 58]. Besides, this link of sensory information influx could possibly be the source for striatal memory processing. The memory and learning system seated on the neostriatum was formerly believed to be strictly segregated from the hippocampal one. Today there is enough evidence, deriving from physiological studies, supporting the idea that these two systems are connected [51, 60]. However, the anatomical evidence of such connections (direct or indirect) has yet to be found. Additionally, connections between the neostriatum and cerebellum have recently been reported [26]. They are supposed to regulate voluntary movements and involvement in other functions hasn't been tested [35].

The role of the ventral striatum in cognitive functions isn't yet very clear as well. Its afferent connections derive from the prefrontal cortex, temporal cortex, basolateral amygdala, hippocampal formation, midline and intra-laminar thalamic nuclei and ventral tegmental area of the midbrain [43]. Efferent fibres are thence directed through the ventral pallidus to the mediodorsa thalamus and finally back to the prefrontal cortex. Another efferent branch targets the lateral hypothalamus and brainstem motor nuclei [1, 27]. Gordon Mogenson first focused attention on the functions of these circuits by describing the nucleus accumbens as a "limbic-motor interface" and regarding the ventral striatum as a key structure in implementing the transition from motivation to action [36]. Especially, the connections of the nucleus accumbens with the ventral tegmental area are in favor of its involvement in emotional processes, pleasure, reward and behavioral motivation [37, 64].

Conclusion

In a time when controversial results and hypotheses are accumulated about the function and the connections of the neostriatum [57], there is a need for anatomical confirmation of this data in animals and, more importantly, in humans. The exact characterization of striatal connections and a better understanding of their function are necessary for the development of effective treatments for basal ganglia disorders.

The evidence about the role of the neostriatum and its involvement in the emergence of cognitive symptoms in motor disorders could also raise the hypothesis that it can be involved in the pathogenesis of cognitive disorders, like Alzheimer's disease [16]. Previous studies on the neostriatum of demented patients do not fully support this idea, since a minimal number of alterations are observed there [3, 4, 9, 12, 41]. However, future research may unravel this issue and change the way we think about the neostriatum.

References

[1] Alexander, GE; DeLong, MR; Strick, PL. Parallel organization of functionally segregated circuits linking basal ganglia and cortex. *Annual Review of Neuroscience*, 1986 9: 357-381.

[2] Arbuthnott, GW; Ingham, CA; Wickens, JR. Dopamine and synaptic plasticity in the neostriatum. *Journal of Anatomy*, 2000 196: 587-596.

[3] Baloyannis, SJ. Neuropathology of dementia. 1st Edition. Thessaloniki: Stavros *J. Baloyannis*; 1993.

[4] Baloyannis, SJ. New horizons in Alzheimer's disease. 1st Edition. Thessaloniki: Stavros *J. Baloyannis*; 1994.

[5] Bernard, V; Bolam, PJ. Subcellular and subsynaptic distribution of the NR1 subunit of the NMDA receptor in the neostriatum and globus pallidus of the rat: co-localization at synapses with the GluR2/3 subunit of the AMPA receptor. *European Journal of Neuroscience,* 1998 10: 3721–3736.

[6] Bolam, JP; Hanley JJ; Booth PCA; Bevan MD. Synaptic organisation of the basal ganglia. *J. Anat.*, 2000 196: 527-542.

[7] Braak, H; Braak, E. Neuronal types in the striatum of man. *Cell and Tissue Research*, 1982 227: 319-342.

[8] Braak, H; Braak, E. Neuronal types in the neocortex-dependent lateral territory of the human thalamus. *Anatomy and Embryology*, 1984 169: 61-72.

[9] Braak, H; Braak, E. Alzheimer's disease: striatal amyloid deposits and neurofibrillary changes. *Journal of Neuropathology and Experimental Neurology*, 1990 49: 215-224.

[10] Brown, VJ; Desimone, R; Mishkin, M. Responses of cell in the tail of the caudate nucleus during visual discrimination learning. *Journal of Neurophysiology*, 1995 74: 1083–1094.

[11] Calabresi, P; De Murtas, M; Bernardi, G. The neostriatum beyond the motor function: experimental and clinical evidence. *Neuroscience*, 1997 78 1: 39-60.

[12] Castellani, RJ; Lee, HG; Zhu, X; Nunomura, A; Perry, G; Smith, MA. Neuropathology of Alzheimer disease: pathognomonic but not pathogenic. *Acta Neuropathologica*, 2006 111: 503-509.

[13] DiFiglia, M; Carey, J. Large neurons in the primate neostriatum examined with the combined Golgi-electron microscopic method. *The Journal of Comparative Neurology*, 1986 244: 36-52.

[14] DiFiglia, M. Synaptic organization of cholinergic neurons in the monkey neostriatum. *The Journal of Comparative Neurology*, 1987 255: 245-258.

[15] Fujiyama, F; Kuramoto, E; Okamoto, K; Hioki, H; Furuta, T; Zhou, L; Nomura, S; Kaneko, T. Presynaptic localization of an AMPA-type glutamate receptor in corticostriatal and thalamostriatal axon terminals. *European Journal of Neuroscience*, 2004 20: 3322–3330.

[16] Gearing, M; Wilson, RW; Unger, ER; Shelton, ER; Chan, HW; Masters, CL; Beyreuther, K; Mirra, SS. Amyloid precursor protein (APP) in the striatum in Alzheimer's disease: an immunohistochemical study. *Journal of Neuropathology and Experimental Neurology*, 1993 52: 22-30.

[17] Gerfen, CR. The neostriatal mosaic: compartmentalization of corticostriatal input and striatonigral output system. *Nature*, 1984 311: 461–464.

[18] Gerfen, CR. The neostriatal mosaic. I. Compartmental organization of projections from the striatum to the substantia nigra in the rat. J. comp. Neurol., 1985 237: 176–194.

[19] Gimenez-Amaya, JM; Graybiel, AM. Compartmental origins of striatopallidal projection in the primate. *Neuroscience* 1990 34: 111–126.

[20] Gold, JI; Shadlen, MN. Neural computations that underlie decisions about sensory stimuli. *Trends in Cognitive Sciences*, 2001 5, 10–16.

[21] Graveland, GA; Williams, RS; Difiglia, M. A Golgi study of the human neostriatum: Neurons and afferent fibers. *The Journal of Comparative Neurology*, 1985 234, 317-333.

[22] Graybiel, AG. Building action repertoires: memory and learning functions of the basal ganglia. *Current Opinion in Neurobiology*, 1995 5: 733–741.

[23] Graybiel, AM; Ragsdale, CW. Histochemically distinct compartments in the striatum of human, monkey, and cat demonstrated by acetylthiocholinesterase staining. *Proceedings of the National Academy of Sciences of the United States of America*, November 1978 75 11, 5723-5726.

[24] Haber, S.N. The primate basal ganglia: parallel and integrative networks. *Journal of Chemical Neuroanatomy*, 2003 26: 317–330.

[25] Holt, DJ; Graybiel, AM; Saper, CB. Neurochemical architecture of the human striatum. *The Journal of Comparative Neurology*, 1997 384: 1-25.

[26] Hoshi, E; Tremblay, L; Feger, J; Carras, PL; Strick, PL. The cerebellum communicates with the basal ganglia. *Nature Neuroscience*, 2005 8 11: 1491–1493.

[27] Kelley, AE; Balbo, BA; Pratt, WE; Will, MJ. Corticostriatal hypothalamic circuitry and food motivation: integration of energy, action and reward. *Physiology and Behavior*, 2005 86, 773-795.

[28] Kemp, JM; Powell, TPS. The site of termination of afferent fibres in the caudate nucleus. Philosophical Transactions of the Royal Society of London, 1971 b 262: 413-427.

[29] Kimura, M; Minamimoto, T; Matsumoto, N; Hori, Y. Monitoring and switching of cortico-basal ganglia loop functions by the thalamo-striatal system. *Neuroscience Research,* 2004 48: 355–360.

[30] Kuppenbender, KD; Standaert, DG; Feuerstein, TJ; Penney, JB; Young, AB; Landwehrmeyer, BG. Expression of NMDA receptor subunit mRNAs in neurochemically identified projection and interneurons in the human striatum. *The Journal of Comparative Neurology*, 2000 419: 407-421.

[31] Kuramoto, E; Fujiyama, F; Unzai, T; Nakamura, K; Hioki, H; Furuta, T; Shigemoto, R; Ferraguti, F; Kaneko, T. Metabotropic glutamate receptor 4-immunopositive terminals of medium-sized spiny neurons selectively form synapses with cholinergic interneurons in the rat neostriatum. *The Journal of Comparative Neurology* 2007 500: 908-922.

[32] Leh, SE; Ptito, A; Chakravarty, MM; Strafella, AP. Fronto-striatal connections in the human brain: A probabilistic diffusion tractography study. *Neuroscience Letters*, 2007 419: 113-118.

[33] Maddox, WT; Filoteo, JV. Striatal contributions to category learning: quantitative modeling of simple linear and complex non-linear rule learning in patients with Parkinson's disease. *Journal of the International Neuropsychological Society*, 2001 7: 710–727.

[34] McHaffie, JG; Stanford, TR; Stein, BE; Coizet, V; Redgrave, P. Subcortical loops through the basal ganglia. *Trends in Neurosciences*, August 2005 28 8: 401-407.

[35] Middleton, FA; Strick, PL. Basal ganglia and cerebellar loops: motor and cognitive circuits. *Brain Research Reviews*, 2000 31: 236–250.

[36] Mogenson, GJ. Limbic-motor integration. *Progress in Psychobiology and Physiological Psychology,* 1987 12: 117–170.

[37] Nakamura, K; Hikosaka, O. Role of Dopamine in the primate caudate nucleus in reward modulation of saccades. *The Journal of Neurosciences*, 2006 26 26: 5360-5369.

[38] Nakano, K; Kayahara, T; Tsutsumi, T; Ushiro, H. Neural circuits and functional organization of the striatum. *Journal of Neurology*, 2000 247 (Suppl 5): 1-15.

[39] Nakano, K. Neural circuits and topographic organization of the basal ganglia and related regions. *Brain and Development*, 2000 22: 5-16.

[40] Onn, SP; Berger, TW; Grace, AA. Identification and characterization of striatal cell subtypes using in vivo intracellular recording and dye labeling in rats. *Synapse*, 1994 16: 231-254.

[41] Oyanagi, K; Takahashi, H; Wakabayashi, K; Ikuta, F. Large neurons in the neostriatum in Alzheimer's disease and progressive supranuclear palsy: a topographic, histologic and ultrastructural investigation. *Brain Research*, 1991 544: 221-226.

[42] Parent, A; Hazrati, LN. Functional anatomy of the basal ganglia. 1. The cortico–basal ganglia–thalamo–cortical loop. *Brain Research Reviews*, 1995 20: 91-127.

[43] Pennartz, CM; Kitai, ST. Hippocampal inputs to identified neurons in an in vitro slice preparation of the rat nucleus accumbens: evidence for feed-forward inhibition. *Journal of Neuroscience* 1991 11: 2838-2847.

[44] Rafols, JA; Cheng, HW; McNeill, TH. Golgi study of the mouse striatum: Age related dendritic changes in different neuronal populations. *The Journal of Comparative Neurology*, 1989 279: 212-227.

[45] Ragsdale, CW; Graybiel, AM. The fronto-striatal projection in the cat and monkey and its relationship to inhomogeneities established by acetylcholinesterase histochemistry. *Brain Research*, 1981 208: 259–266.

[46] Ragsdale, CW; Graybiel, AM. A simple ordering of neocortical areas established by compartmental organization of their striatal projections. *Proceedings of the National Academy of Sciences of the United States of America*, 1990 87: 6196–6199.

[47] Redgrave, P. et al. The basal ganglia: a vertebrate solution to the selection problem? *Neuroscience*, 1999 89: 1009–1023.

[48] Roberts, RC; Conley, R; Kung, L; Peretti, FJ; Chute, DJ. Reduced striatal spine size in schizophrenia: a postmortem ultrastructural study. *Clinical Neurosciences and Neuropathology*, 1996 7: 1214-1218.

[49] Roberts, RC; Roche, JK; Conley, R. Synaptic differences in the postmortem striatum of subjects with schizophrenia: a stereological ultrastructural analysis. *Synapse*, 2005 56: 185-197.

[50] Rosenkilde, CE. Functional heterogeneity of the prefrontal cortex: a review. *Behavioral and Neural Biology*, 1979 25: 301-345.

[51] Rossato, JI; Zinn, CG; Furini, C; Bevilaqua, LRM; Medina, JH; Cammarota, M; Izquierdo, I. A link between the hippocampal and the striatal memory systems of the brain. *Annals of the Brazilian Academy of Sciences*, 2006 78 3: 515-523.

[52] Seger, CA; Cincotta, CM. The roles of the caudate nucleus in human classification learning. *The Journal of Neuroscience*, 2005 25 11: 2941-2951.

[53] Smith, Y; Raju, DV; Pare, JF; Sidibe, M. The thalamostriatal system: a highly specific network of the basal ganglia circuitry. *Trends in Neurosciences*, 2004 27: 520–527.

[54] Smith, Y; Bevan, MD; Shink, E; Bolam, JP. Microcircuitry of the direct and indirect pathways of the basal ganglia. *Neuroscience*, 1998 86: 353–387.

[55] Smith, AD; Bolam, JP. The neural network of the basal ganglia as revealed by the study of synaptic connections of identified neurones. *Trends in Neurosciences*, 1990 13: 259-265.

[56] Stephenson, FA. Structure and trafficking of NMDA and GABAA receptors. *Biochemical Society Transactions*, 2006 34 5: 877-881.

[57] Suvorov, NF. Basal ganglia: structure and function. *Neuroscience and Behavioral Physiology*, 1998 28 3: 219-223.

[58] Takada, M; Itoh, K; Yasui, Y; Sugimoto, T; Mizuno, N. Topographical projections from the posterior thalamic regions to the striatum in the cat, with reference to possible tecto–thalamo–striatal connections. *Experimental Brain Research*, 1985 60: 385–396.

[59] Takagi, M; Somogyi, P; Smith, AD. Aspiny neurons and their local axons in the neostriatum of the rat: a correlated light and electron microscopic study of Golgi impregnated material. *Journal of Neurocytology*, 1984 13, 239-265.

[60] Teng, E; Stefanacci, L; Squire, LR; Zola, SM. Contrasting effects on discrimination learning after hippocampal lesions and conjoint hippocampal-caudate lesions in monkeys. *The Journal of Neuroscience*, 2000 20: 3853–3863.

[61] Tepper, JM; Koos, T; Wilson, CJ. GABAergic microcircuits in the neostriatum. *Trends in Neurosciences*, November 2004 27 11: 662-669.

[62] Tolkunov, BF; Orlov, AA; Afanas'ev, SV. Model of the function of the neostriatum and its neuronal activity during behavior in monkeys. *Neuroscience and Behavioral Physiology,* 1997 27 1: 68-74.

[63] Toni, I; Passingham, RE. Prefrontal-basal ganglia pathways are involved in the learning of arbitrary visuomotor associations: a PET study. *Experimental Brain Research*, 1999 127: 19–32.

[64] Tremblay, L; Hollerman, JR; Shultz, W. Modifications of reward expectation-related neuronal activity during learning in primate striatum. *Journal of Neurophysiology*, 1998 80: 964–977.

[65] Wise, SP; Murray, EA; Gerfew, CR. The frontal cortex-basal ganglia system in primates. *Critical Reviews in Neurobiology*, 1996 10: 317–356.

[66] Yelnik, J; Francois, C; Percheron, G; Tande, D. Morphological taxonomy of the neurons of the primate striatum. *The Journal of Comparative Neurology*, 1991 313: 273-294.

[67] Yelnik, J. Functional anatomy of the basal ganglia. *Movement Disorders*, 2002 17 3: 15-21.

[68] Yeterian, EH; Pandya, DN. Prefrontostriatal connections in relation to cortical architectonic organization in Rhesus Monkeys. *The Journal of Comparative Neurology*, 1991 312: 43-67.

Index

B

C

D

E

F

G

H

I

J

K

L

M

N

T

U

V

W

X

Y

Z